IMMUNOBIOLOGY OF PROTEINS AND PEPTIDES V

VACCINES

Mechanisms, Design, and Applications

Recent Volumes in this Series

Volume 247A
KININS V, Part A
Edited by Keishi Abe, Hiroshi Moriya, and Setsuro Fujii

Volume 247B
KININS V, Part B
Edited by Keishi Abe, Hiroshi Moriya, and Setsuro Fujii

Volume 248
OXYGEN TRANSPORT TO TISSUE XI
Edited by Karel Rakusan, George P. Biro, Thomas K. Goldstick, and Zdenek Turek

Volume 249
MINERAL ABSORPTION IN THE MONOGASTRIC GI TRACT:
Chemical, Nutritional, and Physiological Aspects
Edited by Frederick R. Dintzis and Joseph A. Laszlo

Volume 250
PROGRESS IN POLYAMINE RESEARCH:
Novel Biochemical, Pharmacological, and Clinical Aspects
Edited by Vincenzo Zappia and Anthony E. Pegg

Volume 251
IMMUNOBIOLOGY OF PROTEINS AND PEPTIDES V: Vaccines—Mechanisms, Design, and Applications
Edited by M. Zouhair Atassi

Volume 252
DRUGS, SYSTEMIC DISEASES, AND THE KIDNEY
Edited by Alberto Amerio, Pasquale Coratelli, Vito M. Campese, and Shaul G. Massry

Volume 253A
PURINE AND PYRIMIDINE METABOLISM IN MAN VI, Part A:
Clinical and Molecular Biology
Edited by Kiyonobu Mikanagi, Kusuki Nishioka, and William N. Kelley

Volume 253B
PURINE AND PYRIMIDINE METABOLISM IN MAN VI, Part B:
Basic Research and Experimental Biology
Edited by Kiyonobu Mikanagi, Kusuki Nishioka, and William N. Kelley

---

A Continuation Order Plan is available for this series. A continuation order will bring delivery of each new volume immediately upon publication. Volumes are billed only upon actual shipment. For further information please contact the publisher.

IMMUNOBIOLOGY OF PROTEINS AND PEPTIDES V

VACCINES

Mechanisms, Design, and Applications

Edited by

M. Zouhair Atassi

Baylor College of Medicine
Houston, Texas

Springer Science+Business Media, LLC

Library of Congress Cataloging in Publication Data

International Symposium on the Immunobiology of Proteins and Peptides: Vaccines—Mechanisms, Design, and Applications (1989: Alberta)

Immunobiology of proteins and peptides V: vaccines: mechanisms, design, and applications / edited by M. Zouhair Atassi.

p. cm.

"Proceedings of the Fifth International Symposium on the Immunobiology of Proteins and Peptides: Vaccines—Mechanisms, Design, and Applications, held October 2-5, 1988, in Alberta, Canada"—T.p. verso.

"Major sponsors of the symposium, Praxis, Biologics . . . [et al.]"—T.p. verso.

Includes bibliographies and index.

ISBN 978-1-4757-2048-8 ISBN 978-1-4757-2046-4 (eBook)

DOI 10.1007/978-1-4757-2046-4

1. Vaccines—Congresses. 2. Synthetic vaccines—Congresses. I. Atassi, M. Z. II. Praxis Biologics. III. Title.

[DNLM: 1. Peptides—immunology—congresses. 2. Proteins—immunology—congresses. 3. Vaccines—congresses. QW 805 I624i 1988]

QR189.I637 1989

615′.372—dc20

DNLM/DLC 89-16145

for Library of Congress CIP

Proceedings of the Fifth International Symposium on the Immunobiology of Proteins and Peptides: Vaccines—Mechanisms, Design, and Applications, held October 2-5, 1988, in Alberta, Canada

Originally published by Plenum Press, New York in 1989.

A Division of Plenum Publishing Corporation
233 Spring Street, New York, N.Y. 10013

SCIENTIFIC COUNCIL OF THE SYMPOSIUM

M. Z. Atassi, President
H. L. Bachrach
D. A. Befus
E. Benjamini
W. Hockmeyer
A. Sehon
G. S. Bixler, Meeting Secretary
N. J. Calvanico, General Secretary

MAJOR SPONSOR OF THE SYMPOSIUM

Praxis Biologics
Bristol-Myers
United States Army Medical Research and Development Command
Government of Canada
Province of Alberta

The following organizations also contributed to the Symposium

Smith-Kline and French Laboratories
Amgen

PREFACE

This symposium series was established in 1976 for the purpose of bringing together, once every two or three years, active investigators in the forefront of contemporary immunology, to present their findings, discuss their significance in the light of current concepts and identify important new directions of investigation. The founding of the symposium was stimulated by the achievement of major breakthroughs in the understanding of the immune recognition of proteins and peptides. We believe that these breakthroughs will lead to the creation of a new generation of peptide reagents which should have enormous potential in biological, therapeutic and basic applications. This anticipated explosion has since occurred and many applications of these peptides are now being realized.

The fifth symposium was devoted to the field of vaccines against viral, bacterial and parasitic agents. Many important papers dealt with designs based on whole particles, on subunits or envelope proteins, or on selected active regions of envelope proteins. To achieve an intelligent and effective design of a vaccine, it is crucial to know a great deal about the immunology of infection and about the mechanism of action of the intended vaccine. Therefore many important papers dealing with the details of the humoral and cellular immune responses against the infective agents were presented. How to manipulate the immune system, either to maximize a required antibody or T-cell response or to reduce its magnitude is often desired in the design of an immunological defense strategy. What is the best adjuvant to use for a particular requirement? How to achieve tolerance to certain cases? Can we manipulate regulatory idiotopes? All these are clearly important questions that need to be appreciated in the design of vaccines. I hope that this publication will serve to integrate and correlate the current knowledge of these systems with new discoveries in immunology and to identify the most promising new directions for future investigations.

Conferences of this kind require a considerable amount of funding. I should like to express, on behalf of the organization, our gratitude to our sponsors. In particular, I would like to thank the government of Canada and the Province of Alberta, the U.S. Army Medical Research and Development Command, Bristol-Myers and Praxis for their strong support. I would especially like to thank the authors for the prompt submission of their papers.

M. Z. Atassi

CONTENTS

DEVELOPMENTS TOWARD A RECOMBINANT PERTUSSIS VACCINE

W. Neal Burnette[1], David W. Whiteley[1], Vernon L. Mar[1], Drusilla L. Burns[2], Harvey R. Kaslow[3], Witold Cieplak[4], Jerry M. Keith[4], and Timothy D. Bartley[1]

[1]Amgen Inc., Thousand Oaks, CA 91320; [2]CBER, FDA, Bethesda, MD 20892; [3]Dept. Physiol. Biophys., USC School of Medicine, Los Angeles, CA 90033; [4]NIAID Rocky Mountain Laboratories, Hamilton, MT 59840

INTRODUCTION

Vaccines produced from chemically "inactivated" *Bordetella pertussis* have been remarkably effective in diminishing the incidence of whooping cough in the developed world (Hinman and Koplan, 1984). The frequency of adverse vaccine reactions attributed to pertussis immunization, however, has led to a reassessment of the vaccine material and its components (Hinman and Onorato, 1987). In an effort to provide a less reactogenic vaccine, we have focused on a major virulence factor of the Gram-negative etiologic agent. Pertussis toxin (PTX), also known as lymphocytosis promotion factor (LPF), islet activating protein (IAP), histamine sensitization factor (HSF), and pertussigen (Munoz, 1985), has been shown to be both an important mediator of pertussis disease (Pittman, 1979) and an effective protective element in acellular vaccines that have reduced reactogenicity in pediatric vaccine recipients (Manclark and Cowell, 1984; Robinson *et al*., 1985; Kallings *et al*.,1988).

PTX is a hexameric exotoxin composed of five distinct polypeptide subunits (Tamura *et al*., 1982). Subunit S1, or A protomer, possesses an adenosine diphosphate (ADP)-ribosyltransferase activity (Katada *et al*., 1983; Moss *et al*., 1983) responsible for toxic effects that follow infection with *B. pertussis*. Like the A subunits of cholera toxin and *Escherichia coli* heat-labile enterotoxin, the S1 subunit catalyzes the hydrolysis of nicotinamide adenine dinucleotide (NAD) and transfer of ADP-ribose to proteins regulating the adenylate cyclase complex, resulting in impairment of cell membrane signal transduction (Katada and Ui, 1982; Okajima, *et al*., 1985; Gilman, 1987; Neer and Clapham, 1988). The

B oligomer is a dissociable structural entity of the toxin molecule, comprised of subunits S2, S3, S4 (2 mol/mol toxin), and S5 (Katada et al., 1983). The role of the B oligomer appears to involve transport of the S1 subunit and cell receptor recognition (Tamura et al., 1983; Burns et al., 1987; Brennan et al., 1988).

In addition to the potential adverse effects of PTX in current whole cell B. pertussis, a significant concern has been the presence of other bacterial components capable of eliciting toxic effects. In this regard, B. pertussis lipopolysaccharide endotoxin, a tracheal cytotoxin, and perhaps two adenylate cyclase toxins have been identified (Hewlett et al., 1976; LeDur et al., 1978; LeDur et al., 1980; Goldman et al., 1982; Wardlaw and Parton, 1983; Weiss and Hewlett, 1986; Rosenthal et al., 1987; Rogel et al., 1988). Our goal, therefore, has been to assemble a genetically-inactivated PTX molecule in vitro from subunits produced by recombinant DNA means in a heterologous host; we would utilize an S1 subunit lacking substantial ADP-ribosyltransferase activity yet preserving both its potential to associate with the subunits of the B oligomer and contribute to the formation of immunoprotective antigenic determinants.

RESULTS

We have molecularly subcloned and expressed each of the PTX operon cistronic elements to high levels in E. coli (Burnette et al., 1988a) and identified a region (Y_8-P_{15}) of the 234-amino-acid S1 subunit that is necessary for its enzymatic activity and for formation of its sole neutralizing epitope (Cieplak et al., 1988). By rational site-directed mutagenesis of the S1 coding element, we then produced a recombinant mutant analog of this protein (Burnette et al., 1988b). This S1 analog rS1/1-4, containing a conservative basic amino acid substitution of lysine for arginine at residue 9 (R_9->K), lacks substantial ADP-ribosyltransferase activity ($\leq$ 0.1% of native) but retains the protective antigenic determinant (Sato et al., 1984; Sato et al., 1987).

An important characteristic of S1 analog rS1/1-4 is that the substitution of lysine for arginine 9 has no major impact on the predicted secondary structure of this 234-residue polypeptide. In Figure 1, we have superimposed diagrams of predicted secondary structure (Devereux et al., 1984) of native S1 and the recombinant analog. No discernible differences in alpha helical, beta sheet, beta bend, or coil structure could be detected between these S1 species by Chou-Fasman analysis (Chou and Fasman, 1978); as expected, slight differences in hydropathicity occur surrounding the substituted residue (not shown). In contrast, other residue substitutions for arginine 9 and substitutions for residues in the critical area of Y_8-P_{15} (Cieplak et al., 1988) occasionally had profound effects on predicted structure, particularly with regard to the beta bend regions just downstream of this region.

Utilizing site-directed mutagenesis coupled with sulfhydryl alkylation, we have also studied the effects of the two cysteine residues (at positions 41 and 200) on

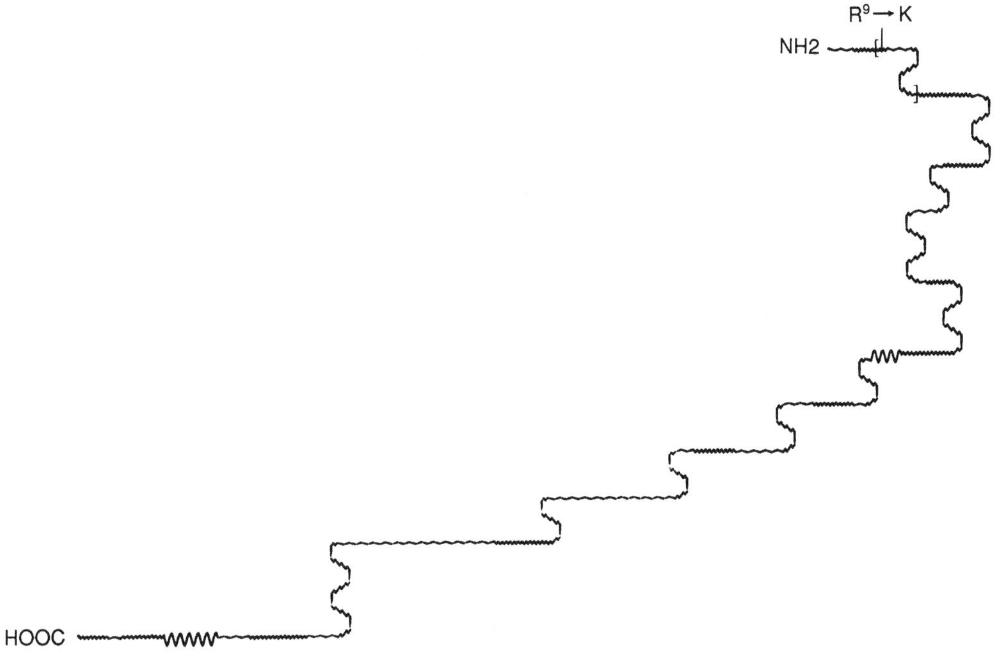

Figure 1. Predicted secondary structure of PTX S1 subunit and its inactive analog. A graphic representation of predicted secondary structure (Chou and Fasman, 1978) for the mature polypeptide sequence of S1 and the inactive recombinant analog rS1/1-4 (Burnette et al., 1988b) was prepared utilizing the Plotstructure program of the University of Wisconsin Genetics Computer Group Sequence Analysis Software Package (Devereux et al., 1984). The brackets indicate the critical S1 region of Y_8-P_{15}. The substitution of lysine for arginine 9 (R_9->K) induced no perceivable changes in this predicted structure, suggesting that this analog would likely acquire a tertiary structure identical to the S1 protein of native amino acid sequence.

enzymatic activity of the S1 subunit (H. Kaslow et al., submitted for publication). These two amino acids, involved in a disulfide pairing, are critical for maintenance of S1 tertiary domains associated with B oligomer interactions (Tamura et al., 1982); reduction of this disulfide bond results in activation of the NAD-dependent ADP-ribosyltransferase and concomitantly permits inactivation of the S1 subunit by sulfhydryl-alkylating agents (Kaslow and Lesikar, 1987). By creating residue substitutions at each cysteine moiety, we have established that alkylation of cysteine 41, but not cysteine 200, is responsible for the inactivation.

We have further demonstrated that NAD retards the alkylation of cysteine 41, but that the sulfhydryl group of this residue is not required for enzyme activity. We are currently investigating the consequences of mutations of residues surrounding cysteine 41.

The *in vitro* assembly of highly complex quaternary structures from recombinant proteins is a developing technology (Hardy *et al.*, 1988). However, Tamura *et al.* (1982) had previously observed that purified native PTX can be reversibly dissociated into oligomeric and monomeric species under the influence of urea. After purification of oxidized and folded recombinant S1 of native amino acid sequence and the enzymatically-inactive rS1/1-4, each species could be shown by electrophoresis in non-reducing, non-denaturing polyacrylamide gels to specifically associate with native B oligomer from *B. pertussis* to form holotoxin molecules physically indistinguishable from native PTX (T. Bartley *et al.*, submitted for publication).

PTX creates a morphologic "clustering" of cultured Chinese hamster ovary (CHO) cells (Hewlett *et al.*, 1983; Burns *et al.*, 1987), a convenient *in vitro* biological assay for holotoxin and the residual toxicity of "inactivated" vaccine materials. This cytotoxic effect has been correlated with the ADP-ribosyltransferase activity of the S1 subunit (Burns *et al.*, 1987), but requires its association with B oligomer to target the plasma membrane (Burns *et al.*, 1987; Brennan *et al.*, 1988). Employing an insertion mutation of the S1 cistronic element (codons for VDGS inserted in-frame between codons for Y_{141} and V_{142}), Black *et al.* (1988) were able to obtain assembly and secretion of a holotoxin from *B. pertussis* that possessed approximately 1% of native toxin CHO cell clustering activity. Each of our semi-recombinant holotoxin species, formed with native B oligomer and either recombinant S1 of native sequence (rS1/1) or analog sequence (rS1/1-4), were applied to CHO cells and assessed for their relative ability to produce the clustered morphology. Holotoxin composed of rS1/1 possessed specific clustering activity essentially identical to that of native PTX, whereas holotoxin comprised of rS1/1-4 had activity $\leq$1% of native toxin (T. Bartley *et al.*, submitted for publication). These *in vitro* experiments confirm the findings of Black *et al.* (1988) and directly support the hypothesis that the ADP-ribosyltransferase activity of the S1 subunit is responsible for the cytopathic effects of PTX (Burns *et al.*, 1987).

CONCLUSIONS

The results obtained here clearly suggest that it will be possible to assemble a totally recombinant holotoxin *in vitro* from *E. coli*-derived protein subunits. We are currently investigating conditions under which the recombinant B oligomer subunits may be associated into this structural moiety to be combined with our analog S1 subunit. Selective site-specific mutagenesis has permitted us to study S1 subunit structure-function relationships; in this regard, we have unambiguously confirmed the coupling between PTX enzyme activity and *in vitro* cytotoxicity and the absence of sulfhydryl participation in catalysis. Employing the same molecular genetic approach to the other subunit cistronic elements will enable a thorough investigation of primary features involved in the tertiary and quaternary organization of this complex protein toxin. But of greater significance, an efficacious whooping cough vaccine component, free of the severe reactogenicity associated with prevailing pertussis vaccines, may be the most meaningful corollary of these investigations.

REFERENCES

Black, W.J., Munoz, J.J., Peacock, M.G., Schad, P.A., Cowell, J.L., Burchall, J.J., Lim, M., Kent, A., Steinman, L., and Falkow, S., 1988, ADP-ribosyltransferase activity of pertussis toxin and immunomodulation by Bordetella pertussis, Science 240:656.

Brennan, M.J., David, J.L., Kenimer, J.G., and Manclark, C.R., 1988, Lectin-like binding of pertussis toxin to a 165-kilodalton Chinese hamster ovary cell glycoprotein, J. Biol. Chem. 263:4895.

Burnette, W.N., Mar, V.L., Cieplak, W., Morris, C.F., Kaljot, K.T., Marchitto, K.S., Sachdev, R.K., Locht, C., and Keith, J.M., 1988a, Direct expression of Bordetella pertussis toxin subunits to high levels in Escherichia coli, Bio/Technology 6:699.

Burnette, W.N., Cieplak, W., Mar, V.L., Kaljot, K.T., Sato, H., and Keith, J.M., 1988b, Pertussis toxin S1 mutant with reduced enzyme activity and a conserved protective epitope, Science 242:72.

Burns, D.L., Kenimer, J.G., and Manclark, C.R., 1987, Role of the A subunit of pertussis toxin in alteration of Chinese hamster ovary cell morphology, Infect. Immun. 55:24.

Chou, P.Y. and Fasman, G.D., 1978, Prediction of the secondary structure of proteins from their amino acid sequence, Adv. Enzymol. 47:45.

Cieplak, W., Burnette, W.N., Mar, V.L., Kaljot, K.T., Morris, C.F., Chen, K.K., Sato, H. and Keith, J.M., 1988, Identification of a region in the S1 subunit of pertussis toxin that is required for enzymatic activity and that contributes to the formation of a neutralizing antigenic determinant, Proc. Natl. Acad. Sci. U.S.A. 85:4667.

Devereux, J., Haeberli, P., and Smithies, O., 1984, A comprehensive set of sequence analysis programs for the VAX, Nucl. Acids Res. 12:387.

Gilman, A.G., 1987, G proteins: transducers of receptor-generated signals, Annu. Rev. Biochem. 56:615.

Goldman, W.E., Klapper, D.G., and Baseman, J.B., 1982, Detection, isolation and analysis of a released Bordetella pertussis product toxic to cultured tracheal cells, Infect. Immun. 36:782.

Hardy, S.J.S., Holmgren, J., Johansson, S., Sanchez, J., and Hirst, T.R., 1988, Coordinated assembly of multisubunit proteins: oligomerization of bacterial enterotoxins in vivo and in vitro, Proc. Natl. Acad. Sci. U.S.A. 85:7109.

Hewlett, E.L., Sauer, K.T., Myers, G.A., Cowell, J.L., and Guerrant, R.L., 1983, Induction of a novel morphological response in Chinese hamster ovary cells by pertussis toxin, Infect. Immun. 40:1198.

Hewlett, E.L., Urban, M.A., Manclark, C.R. and Wolff, J., 1976, Extracytoplasmic adenylate cyclase of Bordetella pertussis, Proc. Natl. Acad. Sci. U.S.A. 73:1926.

Hinman, A.R. and Koplan, J.P., 1984, Pertussis and pertussis vaccine: reanalysis of benefits, risks, and costs, J. Amer. Med. Assoc. 251:3109.

Hinman, A.R. and Onorato, I.M., 1987, Acellular pertussis vaccines, Pediatr. Infect. Dis. J. 6:341.

Kallings, L.O., and Ad Hoc Group for the Study of Pertussis Vaccines, 1988, Placebo-controlled trial of two acellular pertussis vaccines in Sweden - protective efficacy and adverse events, Lancet i:955.

Kaslow, H.R. and Lesikar, D.D., 1987, Sulfhydryl-alkylating reagents inactivate the NAD glycohydrolase of pertussis toxin, Biochemistry 26:3297.

Katada, T., Tamura, M., and Ui, M., 1983, The A protomer of islet-activating protein, pertussis toxin, is an active peptide catalyzing ADP-ribosylation of a membrane protein, Arch. Biochem. Biophys. 224:290.

Katada, T. and Ui, M., 1982, Direct modification of the membrane adenylate cyclase system by islet-activating protein due to ADP-ribosylation of a membrane protein, Proc. Natl. Acad. Sci. U.S.A. 79:3129.

LeDur, A., Caroff, M., Chaby, R. and Szabo, L., 1978, A novel type of endotoxin structure present in Bordetella pertussis, Eur. J. Biochem. 84:579.

LeDur, A., Chaby, R. and Szabo, L., 1980, Isolation of two protein-free and chemically different lipopolysaccharides from Bordetella pertussis phenol-extracted endotoxin, J. Bacteriol. 143:78.

Manclark, C.R. and Cowell, J.L., 1984, Pertussis, in: "Bacterial Vaccines," R. Germanier, ed., Academic Press, Orlando.

Moss, J., Stanley, S.J., Burns, D.L., Hsia, J.A., Yost, D.A., Myers, G.A., and Hewlett, E.L., 1983, Activation by thiol of the latent NAD glycohydrolase and ADP-ribosyltransferase activities of Bordetella pertussis (islet-activating protein), J. Biol. Chem. 258:11879.

Munoz, J.J., 1985, Biological activities of pertussigen (pertussis toxin), in: "Pertussis Toxin," R.D. Sekura, J. Moss, and M. Vaughn, eds., Academic Press, Orlando.

Neer, E.J. and Clapham, D.E., 1988, Roles of G protein subunits in transmembrane signalling, Nature 333:129.

Okajima, F., Katada, T., and Ui, M., 1985, Coupling of the guanine nucleotide regulatory protein to chemotactic peptide receptors in neutrophil membranes and its uncoupling by islet-activating protein, pertussis toxin: a possible role of the toxin substrate in Ca^{+2}-mobilizing receptor mediated signal transduction, J. Biol. Chem. 260:6761.

Pittman, M., 1979, The cause of the harmful effects and prolonged immunity of whooping cough. A hypothesis, Rev. Infect. Dis. 1:401.

Robinson, A., Irons, L.I., and Ashworth, L.A.E., 1985, Pertussis vaccine: present status and future prospects, Vaccine 3:11.

Rogel, A., Farfel, Z., Goldschmidt, S., Shiloach, J., and Hanski, E., 1983, Bordetalla pertussis adenylate cyclase: identification of multiple forms of the enzyme by antibodies, J. Biol. Chem. 263:13310.
Rosenthal, R.S., Nogami, W., Cookson, B.T., Goldman, W.E., and Folkening, W.J., 1987, Major fragment of soluble peptideoglycan released from growing Bordetella pertussis is tracheal cytotoxin, Infect. Immun. 55:2117.
Sato, H., Ito, A., Chiba, J., and Sato, Y., 1984. Monoclonal antibody against pertussis toxin: effect on toxin activity and pertussis infections, Infect. Immun. 46:422.
Sato, H., Sato Y., Ito, A. and Ohishi, I., 1987, Effect of monoclonal antibody to pertussis toxin on toxin activity, Infect. Immun. 55:909.
Tamura, M., Nogimori, K., Yajima, M., Ase, K., and Ui, M., 1983, A role of the B-oligomer moiety of islet-activating protein, pertussis toxin, in development of the biological effects on intact cells, J. Biol. Chem. 258:6756.
Tamura, M., Nogimori, K., Murai, S., Yajima, M., Ito, K., Katada, T., Ui, M., and Ishii, S., 1982, Subunit structure of islet-activating protein, pertussis toxin, in conformity with the A-B model, Biochemistry 21:5516.
Wardlaw, A.C. and Parton, R., 1983, Bordetella pertussis toxins, Pharmacol. Ther. 19:1.
Weiss, A.A. and Hewlett, E.L. 1986, Virulence factors of Bordetella pertussis, Annu. Rev. Microbiol. 40:661.

PROSPECTS AND PROBLEMS IN SYNTHETIC VACCINE DEVELOPMENT: WHAT DID WE LEARN FROM THE CHOLERA TOXIN SYSTEM?

Chaim O. Jacob

Department of Microbiology and Immunology
Stanford University School of Medicine
D-345 Fairchild Bldg., Stanford, California, 94305

INTRODUCTION

The cholera toxin is a 84-kilodalton protein composed of two noncovalently bound subunits, A and B. The A subunit is the proenzyme form of an ADP-ribosyltransferase that is responsible for the biological activity of the toxin through its ability to activate target cell adenylate cyclase. The B subunit is the immunologically dominant region consisting of five identical, noncovalently associated B chains composed of 103 amino acids that are responsible for binding the toxin to target cell membrane receptors containing the oligosaccharide of the GM1 ganglioside. Neutralizing antibodies raised against the holotoxin react mainly with the B subunit, and antibodies against the B subunit are capable of neutralizing the biological activity of the intact toxin.

In the last several years, I have used the cholera toxin B subunit system as a model to study the feasibility and the constraints of the use of short synthetic peptides as the basis for vaccine development. In the present paper I am intending to summarize briefly some of the topics addressed and studies and try to draw some conclusions that are relevant to the general question of synthetic vaccine development.

The following are the topics to be addressed here: 1) the principles underlying peptide antigenicity or what is the most effective peptide? 2) the choice of a carrier and the carrier effect, 3) the choice of adjuvant, and 4) combining the synthetic peptide approach with genetic engineering techniques.

Epitope Analysis

The aim of epitope analysis is to identify those fragments of the protein which will preferentially induce the expected type of immune response. Based on analyses of several proteins such as myoglobin, it was suggested that amino acid residues which form the "corners" of the molecule may induce cross-reactive immune response with the native protein. A second concept was that a fragment of the protein composed of "hydrophylic" amino acids are likely to be exposed on the surface of the protein molecule and thus would be highly antigenic. Another concept introduced was that antibodies to the regions of high mobility reacted with peptide analogues because those peptides could assume in solution the conformation of the mobile segments (Westhof *et al.*, 1985). To locate the regions relevant for antigenic determinants, we have considered results obtained from chemical cleavage of the protein by cyanogen bromide, chemical modification of several amino acid residues such as Arginine 35 and tryptophan at position 88, and the concepts delineated above.

We have studied the following nine peptides: CTP1 (residues 8-20); CTP2 (residues 30-42); CTP2a (residues 27-36); CTP2b (residues 21-36); CTP3 (residues 50-64); CTP3a (residues 45-64); CTP4 (residues 69-85); CTP5 (residues 75-85); and CTP6 (residues (83-97), covering about 80% of the B subunit of cholera toxin sequence, some of them overlapping.

The peptides were synthesized by solid-phase methodology, coupled originally to tetanus toxiod as a carrier and injected in complete Freund's adjuvant to achieve immune response. Of mose interest among these peptides were CTP1 (residues 8-20) and CTP3 (residues 50-64). Antisera against these two peptides cross-reacted with the intact toxin to a level very similar to that of the homologous peptide-antipeptide reaction, and also exerted significant inhibiiton of the biological acitvities of cholera toxin. Thus, they were able to inhibit the cholera toxin induced cyclic AMP production in various cells to an extent of 60-70% and to neutralize significantly the biological activity of cholera toxin as followed by skin vascular permeability assay and by fluid accumulation in ligated small intestinal loops (Jacob *et al.*, 1984).

It is of interest that peptide CTP5 (residues 75-85) was incapable of inducing any cross-reactivity despite the fact that it contains the area of the B subunit (residues 79-84) which has the highest local average hydrophilicity and contains a tyrosine residue which was reported to have a significant role for antigenicity (Sela and Aron, 1960). Similarly, CTP2 (residues 30-42) which was synthesized, since it was suggested based on chemical modification experiments, that arginine 35 or the region surrounding it, is involved in antibody and receptor binding activity (Duffy and Lai, 1979) induced antibodies that were unable to show any cross-reactivity with the native toxin.

CTP6 (residues 83-97), consisting of the region surrounding tryptophan at position 88, synthesized because chemical modification experiments suggested that this residue is near or within the receptor binding site of B subunit (DeWolf *et al.*, 1981) gave rise to antisera of the highest titer of homologous reaction, but showed only slight reactivity with the intact protein. Moreover, antibodies to CTP6 had no effect on the GM1 binding of cholera toxin B subunit (Finkelstein *et al.*, 1987).

On the other hand, CTP3 induced antibodies which, though of lower (than CTP6) absolute titer, gave a very strong cross-reactivity with the intact toxin and neutralized its biological activities. We have some evidence that CTP3 induces antibodies against the GM1 binding site or near it, and exerts its effect by interaction with the B subunit and blocking its attachment to the GM1 receptor (Finkelstein *et al.*, 1987).

It is of interest that elongation of a peptide may affect the reactivity of the antibodies elicited by it in both positive and negative directions (Jacob *et al.*, 1986). It may convert it from one that is incapable of provoking antibodies cross-reactive with the parent protein to one which induces such cross-reactive antibodies. Thus, elongation of CTP2a (residues 27-36) by six amino acids converted it to one which elicited cross-reactive antibody, CTP2b (residues 21-36). Similarly elongation of CTP5 by six amino acids resulted in CTP4 (residues 69-85) which elicited cross-reacted antibodies. Antibodies to a peptide will cross-react with the parent protein if they either possess in free solution a conformation similar to the one within the native protein or if they can easily transconform into such conformation. Thus, if the protein segment is more flexible, the chance of cross-reaction is higher, even though there will be cases when a small peptide is already capable of having a relatively rigid conformation similar to the one it possesses within a native protein. It is obvious for the above reasons that it is not realistic to expect that antibodies to any peptide corresponding to a sequence segment within a native protein will cross-react with that protein. On the other hand, elongation of a peptide, which already leads to cross-reactive antibodies, by addition of amino acid residues, does not necessarily augment such cross-reactivity. As a matter of fact, judging from cross-reactions with toxins immunologically related to cholera toxin, such cross-reactivities may diminish. Consequently, elongation of CTP3 by five amino acids, CTP3a (residues 45-64), induced antibodies which showed cross-reactivity of a definitely lower level with porcine-labile toxin of *E. coli* (P-LT) and vibrio-mimicus toxin than antibodies to the shorter CTP3 peptide. These results are in accord with the observation that antibodies against CTP3 reacted with human-labile toxin (H-LT) of 30 strains of pathogenic *E. coli*. In the case of 9 of these strains, anti-CTP3 showed even higher reactivity than anti-whole cholera toxin (Jacob *et al.*, 1984). It is therefore possible that a shorter peptide which is still cross-reactive, may give a better reactivity with a series of proteins related immunolgically to the parent protein than the cross-reaction between the complete native proteins themselves. This argument may constitute an important consideration in the evaluation of synthetic peptides as the basis for potential vaccines.

As for the question, are there any general methods which allow one to predict whether a given peptide segment will generate antibody which will react with the native protein? It seems that at present a single criterion for the identification of antigenic epitopes in native proteins is unlikely to be universally applicable (Skehel and Ada, 1985).

The Choice of Carrier

In the design of a synthetic vaccine it is clear that it is not sufficient to synthesize the correct peptide which may lead to specific antibodies and neutralize the biological activity. It is of equal importance to establish the manner of choice for presenting the desired epitope so as to lead to the most efficent immune response. In most cases, the synthetic peptides, if not large enough, are not by themselves good immunogens even when administered in potent adjuvant.

In the case of cholera toxin systems, various approaches have been used. used. Thus, the use of tetanus toxoid as a protein carrier, cross-linking and thus polymerization of the peptide without an external carrier and the conjugation of dipalmityl side chains, all had comparable effect on the

H-LT	Ala			Ser			Glu			Ser								Tyr		Ile
CT	Thr	Pro	Gln	Asn	Ile^{5}	Thr	Asp	Leu	Cys	Ala^{10}	Glu	Tyr	His	Asn	Thr^{15}	Gln	Ile	His	Thr	Leu^{20}
P-LT	Ala			Thr			Glu			Ser			Arg					Tyr		Ile

H-LT		Asp			Leu						Met							Val		
CT	Asn	Asn	Lys	Ile	Phe^{25}	Ser	Tyr	Thr	Glu	Ser^{30}	Leu	Ala	Gly	Lys	Arg^{35}	Glu	Met	Ala	Ile	Ile^{40}
P-LT		Asp			Leu						Met							Val		

H-LT				Ser																
CT	Thr	Phe	Lys	Asn	Gly^{45}	Ala	Thr	Phe	Gln	Val^{50}	Glu	Val	Pro	Gly	Ser^{55}	Gln	His	Ile	Asp	Ser^{60}
P-LT				Ser		Glu														

H-LT										Asp					Thr					Thr
CT	Gln	Lys	Lys	Ala	Ile^{65}	Glu	Arg	Met	Lys	Asn^{70}	Thr	Leu	Arg	Ile	Ala^{75}	Tyr	Leu	Thr	Glu	Ala^{80}
P-LT										Asp					Thr					Thr

H-LT		Ile	Asp											Asn	Ser					
CT	Lys	Val	Glu	Lys	Leu^{85}	Cys	Val	Trp	Asn	Asn^{90}	Lys	Thr	Pro	His	Ala^{95}	Ile	Ala	Ala	Ile	Ser^{100}
P-LT		Ile	Asp											Asn	Ser					

H-LT		Glu	
CT	Met	Ala	Asn
P-LT		Lys	

Figure 1. Amino acid sequences of the B subunits of cholera toxin (CT), porcine (P) and human heat labile toxins (H-LT). From Jacob, Arnon, and Finkelstein, Infect. Immun. 52:562-567, 1986.

level of either the homologous anti-peptide response or the cross-reactivity with intact cholera toxin (Jacob *et al.*, 1985).

When a macromolecular carrier is being used, the result obtained depends on its nature. Thus, the branched polypeptide A--L, multipoly (DL-alanyl)--poly(L-lysine) was not an efficient carrier and did not lead to significant antibody production against several cholera toxin peptides, despite that the same A--L carrier proved to be an efficient carrier for synthetic peptides in other systems (Aduibert *et al.*, 1982).

These results indicate that the suitability of a carrier is correlated with the characteristics and specificity of the antigenic determinant coupled to it.

Attachment of palmitoyl groups to the E-amino function of lysine residues was sufficient to convert small peptides into good immunogens (Hopp, 1984). We used this approach by attaching bis-palmitoyl-lysine to peptides CTP1 and CTP3. The addition of this lipid moiety was sufficient for the induction of antibodies against the intact cholera toxin (Jacob *et al.*, 1985). Hopp reported that the dipalmityl conjugates of hepatitis B peptides were even more effective that KLH conjugates (Hopp, 1984). However, in our system protein carriers such as tetanus toxoid showed higher efficacy (e.g. in the case of CTP1).

Polymerization seems to be a decisive parameter in endowing an antigen with immunogenic properties. In our study CTP3 and CTP6 were polymerized using glutaraldehyde. In the case of CTP3, the polymer was as effective as the conjugate with tetnus toxoid. A possible advantage of polymerization is that it can be readily applied when there is more than one peptide. In such a case the resultant polymer leads to efficient immunization against both peptides, to an extent similar to that induced when the two peptides are attached to a macromolecular carrier.

It is of interest that when CTP1 and CTP3 were attached to the same protein carrier molecule, the effectivity of the immune response was considerably lower than when each peptide was attached separately. The reason for this phenomenon is not clear, but it has been observed in an other system as well (Shapira, personal communication).

Another parameter addressed is the effect of prior exposure to the carrier on the immune response towards epitopes attached to it. Thus, previous exposure to tetanus toxoid lead to suppression of the immune response against synthetic epitopes of cholera toxin conjugated to it (Jacob *et al.*, 1985). Such a phenomenon was described first for a hapten by Herzenberg (Herzenberg *et al.*, 1980) and then demonstrated for several peptide systems including the cholera toxin. This argument may be of special significance when one considers tetanus toxoid as a carrier for human vaccines, in view of its worldwide use for anti-tetanus vaccination.

As the peptide epitopes vary in amino acid composition and sequence, in size, and conformation, obviously one type of carrier is unlikely to be universally applicable. Moreover, the methods of binding or cross-linking the epitope with the carrier (not discusssed here) will also play a role in the engineering of the synthetic vaccine.

The Choice of Adjuvant

The effectivity of adjuvants other than complete Freund's adjuvant was studied using the peptide CTP3. These include the synthetic adjuvant

N-acetyl-muramyl-L-alanyl-Δ-isoglutamine (MDP, for muramyl dipeptide) and two derivatives of it, namely threonyl-MDP and MDP-lysine. Another adjuvant (Siddiqui *et al.*, 1981), Avridine (Pfizer), CP-20,961 (N,N,dioctadecyl-N',N'-bis(2-hydroxyethyl-propanediamine), a synthetic lipoidal amine originally characterized as an interferon inducer was also tested.

Both MDP and Avridine were found ineffective in inducing anti-peptide antibodies when used in aqueous medium. No difference was observed between MDP or threonyl-MDP. However, when MDP-lys was covalently conjugated to the peptide-carrier conjugate, highly significant antibody titers were obtained. In this context no significant differences in the antibody level were obtained when tetanus toxoid or the synthetic T,G-(A--L) were used as the carrier. Thus, totally synthetic vaccine with built-in adjuvanticity, namely CTP3-T,G-(A--L)-Lys-MDP could be constructed and successfully used to induce antibody production, when administered in physiological medium (Jacob *et al.*, 1986).

Combined Use of Synthetic Peptides and Recombinant DNA

We attempted to bridge the synthetic and recombinant DNA approaches by inserting a short oligonucleotide encoding for a small peptide and to poexpress that peptide in bacteria other than *V. cholerae*. The starting point in our studies was the identification of a relatively short epitope in the protein that could be encoded in a relatively short oligonucleotide. The potential advantage of this strategy lies in the ability to induce only relevant immune response without resorting to the intact protein.

Initially (Jacob *et al.*, 1985) expressed CTP3 peptide as a b-galactosidase fusion protein. Immunization with the fusion protein did not lead to significant titer of antibodies recognizing cholera toxin. However, when followed by a booster injection of a minute amount of cholera toxin, too small to provide any immune response by itself, it led to a substantial level of neutralizing antibodies (Jacob *et al.*, 1985). Using a recently described lam B "exposition" vector (Charbit *et al.*, 1985), we have constructed recombinant plasmids which express hybrid CTP3 polypeptides in *E. coli*. Expressed hybrid CTP3 polypeptides become major surface constituents upon induction with IPTG at about 10^6 copies per cell (Bavoil *et al.*, unpublished data). Immunization of rabbits with the gel-purified hybrid polypeptide elicited high titers of anti-CTP3 serum antibodies. Most recently we inserted an oligonucleotide specifying CTP3 into the hypervariable region of a cloned *Salmonella flagellin* gene and demonstrated both normal function of the chimeric flagellin and expression of the epitope at the surface of the flagella (Newton *et al.*, 1989). Parenteral administration to mice of a live vaccine strain of *Salmonella* expressing the CTP3-chimeric flagellin gene evoked antibody production reactive with cholera toxin as strongly as with the homologous CTP3.

The availability of synthetic peptides for identification of epitopes in proteins of pathogens, of techniques for the insertion of corresponding oligonucleotides into suitable exposition vectors and of safe effective live *Salmonella* should now allow construction of vaccines likely to be of value for protection agianst infections.

Acknowledgements

I wish to thank my teachers, Professors Ruth Arnon and Michael Sela with whom a large part of these studies were done. Many collaborators have contributed to the data cited in this article, notably, Drs. A. Zamir, M. Leitner, D. Solomon, M. Pines, R. A. Finkelstein, S. Grossfeld, P. Bavoil, S. Newton, and B. A. D. Stocker.

References

Audibert, F., Jolivet, M., Chedid, L., Arnon, R., and Sela, M. (1982), Successful immunization with a totally synthetic diphteria vaccine, Proc. Natl. Acad. Sci. U.S.A. 79:5042-5046.

Bavoil, P., Newton, S., Stocker, B. A. D., and Jacob, C. O. (unpublished data).

Charbit, A., Boulain, J. C., Ryter, A., and Hofnung, M. (1986), Probing the topology of a bacterial membrane protein by genetic insertion of a foreign epitope; expression at the ell surface, EMBO J. 5:3029-3039.

DeWolf, M. J. S., Fridkin, M., Epstein, M., and Kohn, L. D. (1981), Structure-function studies of cholera toxin and its A and B promoters; modification of tryptophan residues, J. Biol. Chem. 256:5481-5488.

Duffy, L. K. and Lai, C. Y. (1979), Involvement of arginine residues in the binding site of cholera toxin subunit B, Biochem. Biophys. Res. Commun. 91:1005-1010.

Finkelstein, R. A., Burks, M. F., Zupan, A., Dallas, W. S., Jacob, C. O., and Ludwig, D. S. (1987), Antigenic determinants of the cholera/coli family of enterotoxins, Rev. Infect. Dis. 9:S490-S502.

Herzenberg, L. A., Tokuhira, T., and Herzenberg, L. H. (1980), Carrier-priming leads to hapten-specific suppression, Nature 285:664-667.

Hopp, T. P. (1984), Immunogenicity of a synthetic HBsAg peptide: Enhancement by conjuation to a fatty acid carrier, Mol. Immunol. 21:13-16.

Jacob, C. O., Harari, I., Arnon, R., and Sela, M. (1986), Antibodies to cholera toxin synthetic peptides of increasing size and their reactivity with related toxins, Vaccine 4:9-98.

Jacob, C. O., Pimes, M., and Arnon, R. (1984), Neutralization of heat-labile toxin of *E. coli* by antibodies to synthetic peptides derived from B subunit of cholera toxin, EMBO J. 3:2889-2893.

Jacob, C. O., Arnon, R., and Sela, M. (1985), Effect of carrier on the immunogenic capacity of synthetic cholera vaccine, Mol. Immunol. 22:1333-1339.

Jacob, C. O., Arnon, R., and Sela, M. (1986), Anti-cholera response elicited in aqueous solution by a completely synthetic antigen with built-in adjuvanticity, Immunol. Letters 14:43-48.

Jacob, C. O., Sela, M., Pimes, M., Hurwitz, S., and Arnon, R. (1984), Both cholera toxin-induced adenylate cyclase activation and cholera toxin induced biological activity are inhibited by antibodies against related synthetic peptides, Proc. Natl. Acad. Sci. U.S.A. 81:7893-7896.

Jacob, C. O., Leitner, M., Zamir, A., Salomon, D., and Arnon, R. (1985), Priming immunization against cholera toxin and *E. coli* heat-labile toxin by a cholera toxin short peptide-β-galactosidase hybrid synthesized in *E. coli*, EMBO J. 4:3339-3343.

Newton, S. M. C., Jacob, C. O., and Stocker, B. A. D. ((1989), Immune response to cholera toxin epitope inserted in *Salmonella flagellin*, Science (in press).

Sela, M., and Arnon, R. (1960), Studies on the chemical basis of the antigenicity of proteins. The role of rigidity in the antigenicity of polypeptidyl gelatins, Biochem. J. 77:394-399.

Shapira, M., personal communication.

Siddiqui, W. A., Kan, S. C., Kramer, K., Case, S., Palmer, K. (1981), Use of a synthetic adjuvant in an effective vaccination of monkeys against malaria, Nature 289:64-66.
Skehel, J. J., and Ada, G. L. (1985), Are peptides good antigens? Nature 316:764-765.
Westhof, E., Altschuk, D., Moras, D., Bloomer, A. C., Mondragon, A., Klug, A., and Van Regenmortel, M. H. V. (1984), Correlation between segmental mobility and the location of antigenic determinants in proteins, Nature 311:123-126.

NEW CONCEPTS IN VACCINES FOR MUCOSAL NON-ENTERIC HUMAN BACTERIAL PATHOGENS

M. A. J. Westerink, A. A. Campagnari, M. B. Nelson, T. F. Murphy, and M. A. Apicella

Division of Infectious Disease
School of Medicine
State University of New York at Buffalo
462 Grider Street
Buffalo, New York 14215

INTRODUCTION

Since the introduction of vaccination in the seventeenth century, the prevention of infectious disease by immunologic methods has been a widely pursued goal of medical practitioners. Until the mid-twentieth century vaccines consisted primarily of inactivated intact organisms or extracts of intact organisms. In 1962, over 100 such vaccine products were available for use in the United States. Many of these products were associated with significant local and systemic toxicity and relatively few could claim efficacy in appropriately designed clinical trials. With a reassessment in the federal drug regulatory mandates after the thalidomide disaster in the 1960's, many products with no proven value were removed from the market. Along with the changes occurring in control of product distribution, work in understanding the pathogenesis of human infection, the nature of the microbial surface, and the scope of the human immune response led to changes in vaccine products which reflected the search for safe and efficacious vaccine products. Over the past decade with application of improved chemical and chromatographic methods combined with the introduction of molecular genetics technology, a wide variety of new vaccines is becoming available.

In this review, the vaccine efforts undertaken against three different non-enteric gram negative human pathogens, *Neisseria meningitidis, Neisseria gonorrhoeae and Haemophilus. influenzae* will be detailed. These microbes have adapted well to their association with the human host and are not found anywhere in nature except in association with human mucosal surfaces. They have evolved redundant systems for survival and have shown the ability to undergo shifts in antigenic expression of principle virulence features further insuring their survival in the face of the human immune response.

The Microbes and Disease

The three species described above are capable of initiating a wide range of disease expression. The meningococcus and *H. influenzae* type b can exist as commensals on the human nasopharyngeal mucosa during the asymptomatic carrier state. Both can cause meningitis and sepsis and the meningococcus can do so in epidemic fashion. The gonococcus, though closely related to the meningococcus and a member of the same genera, is rarely invasive but can inflict considerable morbidity and complications such as infertility on infected populations. A brief review of each organism's disease potential and important structural features is given below.

Neisseria meningitidis (The meningococcus): This organism is the cause of epidemic bacterial meningitis. Until the advent of antibiotics, such epidemics were common in the developed world. Presently sporadic outbreaks and endemic cases occur in the USA and Europe, but large epidemics still occur in Africa, Asia and South America. It is estimated that since 1950, one million people have died in Africa due to this infection (1). The invasive meningococci express a series of carbohydrate capsules. At least 13 different serotypes have been identified (2). Organisms of the four capsular serotypes A, B, C and Y are responsible for the majority of human disease. The principle group at risk for meningococcal infection are children below the age of two; however, during epidemics incidence increases in all age groups. While the capsular polysaccharide is critical for invasion, other bacterial surface structures are important in virulence and pathogenesis of disease. The meningococcus contains a principle surface glycolipid, the lipooligosaccharide (LOS), which is an important virulence factor. It is the major toxin associated with the organism and is the cause of the vascular collapse seen in this infection (3). Like other gram-negative bacteria, the meningococcus has adhesive organelles called fimbriae which serve as attachment factors during mucosal colonization (4). Numerous proteins are present in the outer membrane of the meningococcus. Some serve as channels for nutrients (porins), others act as receptors for important cations (iron repressible proteins) and others as structural proteins involved in maintaining the integrity of the cell surface. The human immune response to the meningococcus is directed against all of these antigens and is a function of the age of the individual infected. Young children do not respond to polysaccharides, yet will evoke a brisk anti-protein and anti-LOS response while adults respond equally well to all components of the bacterial surface with the exception of the capsule from serogroup B meningococci (5,6). This capsule consists of N-acetyl neuraminic acid in a beta 2-8 ketosidic linkage which bears a strong immunologic homology to human glycoproteins (7).

Haemophilus influenzae type b (Hib): This organism is the most common cause of bacterial meningitis in children in the

United States and is the most common cause of acquired mental retardation. While the organism rarely causes epidemic infections, approximately one child in 250 will become seriously infected by it before the age of two. Like the meningococcus, Hib is encapsulated. Unlike the meningococcus, only a single capsular type is responsible for almost all invasive disease. In addition, serious infection in children above the age of four and adults is unusual. Haemophilus has an LOS with toxicities similar to the meningococcus and fimbriae, porin and structural proteins in its outer membrane (8,9,10). The human immune response has been well defined. Children below age two do not respond to the type b capsule even after natural infection. The response to outer membrane proteins and LOS appears to occur as young as three months. The role of these antibodies in protection has not been clearly defined.

N. gonorrhoeae (the gonococcus): Unlike the first two pathogens described, the gonococcus rarely causes invasive disease. The principle serious complication of illness caused by the gonococcus is fallopian tube involvement and the potential infertility which can ensue. In males, infection is symptomatic 98% of the time, while in infected females 50-75% are relatively asymptomatic. As will be pointed out in this chapter, the organism has evolved a number of unique mechanisms to evade host defenses. These include a remarkable degree of antigenic variability in a number of important outer membrane structures including pilin and protein I (11,12). Certain surface proteins evoke antibodies which interfere with the bactericidal effect of antibodies directed against other surface structures (13). The organism expresses surface carbohydrate antigens which mimic self since the terminal tetrasaccharide on its principle surface glycolipid is identical to the human I erythrocyte antigen (14). A polyphosphate matrix surrounds the gonococcus rather than the saccharide residues traditionally found on pathogenic encapsulated bacteria (15). The role of this structure in pathogenesis is unclear. Because its principle sites of infection are the mucosal surface, systemic immunity may only play a minor role in disease prevention. Thus the reliable evocation of immune responses at the secretory immune level are probably crucial to successful vaccine development.

Surface structures of gram-negative non-enteric pathogens which are potential vaccine candidates

Capsular polysaccharides: The capsular polysaccharides have been used as vaccines since the early 1940's. These antigens are serotypic in nature and confer protection by stimulating the serum antibodies which are capable in the presence of serum complement of causing bacterolysis and opsonization. The serotypic nature of these antigens requires that multivalent vaccines for a single species of some bacteria be developed. The pneumococcal vaccine contains 23 different polysaccharides while there are over 80 different capsular serotypes among pneumococcal strains. There are 13 different meningococcal capsular serotypes and four (A, C, W-135, and Y) have been incorporated into a vaccine (16). It should be stressed that the meningococcal group B polysaccharide is not contained in the current vaccine because the polysaccharide is not immunogenic in human populations (17). Meningococci of this serotype are the most frequent causes of this disease in the developed world. Besides the marked heterogeneity of these capsular antigens, they have a limitation in their application based on the nature of the immune response they evoke. They are thymus independent antigens. When used as immunogens, an IgM response is evoked and switching to IgG does not occur. The response to them may be suppressed, or amplified by T-cells but for the most part this response is independent of T-cell regulation. In this regard, an amnestic response does not occur with reimmunization with the same capsular antigen.

The T-independent response limits the usefulness of these vaccines in children below age 2 (18). Children below this age do not respond to such antigens. This is unfortunate because the majority of serious *Haemophilus influenzae* type b and meningococcal infections occur in children below age 2. Hence, vaccine benefits from these immunogens are not available to the group at greatest risk for infection. In older children and adults the available meningococcal and probably the Haemophilus capsular polysaccharide vaccines induce serum anti-capsular antibodies which are protective against the serotypes to which they have been generated.

The U.S. military developed the current meningococcal vaccines and clearly demonstrated their effectiveness in eradicating epidemic meningococcal infection on American military bases (19). These vaccines have been applied in developed and emerging nations and have resulted in termination of epidemics (20). Unfortunately in children below the age of 2 the vaccines were not effective or of reduced efficacy (21).

To stimulate protective antibodies in this group of young children, it has been proposed that by conjugation of these polysaccharide antigens to proteins, vaccines which would contain a polysaccharide antigen whose immune regulation would be similar to a thymus dependent antigen could be developed. Recently, such a vaccine has been approved for use in the United States. This vaccine is composed of the *H. influenzae* type b polysaccharide conjugated covalently to a portion of the diphtheria toxin. The vaccine requires three immunizations for optimal generation of anti-capsular antibody and appeared to be effective in preventing infections among infants in studies in Finland (22). Other polysaccharide-protein conjugates are currently under development using capsular polysaccharides from the pneumococcus and the meningococcus.

Other modes of generation of anti-capsular antibodies are being considered. Our own laboratory has begun evaluations of the possibility of developing an anti-idiotype based anti-capsular vaccine (23). The initial experiments leading to the development of an anti-idiotype antibody (AB2) which would mimic the meningococcal capsular C polysaccharide necessitated the development of a monoclonal antibody to this polysaccharide. Monoclonal antibody 1E4, which is an IgM isotype, was identified from the fusion of splenocytes after immunization with meningococcal strain MP13. This antibody recognized several different preparations of purified group C capsular polysaccharide in immunodot assay. Western blot procedures with *N. meningitidis* as antigen indicated that the antibody did not recognize any other cell component. The high molecular weight polysaccharide capsule did not migrate into SDS-PAGE gels, in contrast to outer membrane proteins and lipooligosaccharides; thus no bands were visualized on Western blots. The ascitic fluids containing 1E4 were tested in ELISA for recognition of purified capsular polysaccharide. Statistical analysis indicated that reactivity of the monoclonal 1E4 ascitic fluid was significantly greater than that of control through a dilution of 1:100,000 in ELISA. In addition, ELISA studies revealed that the 50% inhibition of 1E4 could be achieved with 10 ugms of MP13 capsular C polysaccharide per ml. Meningococcal A capsular polysaccharide failed to inhibit the 1E4 ELISA at concentrations as high as 200 ugms/ml.

In order to determine the prevalence of the antigenic determinants recognized by 1E4 among other *N. meningitidis* strains, we studied 30 group C isolates, 34 B strains, 6 group A strains, 6 group Y strains, 2 group W-135 strains and 2 group X strains. Reactivity was evaluated by dot assay, inhibition ELISA, slide agglutination and bactericidal assays. Optimal concentrations of the antibody used in each assay had been determined previously by using strain MP13. Monoclonal antibody 1E4 recognized determinants on all group C strains tested and on one of 34 *N. meningitidis* serogroup B strains tested. The antigenic determinant recognized by 1E4 was uniformly absent on the capsules of all group A, Y, W-135, and X isolates examined. Studies with 1E4 in a bactericidal assay using meningococcal strain MP13 demonstrated a prozone effect to a dilution of 1:800. One hundred percent killing was observed up to a dilution of 1:10,000.

These studies demonstrated that monoclonal antibody 1E4 had a specificity for an epitope that was common to the capsular polysaccharide of the meningococcal C strains studied. In addition, the epitope was a target for bactericidal activity. These studies led to the selection of this antibody as the idiotype (AB1) for the development of the anti-idiotype (AB2) which would have the surrogate image of the meningococcal C polysaccharide epitope as its antibody combining site.

Previous studies in our laboratory indicated that the sera of mice used to produce ascites to monoclonal antibodies frequently contained evidence of an antiidiotype (AB2), response to these antibodies. On the basis of these observations we elected to use spleens from ascitic mice in which hybridoma 1E4 had been implanted intraperitoneally as a source of splenocytes for the generation of these AB2 secreting hybridomas. Initially hybridomas from these fusions were screened for recognition of AB1 (1E4) in immunodot assay. Six hybridomas positive in this immunodot assay were subjected to further studies.

The six antiidiotype antibodies which recognized the idiotype AB1, 1E4 in immunodot assay were tested for the ability to inhibit 1E4 binding to the capsular C polysaccharide in ELISA. One of these antibodies designated 6F9, showed greater than 70% inhibition of 1E4 in ELISA on repeated occasions. These studies suggested that monoclonal 6F9, an IgG2B isotype was potentially expressing a surface epitope closely related to or immunologically identical to the meningococcal C polysaccharide.

Antiidiotype antibodies which behave as surrogate antigens should also bind polyclonal antisera raised in different animal species in which it is less likely that idiotypes unrelated to antigen binding would be identical. Indirect immunofluorescence analysis demonstrated that hybridoma 6F9 stained circumferentially with anti meningococcal C capsular typing sera and failed to stain with anti meningococcal A capsular typing sera. It is recognized that these polyclonal typing sera contain antibodies to other meningococcal determinants besides the capsular polysaccharide. However, the reactivity of 6F9 with 1E4 and the ability of 6F9 to inhibit 1E4 binding to meningococcal C polysaccharide in ELISA suggests that the fluorescence seen in these studies had a probability of representing rabbit anti C polysaccharide binding to a surface epitope on the 6F9 hybrid which was a surrogate image of the C polysaccharide.

In a series of experiments which were designed to determine whether 6F9 could bind to anti capsular C polysaccharide from other animal species, the ability of 6F9 to bind human antibodies was studied in a meningococcal C polysaccharide specific inhibition ELISA. Complete inhibition of human antibodies in this ELISA was achieved with 200 ug/ml of 6F9 and 50% inhibition was achieved with 50 ug/ml of 6F9. Similar degrees of inhibition were obtained with capsular polysaccharide.

The ability of 6F9 and meningococcal C polysaccharide to generate anticapsular C polysaccharide immune response in rabbits was studied. The pre-immune sera of all rabbits studied did not contain anti meningococcal C polysaccharide antibodies as measured in ELISA. Similarly, a significant anti-meningococcal C polysaccharide response could not be detected after primary or secondary immunization with 6F9 or meningococcal C polysaccharide. Three rabbits were then immunized with 6F9 KLH, meningococcal C polysaccharide KLH and KLH suspended in Freund's complete adjuvant. At two weeks after immunization, an anti-meningococcal C polysaccharide IgG response was detected in the sera of rabbits injected with meningococcal C polysaccharide and 6F9. Subsequent intravenous booster injections on day 28 of these animals resulted in a significant rise in anti-meningococcal C polysaccharide titer measured on day 33. The rabbit immunized with meningococcal C polysaccharide generated an anti-meningococcal C polysaccharide IgG antibody response which was not significantly greater than that of the KLH control. The serum of the rabbit immunized with 6F9 generated an anti-meningococcal C polysaccharide IgG response which was significantly greater than the KLH control to a titer of 1:50,120. Simultaneous measurements of anti-meningococcal C polysaccharide IgM responses in these sera indicated titers which were not significantly different from those of KLH controls. These results indicated that 6F9 contained an antigenic epitope which mimicked meningococcal C capsular polysaccharide. Coupled to a protein carrier, 6F9 functioned as an effective immunogen and stimulated an anti-meningococcal C polysaccharide antibody response which was significantly greater than the response obtained with conjugated meningococcal C polysaccharide preparation in rabbits.

The functional capacity of 6F9 to act as a surrogate image of meningococcal C polysaccharide was exploited in the production of monoclonal antibody AB3. Splenocytes of Balb/C mice immunized SP2/0-Ag14 intraperitoneally with 6F9 hybridoma cells were fused with plasmacytoma cells. Hybridoma cells were screened for recognition of meningococcal C polysaccharide in immunodot assay. Two monoclonal antibodies, 1F2 and 1G11 both of the IgM isotype, were identified; both reacted with meningococcal C polysaccharide. The reactivity of these AB3 monoclonals were tested in immunodot assay against 12 strains of meningococci group C, 6 strains of group B and 2 strains of group A meningococci. Both AB3 monoclonal antibodies reacted with 12 of 12 group C strains, one (strain 6275) of six group B strains and 0 of two group A strains tested. The ascitic fluids containing 1F2 and 1G11 were bactericidal at dilutions from 1:500 to greater than 1:100,000 for the two strains of group C organisms tested (figure 1).

The reactivities of these AB3 monoclonal antibodies were similar to those of monoclonal antibody AB1, 1E4. The results of these experiments indicate that AB3 monoclonal antibodies induced by immunization with 6F9 (AB2) were bactericidal antibodies with specificities for group C

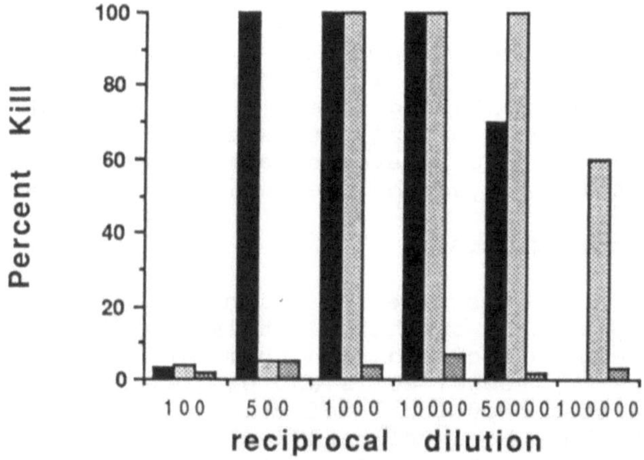

Figure 1. This figure demonstrates the ability of the AB3 monoclonal antibodies 1F3 (■), IG11 (▨) and SP2(▩) to lyse serogroup C strain MP-13 in the presence of complement. As can be seen, SP2 is not bactericidal while 1F3 and 1G11 both are cytolytic to dilutions of 1:50,000.

polysaccharides of the meningococcus identical to the AB1 1E4.

Recently, it has been found that BALB/C mice will respond to 6F9 without conjugation to carriers and without adjuvant. An optimal immunizing dose of 100 ug/animal has been defined. Studies at this dose indicate from the kinetics of the response that 6F9 is acting like a T-dependent antigen. This is based on its ability to cause switching from IgM to IgG and the observation that repeated injections of 6F9 result in increased levels of IgG subclass antibodies to the capsular polysaccharides (figure 2).

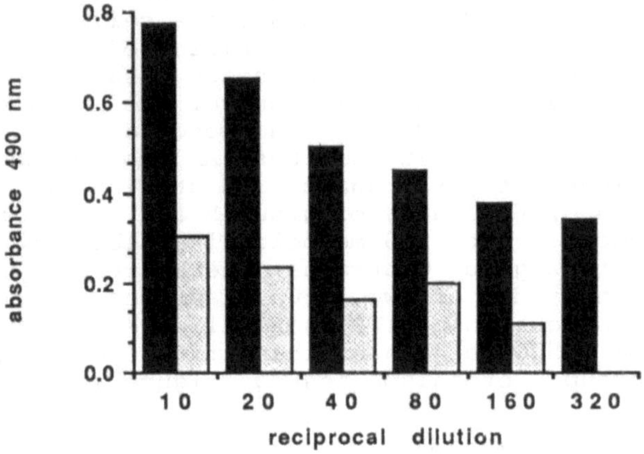

Figure 2. This figure demonstrates an ELISA result for the detection of anti-C polysaccharide IgG after immunization with 100 ugm 6F9 once at day 0 (☐) and twice at day 0 and 28 (■). Animals were sacrificed at day 33. The ELISA wells were coated with 10 ugm/ml capsular C polysaccharide.

At the present time, we are sequencing the heavy and light chain regions of the 6F9 IgG2B gene. If methods to simulate the antibody combining site of these antibodies after sequencing proves fruitful, it is conceivable that peptides which are capable of evoking an antibody response to bacterial capsular polysaccharides will be developed.

Pili (Fimbriae) as Vaccines

Pili are attachment organelles which are composed of self polymerizing protein subunits. It is considered that fimbriae or proteins associated with them mediate attachment to specific sites on mucosal cells and promote colonization and ultimately systemic or local infection (24). The pili of the gonococcus have been well studied immunochemically and physicochemically. The genetics of pilus synthesis have been defined (25,26).

The importance of pilus in pathogenesis was first demonstrated by the studies of Kellog and co-workers. In a group of experimentally infected volunteers, these investigators showed that specific colonial phenotypes, designated T_1 and T_2, were required for infectivity to occur. Random passage of gonococci resulted in loss of these phenotypic characteristics and rendered the organisms not infective. Subsequent studies by Swanson, Kraus and Gotschlich showed that the factor responsible for the T_1 and T_2 colonial phenotype was the piliation of the individual members of the colony (27). Subsequent studies showed that piliated gonococci adhered to a variety of eukaryotic cells and that anti-pilus antibody resulted in inhibition of this adherence (25). Because of the polymerization-depolymerization which can be induced by shifting pH of intact pili, relatively purified preparations of pili could be prepared for use as immunogens (28). These studies demonstrated that pili were antigenically extremely heterogeneous but with some evidence of shared determinants. Vaccine studies in chimpanzees indicated that protection was remarkably strain specific. Physiochemical analysis by Schoolnik showed that pilin, the protein subunit of pili, could be fragmented into three components by trypsin followed by citroconilation. Sequence analysis indicated that the TC-2 contained conserved residues while the TC-3 subunit was the hypervariable portion of the protein.

Analysis of pilin synthesis at the molecular level indicated a unique system for gene movement which explained the antigenic complexity of pilus isolated from different strains and from the same strain. Reversible transitions in pilus expression include both switches in pilin phase (pilus + $\leftrightarrows$ pilus-) and changes in pilin subunit structure (pilin +a $\leftrightarrows$ pilin +b) (26). To accomplish this latter change, DNA identical in sequence to one of the gonococcus genome's multiple partial pilin genes becomes incorporated into the complete pilin gene with attendant eviction of the analogous portion of the completed gene's previously expressed version. The result is a new chimeric pilin structural gene that encodes a corresponding new pilin polypeptide. The gonococcus genome contains over a dozen distinct partial pilin genes as a basic repertoire for synthesis of differing pilin polypeptides. Studies in experimental gonococcal infection indicate that such homologous recombination occurs in vivo and is not merely a product of in vitro growth of this organism. Three dimensional analysis of the structure of gonococcal pili is presently underway. X-ray diffraction studies indicate that the model for the overall peptide fold of a pilin subunit, which is an antiparallel 4-å helix bundle, is similar to tobacco mosaic virus coat protein and myohemerythrin (29).

The adhesion domains on the pilus structure have eluded all forms of analysis until recently. Monoclonal antibody studies by Nicholson and co-workers indicate tha the epitopes for type-specific antibodies lie predominantly in hydrophilic domains that also contain beta turns (30). The epitopes for type specific monoclonal antibodies depend on amino acid changes either in three separate blocks of amino acid sequence in the semi-variable region of pilin, or in discrete regions that lie in the disulphide loop in the hypervariable region of the polypeptide. These studies concluded that antibodies directed against epitopes in both the semi-variable and hypervariable region are able to inhibit adhesion.

The difficulty in defining the adhesion site on pilin has suggested that this site was present on other proteins associated with pilin but not present in the preparation isolated by polymerization-depolymerization methods (31). Using molecular sieve and hydrophobic interaction chromatography purified pili have been isolated and shown to copurify with a consistent array of proteins. Antibody absorption experiments demonstrate that antibodies to these proteins can be absorbed only by piliated gonococci confirming the relationship to piliation. Further studies should define if these proteins have a role in pilus adhesion.

Vaccine studies using pili isolated from a single gonococcal strain were undertaken by the US Army. These studies failed to demonstrate efficacy in the prevention of gonorrhea when the vaccinated population was compared with a placebo vaccinated control group (32). Analysis of the sera of the vaccinated group demonstrated the presence of cross-reacting antibodies to pili. Whether future studies demonstrate that cross-reactive antibodies will be protective remains to be determined.

Outer membrane Proteins: Outer membrane proteins (OMP) have been studied as potential vaccine candidates since the mid-1970's. The results to date have not been successful. Earlier OMP preparations were essentially fragmented outer membrane preparations which contained considerable amounts of lipooligosaccharide. More recently, isolated outer membrane protein preparations have been employed in experimental vaccines. In addition, the isolated proteins tend to be considerably less immunogenic than the intact protein presented in the outer membrane of the bacteria. These factors have stimulated interest in the possiblity of the development of peptide vaccines to exposed relevant regions of these proteins. It is important to note that all exposed sites on outer membrane proteins are not vaccine candidates and that some sites may evoke antibodies which can block bactericidal action of other outer membrane directed antibodies (13). Thus, the importance of choice in site of peptide bears not only on the efficacy of the vaccine but assurances that the site will not evoke antibodies which are detrimental to the host. Recently, studies have demonstrated that several of these mucosal pathogens utilize membrane proteins for iron acquisition (33). Vaccines to these proteins may prevent incorporation of this important nutrient and prevent colonization and disease by these organisms. Studies on vaccines developed to these proteins are under development for these organisms.

A surface exposed epitope on 16,600 molecular weight protein which is highly conserved among *Haemophilus influenzae* type b and nontypable *H. influenzae* strains has been identified. Analysis of sera from humans convalescing from haemophilus bacteremia indicated that antibody to this 16.6 kd protein is present in convalescent human sera (34). Subsequent studies (35) demonstrated that affinity purified human antibody to this 16.6 kd protein, now designated *H.*

influenzae protein P6, was capable of killing nontypable *H. influenzae* strains. Other studies demonstrated the marked degree of conservation of P6 (36). Antibodies raised to isolated P6 in rabbits reacted with P6 from 25 haemophilus strains of diverse origin. In addition, monoclonal antibodies developed to P6 from mice immunized with whole cells of three different strains recognized the P6 epitopes of all 25 strains which were tested. In this study pooled normal human serum contained antibodies to P6. These antibodies reacted with all of the 25 strains which were assayed. Thus, these studies indicated that P6 is a highly conserved antigen on the outer membrane of nontypable *Haemophilus influenzae*.

The gene expressing the *H influenzae* P6 has been cloned and sequenced (37). The gene expressing this protein was cloned as an 800 bp insert into PUC-18. It produced a full length gene product. DNA sequence analysis indicated that this gene contained transcriptional and translational sequences resembling those seen in *E. coli* and a signal sequence characteristic of procaryotic membrane proteins. In addition, the carboxy terminus of this signal sequence shared homology with a common sequence found in bacterial lipoproteins suggesting that P6 is a lipoprotein. The high degree of conservation of this protein, the antigenic conservation of the protein among a wide variety of *H. influenzae* strains, the cloning of the gene expressing this protein coupled with the knowledge of its amino acid sequence have made this protein an attractive vaccine candidate for the prevention of *H. influenzae* infections. Other laboratories have corroborated this work. Other investigators (38) demonstrated that rabbit antiserum to purified P6 protein was specific for a wide variety of *H. influenzae* type b strains isolated from wide geographic areas. In addition, this anti P6 antibody was demonstrated to protect infant rats from challenge. These studies confirmed the value of this protein as a potential vaccine candidate for *Haemophilus influenzae* type b. Subsequent studies from the same laboratory (39), confirmed the sequence previously published (37). In September of 1988, Weinberg and coworkers demonstrated the lipoprotein nature of this protein by showing radiolabelling of this protein utilizing C14 palmitate (40). These studies further confirmed the sequence analysis data which indicated that P6 was a lipoprotein.

Three major outer membrane proteins have been studied as potential vaccine candidates in the gonococcal outer membrane. The porin protein in this organism has been designated PI. This protein shows antigenic variation between different strains and two groups, PIA and PIB, have been defined. These proteins can be further divided by monoclonal antibodies into a number of serovars. Antibodies directed at this protein have been shown to be bactericidal and much vaccine effort has been directed at development of PI based vaccines. The structural genes for this protein have been cloned (41). A group of membrane proteins which mediate intergonococcal adhesion and appear to favor attachment to epithelial cells are the PII proteins. The function of these proteins is not clearly understood. The expression of these proteins can be turned on and off at high frequency and antigenic variation is seen among them. The structural genes for this protein have been cloned (42).

The third protein has been designated PIII. It is present on all strains of gonococci. No antigenic variation has been noted for this protein. The structural gene of PIII has been cloned and sequenced (43). This has shown that this protein contains 236 amino acids. When the predicted se quence was compared to the protein sequences in the Dayhoff data base homology with the Omp A protein of *E.coli* was noted. OmpA contains a proline rich region interrupted by 6 residues of valine or alanine. In PIII there are 5 prolines interspersed with 6 valine/alanine residues. The homology of PIII to OmpA is found only in the portion believed to be periplasmic. If one makes the assumption that the portion of PIII which resembles Omp A is similarly arranged in the outer membrane then only a relatively small portion of PIII (aa 23 to 70) are available to be exposed to the exterior.

Recently it has been shown that blocking IgG with specificity for a common gonococcal surface PIII contributes a major portion of blocking activity in normal and immune serums (13). The relationship between this region which is homologous to OmpA and the blocking IgG is presently speculative. Recent studies suggest that PIV of the meningococcus is structurally and antigenically identical to PIII of the gonococcus. These observations will be important in vaccine design when outer membrane components are the basis of the preparation.

In this paper, we have given an overview of the vaccine developments for three important human bacterial pathogens, the gonococcus, the meningococcus and *H. influenzae* type b.

The advances in molecular biology and carbohydrate and protein chemistry which have occured over the past decade are now being applied by a variety of approaches to the development of new bacterial vaccines. These approaches coupled with a clearer understanding of the pathogenesis of bacterial infections and the human immune response to bacterial antigens will lead to a new generation of safer, more effective vaccine products.

REFERENCES

1. Feldman HA: Meningococcal infections. Adv Intern Med 18:117-140, 1972.

2. Apicella MA: *Neisseria meningitidis*, in The Practices and Principles of Infectious Diseases, edited by; Mandell, G.L., Douglass, R.G., and Bennett, J.E., John wiley and Sons, Inc., New York, NY 1978 and 1983.

3. Giradin E, Grau GE, Dayr J-M, et al. Tumor necrosis factor and interleukin-1 in the serum of children with severe infectious purpura. N. Engl. J. Med., 319:397-400, 1988.

4. DeVoe IW, Gilchrist JE: Piliation and colonial Morphology among laboratory strains of meningococci. J. Clin Microbiol, 7:379-384, 1978.

5. Griffiss JM, Brandt BL, Broud DD, et al. Immune response of infants and children to disseminated infections with *Neisseria meningitidis*. J. Infect. Dis. 150:71, 1984.

6. Wyle FA, Artenstein MS, Brandt BL, et al: Immunologic response of man to group B meningococcal polysaccharide vaccines. J Infect Dis, 126:514-522, 1972.

7. Bhattacharjee AK, Jennings JH, Kenny CP: Characterization of 3-deoxy-D-manno-octulosonic acid as a component of the capsular polysaccharide antigen from *Neisseria meningitidis* serogroup 29E. Biochem Biophys Res Commun, 61:489-493, 1974.

8. Campagnari AA, Gupta MR, Dudas KC, Murphy TF and Apicella MA. Development of a lipooligosaccharide based serotyping system for *Haemophilus influenzae*. Infect. Immun. 54:843-847, 1987.

9. Apicella MA, Shero M, Dudas KC, Stack RR, Klohs W, LaScolea LJ, Murphy TF, Mylotte JM. Fimbriation of Haemophilus species isolated from the respiratory tract of adults. J. Infect. Dis. 150:40-53, 1984.

10. Munson RS, Jr., and Granoff DM. Purification and partial characterization of outer membrane proteins P5 and P6 from *Haemophilus influenzae* type b. Infect. Immun. 49:544-599, 1985.

11. Lambden PR, Robertson JN, and Watt PJ. Biological properties of two distinct pilus types produced by isogenic varients of *N. gonorrhoeae* P9., J Bacteriol, 141:393-396, 1980.

12. Poolman JT, deMarie S, Janen C. Variability of low molecular weight, heat modifiable outer membrane proteins of *Neisseria meningitidis*. Infect. Immun. 30:642-648, 1980.

13. Rice PA, Vayo HE, Tam MR and Blake MS. Immunoglobulin G antibodies directed against protein III block killing of serum resistant *Neisseria gonorrhoeae* by immune serum. J Exp Med, 164:1735-1748, 1986.

14. Mandrell RE, Griffiss JM, Macher BA. Lipooligosaccharide (LOS) of *Neisseria gonorrhoeae* and *Neisseria meningitidis* have components that are immunochemically similar to precursors of human blood group antigens. J. Exp. Med. 168:107-126, 1988.

15. Noegel A, Gotschlich EC. Isolation of a high molecular weight polyphosphate from *Neisseria gonorrhoeae*. J. Exp. Med. 157:2049-60, 1983.

16. Griffiss JM, Brandt BL, Broud DO: Human immune response to varous doses of group Y and W135 meningococcal polysaccharide vaccines. Infect Immun, 37:205-208, 1982.

17. Brandt BL, Artenstein MS, Smith CD: Antibody response to meningococcal polysaccharide vaccines. Infect Immun, 8:590-596, 1973.

18. Lepow ML, Beeler J, Randolph M, Samuelson JS, Hankins WA: Reactogenicity and immunogenicity of a quadrivalent combined meningococcal vaccine in children. J. Infect Dis, 154:1033-1036, 1986.

19. Artenstein MS, Gold R, Zimmerly JG, et al: Prevention of meningococcal disease by group C polysaccharide vaccine. N Eng J Med, 282:417-420, 1970.

20. Wahdan MH, Rizh F, El-Akkad AM, et al: A controlled field trial of a serogroup A meningococcal polysaccharide vaccine. Bull WHO, 48:667-673, 1973.

21. Reingold AL, Hightower AW, Bolan GA, Jones EE, Tiendrebeogo H, Yada A, Phillips C, Adamsbaum C, Ajello GW, Broome CV: Age specific differences in duration of clinical protection after vaccination with meningococcal polysaccharide vaccines. Lancet. 2:114-118, 1985.

22. Eskola J, Kayhty H, Gordon LK, Hovi T, Stenvik M, Ronnberg PR, Kela E, Peltola H. Simultaneous administration of *Haemophilus influenzae* type b capsular polysaccharide-diphtheria toxoid conjugate vaccine with routine diphtheria-tetanus-pertussis and inactivated poliovirus vaccinations of childhood. Pediatr. Infect. Dis. J. 7:480-4, 1988.

23. Westerink MAJ, Campagnari A, Wirth Ma, Apicella MA: Development and characterization of an anti-idiotype antibody to the capsular polysaccharide of *Neisseria meningitidis* serogroup C. Infect Immun, 56:1120-1127, 1988.

24. Swanson J. Studies on gonococcus infection IV. Pili: their role in attachment of gonococci to tissue culture cells. J. Exp. Med.. 127:571, 1973.

25. Hagblom P, Segal E, Bellyard E, So M. Intragenic recombination leads to pilus antigenic variation in *Neisseria gonorrhoeae*. Nature 315:156-158, 1985.

26. Swanson J, Bergstrom S, Barrera O, Robbins K, and Cowin D. Pilus-gonococcal variants. Evidence for multiple forms of piliation control. J. Exp. Med. 162:729-744, 1985.

27. Swanson J, Kraus SJ, Gotschlich EC. Studies on gonococcus infection. I. Pili and zones of adhesion: their relation to gonococcal growth patterns. J. Exp. Med. 134:886-906, 1971.

28. Brinton C. In: Immunobiology of *Neisseria gonorrhoeae*. ASM publications, 1978. Washington, DC, pp. 155-178.

29. Parge HE, McRee DE, Capozza MA., et al. Three dimensional structure of bacterial pili. Antonie van Leeuwenhok 53:447-453, 1987.

30. Nicolson IJ, Perry AC, Virji M, et al. Localization of antibody-binding sites by sequence analysis of cloned pilin genes from *Neisseria gonorrhoeae*. J. Gen. Microbiol. 133:825-833, 1987.

31. Muir LL, Slugnell RA, Davies JK. Proteins that appear to be associated with pili in *Neisseria gonorrhoeae*. Infect. Immun. 56:1743-1747, 1988.

32. Tramont EC, Boslego JW. Pilus vaccines. Vaccine. 3:3-20, 1985.

33. David DW, West EP, McKenna W, Thompson SA and Sparling PF. A pleiotropic iron-uptake mutant of *Neisseria meningitidis* lacks a 70-kilodalton iron-regulated protein. Infect. Immun. 56:977-983, 1988.

34. Murphy TF, Nelson MB, Dudas KC, Mylotte JM and Apicella MA. Identification of an *Haemophilus influenzae* specific epitope on a 16,000 dalton outer membrane protein. J. Infect. Dis. 152:1300-1307, 1985.

35. Murphy TF, Bartos LC, Rice PA, Nelson MB, Dudas KC, Apicella MA. Identification of a 16,600 dalton outer membrane protein on nontypable *Haemophilus influenzae* as a target for human serum bactericidal antibody. J. Clin. Invest. 78:1020-1027, 1986.

36. Murphy TF, Bartos LC, Campagnari AA, Nelson MB and Apicella MA. Antigenic characterization of the P6 protein

of nontypable *Haemophilus influenzae*. Infect. Immun. 54:751-760, 1986.

37. Nelson MB, Apicella MA, Murphy TF, VanKeulen H, Spotila LD and Rekosh D. Cloning and sequencing of *Haemophilus influenzae* outer membrane protein P6. Infect Immun; 56:128-134, 1988.

38. Green BA, Quinn-Dey T, Zlotnick GW. Biologic Activities of antibody to a peptidoglycan-associated lipoprotein of *Haemophilus influenzae* against multiple clinical isolates of *H. influenzae* type b. Infect. Immun. 55:2878-2883, 1987.

39. Deich RA, Metcalf BJ, Finn CW, Farley JE, Green BA. Cloning of genes encoding a 15,000 dalton peptidoglycan-associated outer membrane lipoprotein and an antigenically related 15,000 dalton protein from *Haemophilus influenzae*. J. Bacteriol. 170:489-498, 1988.

40. Weinberg GA, Towler DA, Munson RS: Lipoproteins of *Haemophilus influenzae* type b. J. Bacteriol. 170:4161-4164, 1988.

41. Gotschlich EC, Seiff ME, Blake MS, and Koomey M. Porin protein of *Neisseria gonorrhoeae*: Cloning and gene structure. Proc. Natl. Acad. Sci. USA 84:8135-8139, 1987.

42. Gotschlich EC, Blake MS, Koomey JM, Seif M, Derman A. Cloning of the structural genes of three H8 antigens and protein III of *Neisseria gonorrhoeae*. J. Exp. Med. 164:868-881, 1986.

43. Gotschlich EC, Seiff M, Blake MS. The DNA sequence of the structural gene of gonococcal protein III and the flanking region containing a repetitive sequence. J. Exp. Med. 165:471-482, 1987.

STABLE RECOMBINANT AVIRULENT SALMONELLA VACCINE STRAINS

Roy Curtiss III, Sandra M. Kelly, Paul A. Gulig[1]
and Koji Nakayama[2]

Department of Biology
Washington University
St. Louis, MO 63130

INTRODUCTION

All infectious disease agents have specific mechanisms to colonize, invade, and overcome the host. By use of gene cloning and genetic and biochemical techniques, it is possible to elucidate these mechanisms in biochemical terms as well as to understand the biology by which pathogens accomplish these feats. Once these mechanisms are understood, the information can be used to design novel strategies to immunize the host to prevent infection and disease by the pathogen. Since most pathogens colonize on and invade through a mucosal surface, it is appropriate to develop an immunization strategy that elicits a mucosal immune response as well as production of serum IgG and a cellular immune response.

The use of avirulent *Salmonella* expressing foreign colonization and virulence antigens as a means to elicit such secretory, humoral and cellular immune responses was predicated on four discoveries. First, the delivery of antigens to the gut-associated lymphoid tissue (GALT or Peyer's Patches) leads to a generalized secretory immune response in which secretory IgA (sIgA) bathes all mucosal surfaces and is included in the secretions of

[1]Dept. of Immunology and Medical Microbiology, Box J-266 JHMHC, University of Florida College of Medicine, Gainesville, FL 32610
[2]Dept. of Microbiology, School of Dentistry, Kyushu University, Higashi-ku, Fukuoka 812, Japan

major exocrine glands (Cebra et al., 1976; Bienenstock et al., 1978; Weisz-Carrington et al., 1979; McNabb and Tomasi, 1981; McCaughan and Basten, 1983; LeFever and Joel, 1984; Brandtzaeg, 1988 in press). Second, it is known that Salmonella have the capability following oral inoculation to attach to, invade and persist in the GALT prior to reaching deeper tissues such as the liver and spleen (Carter and Collins, 1974). Third, Salmonella can be attenuated by introduction of a diversity of mutations (Bacon et al., 1950; 1951; Germanier and Furer, 1971;1975; Hoiseth and Stocker, 1981; Curtiss, 1986; Curtiss et al., 1987b; Curtiss and Kelly, 1987; McFarland and Stocker, 1987) without altering the initial tissue tropism to the GALT (Curtiss, 1986; Dougan et al., 1986; Maskell et al., 1986; Curtiss and Kelly, 1987; Dougan et al., 1987; Curtiss et al., 1987b, 1988a, 1988b, 1988d). Fourth, many researchers have contributed information about important colonization and virulence antigens from a diversity of pathogens by use of modern molecular genetic techniques in conjunction with biochemical and immunological analyses and employment of animal models (Macrina, 1984; Curtiss, 1985; Goebel, 1985; Urbaschek, 1987).

Many different mutations have been investigated as to their ability to render Salmonella avirulent yet immunogenic and many of these strains have been used as recipients for genes encoding colonization and virulence antigens from a diversity of pathogens (see Curtiss et al., 1988c). Although many of these recombinant avirulent Salmonella strains elicit secretory, humoral and cellular immune responses against the cloned gene product, many of the constructs do not express cloned genes at very high levels and, more importantly, do not maintain the cloned gene stably when the vaccine strain is administered to the animal host (Curtiss et al., 1987b; 1988a; 1988c). In addition, many of the attenuating mutations previously available have attributes that are not quite optimal for use in vaccine constructs (see Curtiss et al., 1988b, 1988c).

The ideal attenuated vaccine strain should have a number of features. First, it must be completely avirulent and highly immunogenic. This requires a balance that is often difficult to achieve. Second, at least in relation to avirulent Salmonella, it must retain its tropism for the intestine and GALT without causing disease or impairment of normal host physiology and growth. Third, it must have two or more attenuating deletion mutations to preclude loss of the traits by reversion or gene transfer. Fourth, the attenuating phenotype should be unaffected by anything supplied in the diet or by the animal host. Fifth, the system must provide stable high-level expression of the cloned genes in the immunized animal host. Sixth, it must be easy to grow, store, and administer.

We describe in this manuscript two separate means to attenuate Salmonella that work well together and a balanced-lethal host vector system that ensures high-level, stable expression of cloned gene products in vivo.

RESULTS

Attenuation of Salmonella by curing the 100 kb virulence plasmid

Invasive strains of Salmonella possess a 100 kb plasmid that contributes to the virulence of the pathogen (Jones et al., 1982; Nakamura et al., 1985; Barrow et al., 1987; Heffernan et al., 1987). S. typhimurium strains cured of this plasmid have increased LD_{50} values when fed by the oral route, but an unaltered or slightly elevated LD_{50} value by the intraperitoneal route (Pardon et al., 1986; Gulig and Curtiss, 1987; (see TABLE 1). The plasmid-cured strains have an equal ability to the wild-type plasmid-containing strains to attach to, invade and persist in the GALT (see FIGURE 1) but a relatively diminished ability to either reach or more likely to survive in phagocytic cells encountered in the liver and spleen (see FIGURE 2) (Pardon et al., 1986; Gulig and Curtiss, 1987). Because of diminished ability to cause disease and induction of immunity to challenge with virulent strains, Nakamura et al. (1985) were first to suggest the use of such plasmid-cured derivatives as live vaccines. However, the fact that plasmid-cured strains are still able to cause disease symptoms even when administered orally and cause death when administered by other routes (Pardon et al., 1986; Gulig and Curtiss, 1987) necessitates use of additional attenuating mutations to confer complete safety.

Attenuation of Salmonella by deleting genes for adenylate cyclase and cyclic AMP receptor protein

Cyclic AMP synthesized by adenylate cyclase combines with the cyclic AMP receptor protein (CRP) to activate transcription of a large number of bacterial genes and operons for carbohydrate transport and catabolism, glycogen synthesis, amino acid transport and catabolism and synthesis of flagella, fimbriae/pili, and several outer membrane proteins (see Curtiss and Kelly, 1987; Curtiss et al., 1988c). Transposon mutagenesis with Tn10 has been used to construct S. typhimurium strains with deletion mutations in the cya and crp genes, encoding adenylate cyclase and the cyclic AMP receptor protein (Curtiss and Kelly, 1987). These strains are totally avirulent when fed orally to mice at doses up to 10,000 times what would be a lethal dose of the wild-type parental strain (Curtiss and Kelly, 1987; Curtiss et al., 1988a, 1988b, 1988d). These mutants have a generation time

TABLE 1

Mouse LD_{50} values[a] (CFU) for wild-type and 100 kb plasmid-cured <u>S</u>. <u>typhimurium</u>

Strain	100 kb plasmid	LD_{50} (CFU) per oral	intra-peritoneal
SR-11			
χ3306	+	3×10^5	<50
χ3337	-	$>10^8$	<50[b]
χ3338[c]	+	1×10^5	<50
SL1344			
χ3339	+	6×10^4	<50
χ3340	-	$>6 \times 10^8$	400
χ3351[c]	+	$<5 \times 10^4$	<50

[a]Determined for female BALB/c mice by the method of Reed and Muench (1938). From Gulig and Curtiss (1987).
[b]The mean time to death for χ3337 (12.3 days) was greater than that of χ3306 (7.0 days).
[c]χ3337 and χ3340, respectively, reendowed with the virulence plasmid.

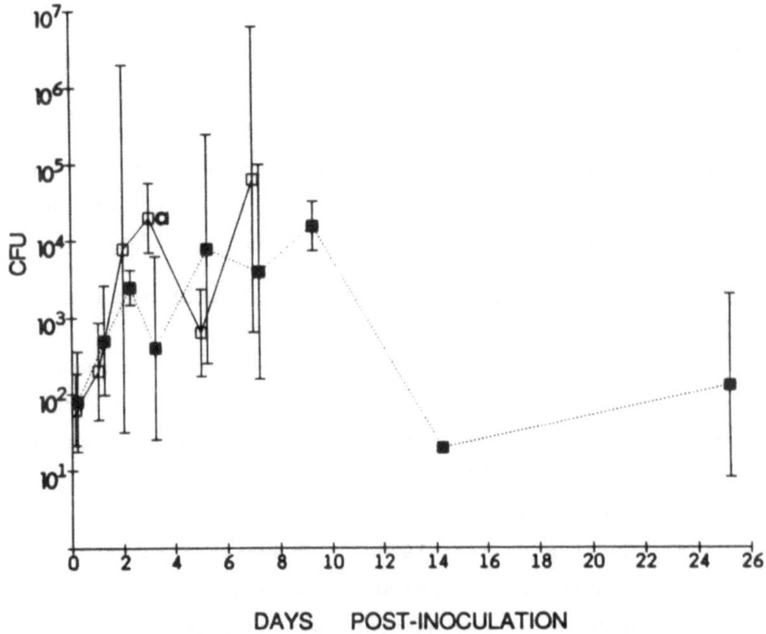

FIGURE 1. Total CFU in Peyer's patches after p.o. inoculation of mice with <u>S</u>. <u>typhimurium</u> SR-11 strains χ3306(pStSR100) (□) and χ3337 (plasmid cured) (■). Results are given as geometric means ± standard deviations for two to seven mice. All mice infected with χ3306 died by 8 days post-inoculation. From Gulig and Curtiss (1987).

in synthetic media almost twice as long as the wild-type and this may serve as an important reason for their attenuated behavior (Curtiss and Kelly, 1987). Orally fed Δcya Δcrp S. typhimurium strains colonize the GALT and the intestine to the same extent as wild-type strains but possess a much reduced ability to either reach or survive after reaching the spleen. The avirulence of the Δcya Δcrp mutations in S. typhimurium has also been demonstrated in one-week old mice that tolerate 10^6 to 10^8 bacteria by the oral route. In addition, repeated oral infection of mice with recovery of bacteria that reach the spleen one week after infection and reinoculation into a subsequent group of mice leads to no alteration in the avirulence properties of the Δcya Δcrp S. typhimurium χ4064 vaccine strain. Indeed, such repeated infection cycles have been conducted eight times with no illness or death of immunized animals.

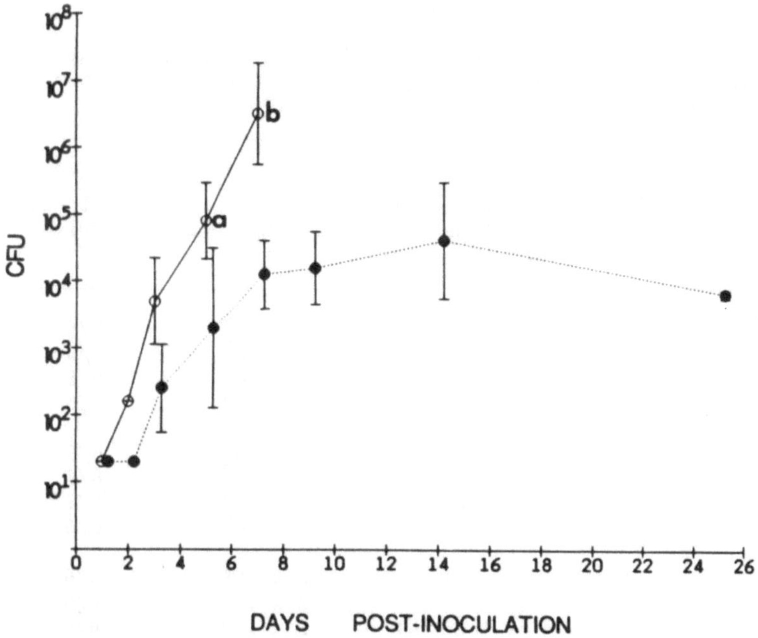

FIGURE 2. Total CFU in spleens after p.o. inoculation of mice with S. typhimurium SR-11 strains χ3306(pStSR100) (○) and χ3337 (plasmid cured) (●). Results are given as geometric means ± standard deviations for two to seven mice. All mice infected with χ3306 died by 8 days postinoculation. From Gulig and Curtiss (1987).

Immunogenicity of Δcya Δcrp S. typhimurium vaccine strains

The avirulent Δcya Δcrp S. typhimurium strains are also highly immunogenic (Curtiss and Kelly, 1987; Curtiss et al., 1988a, 1988b, 1988d). The data in TABLE 2 indicate that oral immunization of mice with 10^7 χ4064 cells on one occasion can induce protective immunity 30 days later to a challenge with up to 2 x 10^9 of fully virulent S. typhimurium SR-11 cells. The data in TABLE 3 indicate that this immunity lasts for a duration of up to four months. TABLE 4 shows that there is a difference in immunogenicity of the plasmid-containing Δcya Δcrp S. typhimurium strain χ4064 compared to the nonisogenic plasmid-cured strain χ4062. This is revealed by challenge of mice, immunized 8 or 15 days earlier, by varying doses of the wild-type virulent S. typhimurium SR-11 parent strain. As is evident from data in TABLE 4, the plasmid-cured strain χ4062 does not induce high-level immunity until 15 days after immunization, but the plasmid-containing χ4064 strain induces a relatively high level of immunity 8 days after immunization. It should be emphasized that the Δcya Δcrp mutations in these two strains are different and thus the effect may be due to the extent of the deletion mutations as well as to the presence or absence of the virulence plasmid.

TABLE 2

Effectiveness of immunization with avirulent S. typhimurium SR-11 Δcya Δcrp plasmid-containing χ4064 to subsequent challenge with wild-type virulent S. typhimurium SR-11 χ3306[a]

Dose of immunizing strain (cfu)	Immunization schedule: No. of doses	Immunization schedule: Time between doses	Wild-type challenge dose (cfu)	Survival[b] live/total
1 x 10^7	1	--	2 x 10^8	5/6
			2 x 10^9	6/6
	3	1 day	2 x 10^8	6/6
			2 x 10^9	6/6
1 x 10^8	1	--	2 x 10^8	6/6
			2 x 10^9	6/6
	3	1 day	2 x 10^8	5/6
			2 x 10^9	6/6
1 x 10^9	1	--	2 x 10^8	6/6
			2 x 10^9	6/6
	3	1 day	2 x 10^8	6/6
			2 x 10^9	6/6

[a]Thirty days after immunization with χ4064, each group of 8-week-old mice immunized with χ4064 was orally challenged with χ3306. The oral LD_{50} of χ3306 for non-immunized mice is 2 x 10^5 cfu (Gulig and Curtiss, 1987). From Curtiss et al. (1988b).
[b]Thirty days after challenge.

TABLE 3

Duration of protection of orally immunized 8-week-old mice[a]

Strain number	Relevant genotype	Virulence plasmid	Immunizing dose (CFU)	Survival after challenge at day: 31	61	91	121
χ4032	Δcya-1	+	1×10^9	4/5	5/5	5/5	5/5
χ4060	Δcya-3	-	9×10^8	4/5	5/5	5/5	5/5
χ4062	Δcya-3 Δcrp-2	-	7×10^8	5/5	5/5	5/5	5/5
χ4064	Δcya-1 Δcrp-1	+	1×10^9	5/5	5/5	5/5	5/5

[a]Survival was monitored for 60 days following oral challenge with 1×10^7 CFU of χ3306.

TABLE 4

Onset of protective immunity following oral immunization of 8-week-old BALB/c mice with 1×10^9 CFU of S. typhimurium SR-11 Δcya Δcrp strains χ4062 (plasmid-cured) and χ4064 (plasmid-containing)[a]

Strain	Challenge dose (CFU)[b]	Time after immunization for challenge	Survival[c] live/total
χ4062	1×10^8	8 days	1/5
χ4062	1×10^9	8 days	0/5
χ4064	1×10^8	8 days	5/5
χ4064	1×10^9	8 days	2/5
χ4062	1×10^8	15 days	5/5
χ4062	1×10^9	15 days	5/5
χ4064	1×10^8	15 days	5/5
χ4064	1×10^9	15 days	5/5

[a]From Curtiss et al., 1988d.
[b]Wild-type virulent S. typhimurium SR-11 χ3306 was administered orally.
[c]Thirty days after challenge.

Stable high-level expression of cloned gene products in Δcya Δcrp S. typhimurium vaccine strains

We have found that Δcya Δcrp S. typhimurium vaccine strains stably maintain and express plasmid cloning vectors containing foreign genes more stably than is true for a number of other avirulent Salmonella strains and for some of the standard E. coli cloning hosts (Curtiss, et al., 1988a,

1988b; Nakayama et al., 1988). Nevertheless, a host-vector system that would ensure 100 percent maintenance of the plasmid cloning vector would be beneficial. Furthermore there is a necessity of eliminating antibiotic resistance markers because of Food and Drug Administration requirements for live vaccines. We have, therefore, developed a novel balanced-lethal host-vector system wherein a deletion mutation is introduced into the chromosome of S. typhimurium to preclude the synthesis of diaminopimelic acid, a unique and essential constituent of the rigid layer of the bacterial cell wall and then use of a wild-type gene on the plasmid cloning vector to complement this deletion (Nakayama et al., 1988). For this purpose, the Streptococcus mutans gene for β-aspartate semialdehyde dehydrogenase which is essential for DAP synthesis has been cloned (Jagusztyn-Krynicka et al., 1982), sequenced (Cardineau and Curtiss, 1987) and used on plasmid cloning vectors to complement a Δasd mutation in the S. typhimurium chromosome (Nakayama et al., 1988). The constructed cloning vector, pYA248, is depicted in FIGURE 3, and possesses in addition to the S. mutans asd^+ gene, a replicon on a BglII fragment obtained from the plasmid p15a, the trc promoter, a multiple cloning site that facilitates insertion of foreign genes to be expressed under the control of the trc promoter and the transcription terminator from the rrnB gene. The EcoRI site in pYA248 is in the same reading frame as is the EcoRI site in λgt11 expression cloning vector therefore facilitating expression of genes cloned and detected in λgt11. pYA248 has been made in a way to facilitate alteration and interchange of various components as cassettes.

Δasd mutants of S. typhimurium are completely avirulent when administered to mice orally, intraperitoneally or intravenously (Curtiss et al., 1987b). Such S. typhimurium Δasd mutants are restored to complete virulence by introduction of the pYA248 cloning vector thus illustrating that loss of the pYA248 vector, which would lead to lysis of cells due to DAPless death, is infrequent (Nakayama et al., 1988).

The E. coli lacZ gene has been cloned into pYA248 to yield the plasmid pYA260. When pYA260 and pYA248 were introduced into various E. coli cloning hosts with a Δasd mutation in the chromosome and S. typhimurium Δasd strain χ3115 and the Δcya Δcrp Δasd vaccine strain χ4072, β-galactosidase is synthesized constitutively and at high levels by cells with pYA260 (TABLE 5). It is evident from the data in TABLE 5 that the Δcya Δcrp Δasd strain χ4072 permits a higher level of β-galactosidase expression than is achieved in other strains.

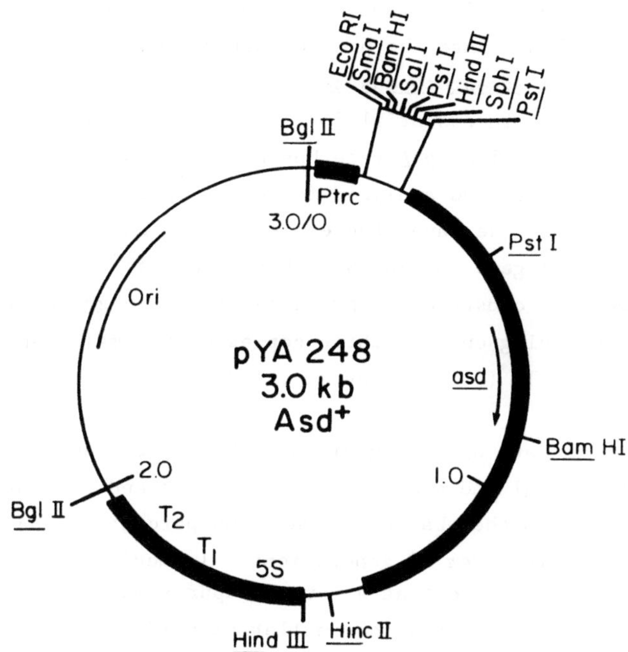

FIGURE 3. Asd+ pYA248 cloning vector from Nakayama et al. (1988).

TABLE 5
β-galactosidase activities[a]

Strain/plasmid	Relevant genotype	β-galactosidase units -IPTG	+IPTG
Escherichia coli K-12			
χ6097(pYA248)	ΔasdA4	0	0
χ6097(pYA260)	ΔasdA4	1555	1687
Salmonella typhimurium SR-11			
χ3115(pYA248)	ΔasdA3	0	0
χ3115(pYA262)	ΔasdA3	6531	6898
χ4072(pYA248)	ΔasdA1 Δ[zhf-4::Tn10] Δcya-1 Δcrp-1	0	0
χ4072(pYA262)	ΔasdA1 Δ[zhf-4::Tn10] Δcya-1 Δcrp-1	8358	7915

[a]β-galactosidase assays were carried out by methods described by Miller (1972).

Genetic information for important colonization antigens from other pathogens have been cloned into the pYA248 vector. Much of our work has been concerned with the Streptococcus sobrinus spaA gene which encodes a major streptococcal protein essential for colonization and virulence (Curtiss et al., 1987a). The spaA gene has been cloned (Holt et al., 1982), partially sequenced and analyzed for major antigenic determinants (Curtiss et al., 1988a). We have, therefore, made a number of constructs of portions of the spaA gene in pYA248 (Nakayama et al., 1988). pYA261, and pYA262 are two such constructs expressing at very high levels portions of the spaA gene including one and three copies of the major antigenic determinant, respectively (Nakayama et al., 1988).

FIGURE 4 presents data on the in vitro stability of χ4072 cells possessing pYA260, pYA261 and pYA262 in L broth with and without diaminopimelic acid. In the absence of DAP, 100 percent of the cells maintain the plasmids for over 50 generations. In medium with DAP the largest plasmid, pYA260, is lost at a rate of approximately 1 percent per bacterium per generation, whereas the smaller plasmids, pYA261 and pYA262, are lost at a lower rate. It should be noted that other types of plasmid cloning vectors in other types of Salmonella are lost at a much more rapid

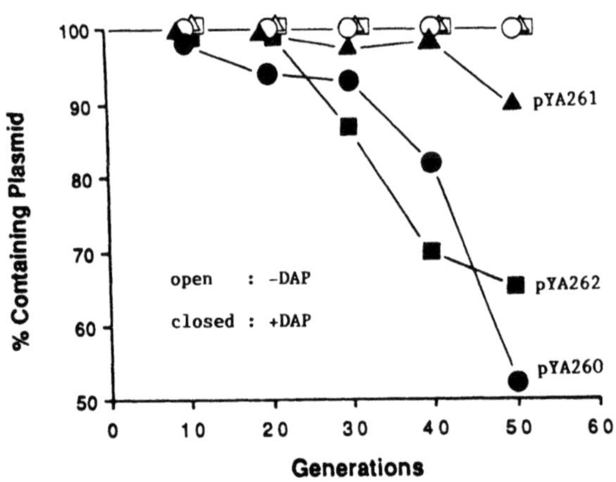

FIGURE 4. Plasmid stability of χ4072(pYA260) in L broth with DAP (●) and without DAP (○), χ4072(pYA261) in L broth with DAP (▲) and without DAP (△), and χ4072(pYA262) in L broth with DAP (■) and without DAP (□). From Nakayama et al. (1988).

rate (Curtiss et al., 1987b, 1988a, 1988c). Of great importance is the data in TABLE 6 which indicate that χ4072 cells expressing the SpaA protein due to the presence of pYA262 are stably maintained <u>in vivo</u> for at least three weeks. In other experiments, like the one providing the data for TABLE 6, all cells so far evaluated by colony immunoblot produced SpaA protein and furthermore were able to grow on medium with or without DAP. It should be emphasized that χ4072 cells harboring pYA262 were recovered from mice by plating on medium with DAP and then tested for production of SpaA protein and ability to grow without DAP.

TABLE 6

In vivo stability of pYA262 specifying SpaA protein in the <u>S</u>. <u>typhimurium</u> SR-11 Δ<u>cya</u> Δ<u>crp</u> Δ<u>asd</u> strain χ4072[a]

Day sacri- ficed	Mouse	Total CFU in Peyer's patches	pYA262[b]		Total CFU in spleen	pYA262[b]	
			Asd^+	$SpaA^+$		Asd^+	$SpaA^+$
1	1	2×10^2	12/12	12/12	N/T	N/T	N/T
	2	6×10^1	4/4	4/4	N/T	N/T	N/T
	3	1×10^2	7/7	7/7	N/T	N/T	N/T
4	1	1×10^4	20/20	20/20	2×10^2	13/13	13/13
	2	2×10^4	20/20	20/20	<10	N/A	N/A
	3	5×10^3	20/20	20/20	<10	N/A	N/A
7	1	2×10^5	20/20	20/20	1×10^3	20/20	20/20
	2	1×10^5	20/20	20/20	5×10^1	3/3	3/3
	3	1×10^5	20/20	20/20	<10	N/A	N/A
14	1	7×10^3	20/20	20/20	1×10^2	11/11	11/11
	2	5×10^3	20/20	20/20	8×10^1	5/5	5/5
	3	<10	N/A	N/A	5×10^2	20/20	20/20
21	1	2×10^3	20/20	20/20	2×10^2	16/16	16/16
	2	1×10^3	20/20	20/20	2×10^1	1/1	1/1
	3	6×10^2	20/20	20/20	<10	N/A	N/A

[a]Eight-week-old BALB/c mice were orally inoculated with 1.2×10^9 χ4072(pYA262) cells on Day 0. From Curtiss et al., 1988d.
[b]Number positive/number tested. N/T = not tested. N/A = none available for testing.

Discussion

There are numerous advantages of the Δ<u>cya</u> Δ<u>crp</u> Δ<u>asd</u> <u>Salmonella</u> vaccine strains expressing genes for important colonization and virulence antigens from other pathogens cloned on the Asd^+ vector. First, the strains are genotypically and phenotypically stable, being unable to revert to the wild-type state with little likelihood of loss of all attenuated mutations by gene transfer from other microorganisms. Second, the strains are completely avirulent and are highly immunogenic. Third, genetic defects leading to the attenuation are not remediable by anything in the diet or anything that can be supplied by the animal host. Fourth, the strains are easy to grow and store. Fifth, there is stable maintenance of recombinant plasmids. Sixth, there is high-level expression of foreign genes. It should also be noted that the occasional loss of the plasmid cloning vector will result in DAPless death with the release of cell contents which should augment the immune response.

The system developed should be useful in analyzing the details of the secretory immune response. Thus, issues of antigen presentation, oral tolerance, and immunomodulation are amenable to investigation using the antigen delivery system developed. It is also possible to use the host vector system constructed to investigate whether putative colonization and virulence antigens from various pathogens can elicit an immune response in a suitable animal to prevent infection and disease by that pathogen. This would constitute a new method to evaluate the relative importance of surface antigens in colonization and virulence. This would be particularly important with pathogens in which introduction of specific genes and other genetic analyses are still not possible. Lastly, it is anticipated that this approach should lead to the development and testing of a number of vaccine constructions with the potential to elicit long-lasting protective immunity due to stimulation of the secretory, humoral and cellular immune responses.

References

Bacon GA, Burrows TW, Yates M (1950) The effects of biochemical mutation on the virulence of Bacterium typhosum: the virulence of mutants. Brit J Exp Pathol 32:714-724

Bacon GA, Burrows TW, Yates M (1951) The effects of biochemical mutation on the virulence of Bacterium typhosum: the loss of virulence of certain mutants. Brit J Exp Pathol 32:85-96

Barrow PA, Simpson JM, Lovell MA, Binns MM (1987) Contribution of Salmonella gallinarum large plasmid toward virulence in fowl typhoid. Infect Immun 55:388-392

Bienenstock J, McDermott M, Befus D, O'Neil LM (1978) A common mucosal immunologic system involving the bronchus, breast, and bowel. Adv Exp Med Biol 107:53-59

Brandtzaeg P (1988) Overview of the Mucosal Immune System. Curr Top Microbiol Immun In press.

Cardineau GA, Curtiss R III (1987) Nucleotide sequence of the asd gene of Streptococcus mutans: identification of the promoter region and evidence for attenuator like sequences preceding the structural gene. J Biol Chem 262:3344-3353

Carter PB, Collins FM (1974) The route of enteric infection in normal mice. J Exp Med 139:1189-1203

Cebra JJ, Gearhart PJ, Kamat R, Robertson SM, Tseng J (1976) Origin and differentiation of lymphocytes involved in the secretory IgA response. Cold Spring Harbor Symp Quant Biol 41:201-221

Curtiss R III (1985) Genetic analysis of Streptococcus mutans virulence. Curr Top Microbiol Immun 118:253-277

Curtiss R III (1986) Genetic analysis of Streptococcus mutans virulence and prospects for an anticaries vaccine. J Dent Res 65:1034-1045

Curtiss R III, Goldschmidt R, Barrett J, Thoren-Gordon M, Salzberg DJ, Murchison HH, Michalek SM (1987a) Genetic analysis of surface proteins essential for virulence of Streptcoccus sobrinus, pp. 212-216 In: Streptoccal Genetics Ferretti JJ, Curtiss R III (eds) American Society for Microbiology, Washington DC

Curtiss R III, Goldschmidt R, Kelly SM, Lyons M, Michalek S, Pastian R, Stein S (1987b) Recombinant avirulent Salmonella for oral immunization to induce mucosal immunity to bacterial pathogens. pp. 261-271 In: Vaccines: New concepts and developments - Proceedings of the Tenth International Convocation on Immunology Kohler H, LoVerde PT (eds) Longman Scientific and Technical, Harlow, Essex, Great Britain

Curtiss R III, Kelly SM (1987) Salmonella typhimurium deletion mutants lacking adenylate cyclase and cyclic AMP receptor protein are avirulent and immunogenic. Infect Immun 55:3035-3043

Curtiss R III, Goldschmidt RM, Fletchall NB, Kelly SM (1988a) Avirulent Salmonella typhimurium Δcya Δcrp oral vaccine strains expressing a streptococcal colonization and virulence antigen. Vaccine 6:155-160

Curtiss R III, Kelly SM, Gulig PA, Gentry-Weeks CR, Galan JE (1988b) Avirulent Salmonella expressing virulence antigens from other pathogens for use as orally-administered vaccines. pp. 311-328 In: Virulence mechanisms of bacterial pathogens. JA Roth (ed) American Society for Microbiology, Washington DC

Curtiss R III, Kelly SM, Gulig PA, Nakayama K (1988c) Selective delivery of antigens by recombinant bacteria. Curr Top Microbiol Immun In press.

Curtiss R III, Nakayama K, Kelly SM (1988d) Recombinant avirulent Salmonella vaccine strains with stable maintenance and high level expression of cloned genes in vivo. In: Immunology and Immunopathology of the Alimentary Canal, NY M. Dekker In press.

Dougan G, Sellwood R, Maskell D, Sweeney K, Liew FY, Beesley J, Hormaeche C (1986) In vivo properties of a cloned K88 adherence antigen determinant. Infect Immun 52:344-347

Dougan G, Hormaeche CE, Maskell DJ (1987a) Live oral Salmonella vaccines: potential use of attenuated strains as carriers of heterologous antigens to the immune system. Parasite Immunology 9:151-160

Germanier R, Furer E (1971) Immunity in experimental salmonellosis. II. Basis for the avirulence and protective capacity of galE mutants of Salmonella typhimurium. Infect Immun 4:663-673

Germanier R, Furer E (1975) Isolation and characterization of GalE mutant Ty21a of Salmonella typhi: a candidate strain for a live, oral typhoid vaccine. J Infect Dis 131:553-558

Goebel W (ed.) (1985) Genetic approaches to microbial pathogenicity. Curr Top Microbiol Immunol. Vol. 118. Springer-Verlag, New York

Gulig PA, Curtiss III R (1987) Plasmid-associated virulence of Salmonella typhimurium. Infect Immun 55:891-2901

Heffernan EJ, Fierer J, Chikami G, Guiney D (1987) Natural history of oral Salmonella dublin infection in BALB/c mice: effect of an 80-kilobase-pair plasmid on virulence. J Infect Dis 155:1254-1259

Hoiseth S, Stocker BAD (1981) Aromatic-dependent Salmonella typhimurium are nonvirulent and effective as live vaccines Nature 291:238-239

Holt RG, Abiko Y, Saito S, Smorawinska M, Hansen JB, Curtiss R III (1982) Streptococcus mutans genes that code for extracellular proteins in Escherichia coli K-12. Infect Immun 38:147-156

Jagusztyn-Krynicka EK, Smorawinska M, Curtiss R III (1982) Expression of Streptococcus mutans aspartate semialdehyde dehydrogenase gene cloned into plasmid BR322. J Gen Microbiol 128:1135-1145

Jones GW, Rabert DK, Svinarich DM, Whitfield HJ (1982) Association of adhesive, invasive, and virulent phenotypes of S. typhimurium with autonomous 60-megadalton plasmids. Infect Immun 38:476-486

LeFever ME, Joel DD (1984) Peyer's patch epithelium: an imperfect barrier pp. 45-56 In: Schiller CM (ed.) Intestinal toxicology. Raven Press, New York

Macrina FL (1984) Molecular cloning of bacterial antigens and virulence determinants. Annu Rev Microbiol 38:193-219

Maskell D, Liew FY, Sweeney K, Dougan G, Hormaeche C (1986) Attenuated Salmonella typhimurium as live oral vaccines and carriers for delivering antigens to the secretory immune system pp. 213-217 In: Vaccines 86. Cold Spring Harbor Laboratory, Cold Spring Harbor

McCaughan G, Basten A (1983) Immune system of the gastrointestinal tract Internal Rev Physiol 28:131-157

McFarland WC, Stocker BAD (1987) Effect of different purine auxotrophic mutations on mouse-virulence of a Vi-positive strain of Salmonella dublin and of two strains of Salmonella typhimurium. Microbial Pathogenesis 3:129-141

McNabb PC, Tomasi TB (1981) Host defense mechanisms at mucosal surfaces. Ann Rev Microbiol 35:477-496

Miller JH (1972) Experiments in Molecular Genetics Cold Spring Harbor Laboratory, Cold Spring Harbor, NY

Nakamura M, Sato S, Ohya T, Suzuki S, Ikeda S, Koeda T (1985) Plasmid-cured Salmonella enteritidis AL1192 as a candidate for a live vaccine. Infect Immun 50:586-587

Nakayama K, Kelly SM, Curtiss III R (1988) Construction of an Asd^+ expression-cloning vector: stable maintenance and high level expression of cloned genes in a Salmonella vaccine strain. Bio/Technology 6:693-697

Pardon P, Popoff MY, Coynault C, Marly J, Miras I (1986) Virulence-associated plasmids of Salmonella serotype typhimurium in experimental murine infection. Ann Microbiol (Paris) 137B:47-60

Reed LJ, Muench H (1938) A simple method of estimating fifty percent endpoints. Am J Hyg 27:493-497

Urbaschek B (1987) Perspectives on bacterial pathogenesis and host defense. In Reviews of Infectious Diseases, Volume 9, Supplement 5. Infectious Diseases Society of America, the University of Chicago Press, Chicago

Weisz-Carrington P, Roux M, McWilliams M, Phillips-Quagliata JM, Lamm ME (1979) Organ and isotype distribution of plasma cells producing specific antibody after oral immunization: Evidence for a generalized secretory immune system. J Immunol 123:1705-1708 (1979)

CYTOTOXIC AND HELPER T-LYMPHOCYTE RESPONSES TO ANTIBODY RECOGNITION REGIONS ON INFLUENZA VIRUS HEMAGGLUTININ

M. Zouhair Atassi[1,2], Jose V. Torres[2] and Philip R. Wyde[2]

[1]Department of Biochemistry and [2]Department of Microbiology/ Immunology, Baylor College of Medicine Houston, Texas, 77030, U.S.A

ABSTRACT

We have previously localized and synthesized twelve antibody recognition sites on influenza virus hemagglutinin (HA). These peptides correspond to exposed surface areas in the 3-D structure of HA. Using intact X31 virus as the immunogen, we have determined the recognition of these synthetic peptides by proliferative T-helper lymphocytes (ThL), delayed type hypersensitivity (DTH), and cytotoxic T-lymphocytes (CTL) responses. The responses to the individual determinants in each of these immune compartments were under separate Ir gene control. Conversely, using the peptides as immunogens, we have determined the ability of various peptide-specific antibodies (in outbred mice) and ThLs (in $H\text{-}2^k$, $H\text{-}2^d$, $H\text{-}2^s$ and $H\text{-}2^b$ mice) to recognize intact virus. Whereas most of the peptides primed the mice for an anti-peptide proliferative ThL response, only very few of these cross-reacted with the virus. The identity of the peptide(s) eliciting virus cross-reactive ThLs varied with the strain. The importance of antibody, ThL, CTL and DTH responses in protection against viral infection and in vaccine design is discussed.

INTRODUCTION

Although much progress has been made in virus vaccine research, viral diseases continue to plague man. There are no vaccines currently available for respiratory syncytial virus or parainfluenza viruses, the two most important respiratory viral pathogens of children under two years of age. No vaccines exist for rhinoviruses. Although not usually fatal, these viruses have a major economic impact and can cause much human misery. Vaccines do exist for influenza viruses but these vaccines are not wholly effective and are not as accepted by the general (or medical) population as they could be. Influenza virus epi- and pandemics occur annually and have enormous economic and medical impact. No effective vaccines exist for herpes or human immunodeficiency viruses. The impact of these viruses on humankind needs no amplification. There are many other virus pathogens for which there are no, or only partially effective vaccines. There is a great need, therefore, for vaccines against these viruses.

X-31 Hemagglutinin Synthetic Surface Peptides

HA 1 Peptides

Peptide	Start	End	Sequence
Peptide 1	23	36	G.T.L.V.K.T.I.T.D.D.Q.I.E.V
Peptide 2	124	134	G.F.T.W.T.G.V.T.Q.N.G
Peptide 3	138	152	A.G.K.R.G.P.G.S.G.F.F.S.R.L.N
Peptide 4	154	167	L.T.K.S.G.S.T.Y.P.V.L.N.V.T
Peptide 5	175	188	D.K.L.Y.I.W.G.V.H.H.P.S.T.N
Peptide 6	183	199	H.H.P.S.T.N.Q.E.Q.T.S.L.Y.V.Q.A.S
Peptide 7	201	218	R.V.T.V.S.T.R.R.S.Q.Q.T.I.I.P.N.I.G
Peptide 8	272	288	A.P.I.D.T.G.I.S.E.G.I.T.P.N.G.S.I
Peptide 9	300	315	I.T.Y.G.A.G.P.K.Y.V.K.Q.N.T.L.K

HA 2 Peptides

Peptide	Start	End	Sequence
Peptide 10	1	11	G.L.F.G.A.I.A.G.F.I.E
Peptide 11	56	68	I.E.K.T.N.E.K.F.H.Q.I.E.K
Peptide 12	68	84	K.E.F.S.E.V.E.G.R.I.Q.D.L.E.K.Y.V

Fig. 1. Covalent structures of surface peptides of the HA molecule synthesized in our laboratory. In the synthetic peptides, the cysteine residues at positions 227, 281, and 305 are replaced with glycine. It is not suggested that the antigenic sites stretch the full length of these peptides but rather that the sites reside within the boundaries of these peptides. HA1 is the large (molecular weight 47,500) and HA2 is the small (molecular weight 28,700) fragment resulting from proteolytic cleavage of HA. Figure is from Atassi and Kurisaki (1984).

Previous attempts to make peptide vaccines against invasive pathogens have not been as effective as initially anticipated. These attempts were based on naive and grossly incomplete understanding of the molecular immune recognition of the virus. This paper describes briefly our studies on the dissection of the submolecular recognition of influenza virus hemagglutinin by various humoral and cellular compartments of the immune system.

Table 1. Binding of human antiviral antibodies to synthetic peptides

Antigen	Influenza antibodies bound (Δ cpm)			
	Human antiserum[a]			Mouse antiserum[b]
	ZA	NB	PA	
Virus X-31	93,425	105,709	110,580	110,650
Peptide 1	0	5,994	843	7,360
Peptide 2	12,006	11,220	8,764	12,340
Peptide 3	17,098	20,920	14,496	10,580
Peptide 4	8,346	1,212	7,946	8,330
Peptide 5	1,240	4,850	0	1,810
Peptide 6	5,902	18,640	9,080	7,730
Peptide 7	9,970	11,790	8,855	11,210
Peptide 8	0	2,470	1,774	5,630
Peptide 9	5,305	11,740	9,622	8,920
Peptide 10[c]	-	-	-	18,850
Peptide 11	3,410	10,175	2,120	7,640
Peptide 12	13,185	11,387	13,860	12,580
Mb (sperm whale)	0	320	0	0
Hb (human)	110	80	75	0
Mb peptide[d]	160	210	0	150
Hb peptide[d]	0	108	95	0

[a]Serum was obtained from individuals with recent influenza A infections and diluted 1:1000 (vol/vol) with PBS/BSA (0.15M NaCl in 0.01 M sodium phosphate buffer, pH 7.2 containing 0.1% BSA). Peptide-protein conjugates were used as plate antigens. Bound antibodies were reacted with 1:1,000 (vol/vol) rabbit anti-human immunoglobulin G (IgG) + IGM followed by treatment with ^{125}I-labeled protein A (2 x 10^5 cpm). The results have been corrected for non-specific binding by BSA (1 to 2.5% of total label) and are the average of triplicate analyses, which varied ± 1.5% or less. Note that the plate assay is semiquantitative only, so the sum of the activities of peptides cannot be compared with activity of intact virus.

[b]Antisera from five outbred mice were mixed and diluted 1:1000 (vol/vol) with PBS-0.1% BSA. Rabbit antimouse IgG + IgM was used.

[c]The binding activity of this peptide has been reported previously (Atassi and Webster, 1983).

[d]BSA conjugates of peptides 1-15 of the α chain of Hb (Kazim and Atassi, 1982) and 121-137 of myoglobin (Bixler and Atassi, 1983, 1984) were used as negative controls.

Table is from Atassi (1985).

Two major components make up the coat of influenza viruses, hemagglutinin (HA) and neuraminidase (NA) (Webster et al., 1982). Of the two molecules, HA appears to be most responsible for virus recognition by the immune system, and is the molecule which binds to a receptor on target cells (and thus is most important in mediating infection) (Webster et al., 1982; White et al., 1982).

Much is known about the makeup of HA. It is expressed on the virus as a trimer. Each subunit is glycosylated at seven sites (Bosch et al.,

1979; Rott, 1979; Potter and Oxford, 1979; Wiley et al., 1977; Wiley et al., 1981) and consists of a large external domain, a small hydrophobic transmembrane region and a small hydrophilic internal segment (Webster et al., 1982: Potter and Oxford, 1979). The molecule undergoes scission at position 329 resulting in the loss of an arginine and the cleavage of the molecule into two chains (HA1 and HA2) linked by a single disulfide bond (Lazarowitz and Choppin, 1975; Skehel and Schild, 1971; Skehel, 1972). This cleavage with its attendant conformational changes in the molecule is an absolute requirement for virus infectivity (Lazarowitz and Choppin, 1975).

Table 2. Differences in anti-site antibody response in sera of influenza patients.

	Antiserum						
Antigen	TD	GB	BA	ZA	NB	PA	MT
X-31	+	+	+	+	+	+	+
Peptide 1	-	+	+	-	+	-	+
Peptide 2	+	+	+	+	+	+	+
Peptide 3	+	+	+	+	+	+	+
Peptide 4	-	+	-	+	+	+	-
Peptide 5	-	+	-	+	+	-	+
Peptide 6	+	+	+	+	+	+	+
Peptide 7	+	+	+	+	+	+	+
Peptide 8	-	-	+	-	+	+	+
Peptide 9	+	+	+	+	+	+	+
Peptide 10	ND[a]	+	+	+	+	+	+
Peptide 11	+	+	+	+	+	+	+
Peptide 12	+	∓	+	+	+	+	-

[a]Not done.

Table is from Atassi (1985).

ANTIGENIC (ANTIBODY RECOGNITION) SITES

Because HA is such a critical viral protein, there has been much interest in trying to localize antigenic sites on the HA for their potential use in a synthetic vaccine preparation (Potter and Oxford, 1979; Skehel, 1972; Ward, 1981; Yewdell et al., 1979; Klenk et al., 1981; Atassi and Kurisaki, 1984; Atassi, 1985, Atassi and Kurisaki, 1986). Previous studies from our laboratories have revealed that protein antigenic sites are invariably situated on surface areas of a protein molecule (Atassi and Kurisaki, 1986; Atassi, 1984). This is regardless of whether the antigenic sites are of the "continuous" type [i.e., occupying a continuous portion of the polypeptide chain such as occcurs with myoglobin (Atassi, 1975), hemoglobin (Kazim and Atassi, 1980, 1982; Yoshioka and Atassi, 1986), cytochrome c (Atassi, 1981) and some of the serum albumin sites (Atassi et al., 1979), or the "discontinuous" type [such as the antigenic sites on lysozyme (Atassi, 1978) and some of the serum albumin sites (Atassi, 1982; Atassi et al., 1979). By synthesizing peptides corresponding to the surface areas of the HA molecule, twelve antigenic sites (Figure 1) (i.e. reacting with anti-virus antibodies) on influenza virus X31 (A/Hong Kong/68 serotype) were localized (Atassi and Kurisaki, 1984).

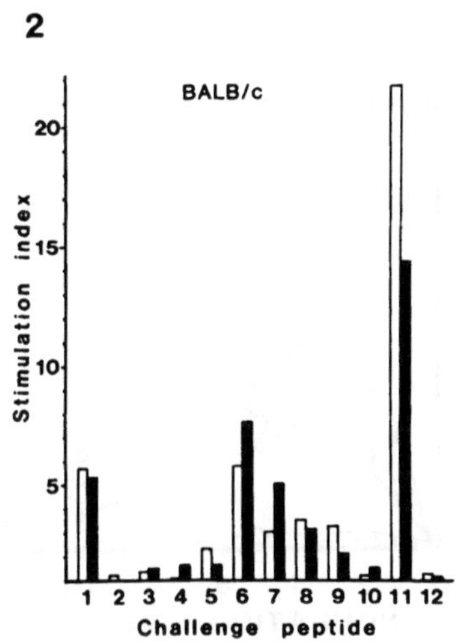

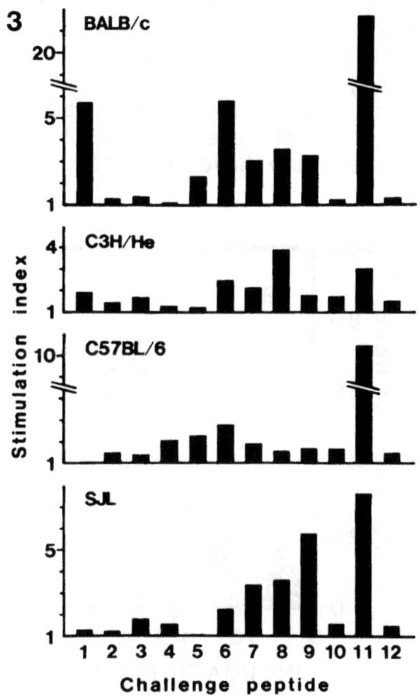

Fig. 2. (left) The T-cell recognition profile of the peptides after immunization with X31 virus in PBS (open bars) or incomplete Freund's adjuvant (solid bars). Balb/c mice were immunized with X31 at a dose 100 HAU/mouse. Results represent responses at optimum *in vitro* challenge doses. Unstimulated cells from immunization in PBS or *in adjuvant* were 1,090 and 1,510 cpm, respectively. Stimulation index is: cpm of stimulated cells/cpm of unstimulated cells. Figure is from Atassi and Kurisaki (1986).

Fig. 3.(right) The peptide T-cell recogntion profile in four mouse strains that had been immunized with an optimum dose of X31 virus. The results, which are expressed in stimulation index (cpm of stimulated cells/cpm of unstimulated cells), summarize the values of maximum responses at the optimum *in vitro* challenge dose for each peptide. Figure is from Atassi and Kurisaki (1986).

Four of these peptides (and possibly five) carry residues that had previously been postulated to reside in sites on the influenza virus coat that are involved in antibody neutralization (Wiley et al., 1981). Thus in HA, residues 133, 143, 144, 145 and 146 were postulated to be what was termed as site A (Wiley et al., 1981); residue 154 and 155 were postulated to fall in what was termed site B; and residues 275 and 278 were postulated to fall in site C. The binding activity to anti-viral neutralizing antibodies of our synthetic HA1 peptides (peptides 124-134, 138-152, 154-167, 183-199 and 272-288) indicated the correctness of the postulations.

It is important to point out that the antigenic sites described (Atassi and Kurisaki, 1984, 1986) were recognized by both human (Table 1) and mouse (Tables 1 and 2) antisera. This is consistent with our previous conclusions that protein antigenic sites are inherent in the protein and are not dependent on the host species in which the antisera are elicited (Atassi and Kazim, 1978; Atassi, 1984). More importantly, it makes it more likely that results obtained in the mouse model will be predictive of what will occur in humans.

There were considerable differences observed in the levels of human anti-viral antibodies (Tables 1 and 2) to some of the twelve peptides tested (Atassi and Kurisaki, 1984). This may be indicative of genetic control, at the level of individual sites (epitopes). Since the antisera used in the mouse studies described above were a mixture from five outbred mice, the differences in the levels of responses to individual sites (peptides) may have been obscured.

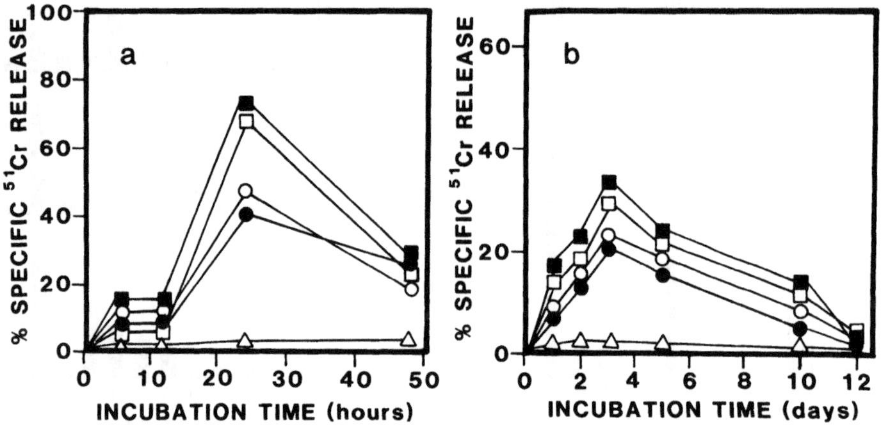

Fig. 4. Determination of optimum incubation time prepare L929 targets coated with virus HA peptide (panel a) or to induce C3H/He cytotoxic T lymphocytes in splenocytes suspension from virus-primed mice using virus HA peptides (panel b). Panel a shows virus-primed cytotoxic T lymphocytes (CTL) lysis of L929 target cells preincubated with representative peptides or virus from 0 to 48 h. Panel b shows determination of the optimum time for peptide incubation with virus-primed CTL in order to obtain a maximum cytotoxic response towards virus-infected targets. Effector to target cell ratio was 50:1. Symbols represent: o, peptide 1 (HA1 23-36); □, peptide 5(HA1 175-188); •, equimolar mixture of peptides 1 to 12; ■, AX-31 virus; Δ, control peptide. Figure is from Torres et al. (1988).

The fact that significantly more than 4 peptides were identified as being able to interact with virus-specific antibodies indicates that the antigenicity of the HA on the virus may be more complex than is presently thought. It also suggested that these twelve sites may be involved in the induction of virus-specific ThL, CTL and DTH responses. This is one area of knowledge that is presently incomplete and one that we hope to investigate.

PROLIFERATIVE T-HELPER LYMPHOCYTE RECOGNITION SITES

Immunization with intact virus (without adjuvant; see Figure 2) into mice of Balb/c (H-2^d), C57/BL6 (H-2^b), C3H/He (H-2^k) and SJL (H-2^s) showed that, of the 12 synthetic antigenic sites elucidated, at least seven and possibly eight, were recognized by virus-primed ThLs (Atassi and Kurisaki, 1984, 1986). The ThL responses were under Ir gene control (Figure 3). These findings are consistent with our previous reports that sites recognized by antibodies (i.e, B cells) can also be recognized by T cells and that the responses to each site are under separate genetic control (Okuda et al., 1979; Bixler and Atassi, 1984; Atassi and Bixler,

1988). Conversely, the profiles of T-cell recognition for myoglobin (Bixler and Atassi, 1984) and lysozyme (Bixler et al., 1984a,b) reveal that in addition to sites that are recognized by both antibodies and T cells, each of these two proteins also contains sites that are recognized only by T cells and for which no detectable antibody responses were found.

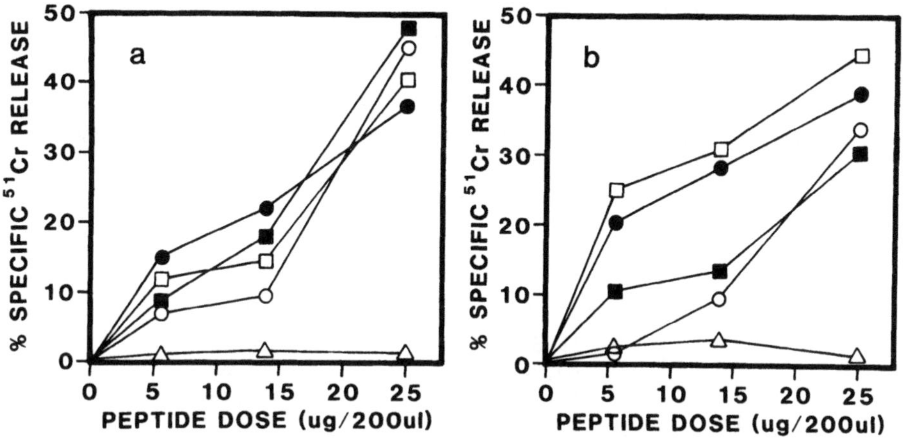

Fig. 5. Examples of dose-response curves for peptide recognition and target lysis by AX-31-primed CTL. Panel a shows that when syngeneic target cells are preincubated with synthetic peptides representing different areas in the influenza virus HA, their recognition and lysis by virus-directed CTL follows dose-response patterns. The cytotoxicity assays were performed as indicated in Section 3. Panel b shows that when virus-primed CTL are preincubated with synthetic HA peptides, recognition and lysis of virus-infected syngeneic target cells is enhanced in a dose-dependent manner. Protocol was the same as in Fig. 1, as described in Section 3. Symbols represent: o, peptide 1 (HA1 23-36); □, peptide 5 (HA1 175-188); ●, peptide 3 (HA1 138-152); ■, peptide (HA1 300-315); Δ, control peptide. Figure is from Torres et al. (1988).

It has been shown by Anderer (Anderer, 1963 and Anderer and Schlumberger, 1965) and Benjamini (Benjamini et al., 1972; Benjamini, 1977; Fearny et al., 1971) that antibodies which neutralize tobacco mosaic virus in vitro, can be obtained by immunization with conjugates of small viral peptides. However, eliciting protective immunity in vivo in a host may require a more complex antigen than a simple peptide. Attempts to develop effective anti-viral vaccines using small synthetic peptides have not been too successful (for recent reviews, see refs Arnon, 1987; Kohler et al., 1988). Such studies have relied entirely on the production of anti-peptide antibodies and not paid much attention to immune recognition and regulation by T cells (Okuda et al., 1979; Bixler and Atassi, 1983; Young and Atassi, 1982; Krco et al., 1984; Krco et al., 1981a,b).

CYTOTOXIC T-LYMPHOCYTE RECOGNITION SITES

Synthetic peptides representing antigenic areas in a molecule can be recognized by protein-specific CTL if they are pre-incubated with syngeneic targets coated with these peptides (Torres et al., 1988; Wabuke-Buenoti et al., 1984). Initial studies have suggested that although HA is the major protein available at the virus surface, it is recognized only by minor populations of CTL from mice primed to this HA (Caton et al., 1982; Braciale, 1979). The major population of CTLs from these mice were thought to recognize virus nucleoprotein antigens (Townsend et al., 1984). However, recently it was shown that a very significant percentage of target cells coated with noninfective virus or HA can be recognized and lysed by CTL from appropriately primed mice (Morrison, et al., 1986; Torres et al., 1988). Furthermore, CTLs from mice primed with virus, express stable cytotoxic activity when propagated in tissue culture (Andrew et al., 1985; Torres et al., 1988).

Table 3. Binding of antipeptide antibodies to intact virus[a]

	Antibody bound (Δ cpm) to:	
Peptide	Immunizing peptide	X-31 virus
1	19,390	18,688
2	20,065	18,320
3	12,530	11,340
4	14,920	11,420
5	8,860	3,190
6	24,230	19,430
7	20,480	18,490
8	13,600	9,840
9	16,390	22,130
10	16,800	15,380
11	22,300	21,910
12	22,660	20,890

[a]Antisera were elicited in three outbred mice for each peptide by immunization with free (noncoupled) peptides. For the plate assay, the peptides were used as protein conjugates. Antipeptide antisera from 60-day postimmunization bleedings were diluted 1:100 to 1:1,000 with PBS-0.1% BSA for assay. The results, corrected for nonspecific binding (1 to 2% of total label) of BSA, are the average of triplicate analyses that differed $\pm$ 1.6% or less. None of these antisera showed any significant binding $P < 0.05$, Student's t test) to Mb or Hb.

Table is from Atassi and Kurisaki (1984).

However, despite their potential use as a peptide vaccine, until recently little information was available concerning the submolecular *areas in the hemagglutinin that are responsible for its immunologic* recognition by cytotoxic T lymphocytes. Recently, we carried out a study (Torres et al., 1988) to determine whether submolecular areas recognized by antibodies and helper T cells are also important in the virus-specific, T-lymphocyte-mediate cytotoxic responses generated towards virus-infected cells. The 12 synthetic peptides representing areas of HA in influenza AX-31 virus, which have previously been shown (Atassi and Kurisaki, 1984, 1986) to bind anti-influenza virus antibodies and provoke proliferation of virus-primed ThLs were tested in two different cytotoxicity assays. It was found (Torres et al., 1988) that some of these peptides, when incubated with appropriate L929 target cells, lysis by cytotoxic T cells from virus primed C3H/He (H-2^k) mice was observed (Figure 4a). In addition, AX-31 primed lymphocytes preincubated with these synthetic peptides (both individually and as an equimolar mixture) exhibited enchanced lysis of virus-infected syngeneic targets (Figure 4b). The cytotoxic responses showed dose-response characteristics (Figure 5) in all cases and, in each of the two assays used, the same patterns of cytotoxic induction were observed (Figure 6). The recognition of the peptides was MHC-restricted since virus-specific cytotoxic T cells from C3H/He mice (H-2^k) lysed L929 (H-2^k) target cells after incubation with peptides or viruses, but did not lyse P815 (H-2^d) targets under the same conditions.

Table 4. Recognition of X31 or peptide by T cells after priming with the individual peptides[a].

	Balb/c		C57/BL6		C3H/He		SJL	
Peptide	Recog. of Prot.	Recog. of X31	Recog. of Pept.	Recog. of X31	Recog. of Pept.	Recog. of X31	Recog. of Pept.	Recog. of X31
1	-	-	-	-	++	+	-	-
2	-	-	-	-	+++	+	-	-
3	-	-	-	-	-	-	-	-
4	-	-	-	-	-	-	-	-
5	-	-	++	+	-	+	-	-
6	++++	++	+	+	++++	+	-	+
7	++++	-	+++	-	++	-	+	-
8	++++	-	-	-	-	-	-	-
9	+++	-	+++	-	+++	-	++	-
10	-	-	+	-	-	-	-	-
11	+++	-	+++	-	-	-	-	-
12	-	-	-	-	-	-	-	-

[a]Symbols denote the following cpm values: (+) 3,000-10,000; (++) 10,000-25,000; (+++) 25,000-50,000; (++++) 50,000.

Table is from Atassi and Kurisaki (1986).

SITES OF DELAYED TYPE HYPERSENSITIVITY

To measure the delayed type hypersensitivity (DTH) response, C3H/He mice were primed with live AX31 virus and challenged in the footpad with the individual synthetic peptides. Swelling after 48 hours was recorded and the results are shown in Figure 7. Some of the peptides elicited very good DTH responses in virus-primed mice. Specifically, peptides 3, 5, 9 and 12 were responsible for net swelling responses higher than 50% in a 48 hour period. Clearly, therefore, several peptides representing surface areas of HA are able to activate DTH effecors primed by the intact virus. In this case, most of the areas which are positive for DTH coincide with those responsible for CTL responses in the C3H/He mice. Peptides 3, 9, and 12 are both effective in stimulating CTL and DTH responses originally directed to the intact virus. Peptide 6 (183-199) is effective in T-cell proliferation as well as the DTH response. Since both assays are implicated in the detection of helper T-cell activation, this area should be a strong helper T-cell epitope in HA.

THE SYNTHETIC PEPTIDES AS IMMUNOGENS

Antisera to the peptides raised in outbred mice, were found to bind to intact virus (Table 3). In most cases, these antibodies did not bind disrupted virus. The ThL responses were studied in four mouse strains (Atassi and Kurisaki, 1984, 1986). In the four mouse strains (Balb/c, C57/BL6, C3H/He and SJL), several of the peptides (as already mentioned) stimulated _in vitro_ proliferative responses of T helper cells from virus-primed mice. Conversely, priming with the peptides (Atassi and Kurisaki, 1986), individually (Table 4) or in equimolar mixture (Figure 8), produced in most cases T cells that recognized the peptide _in vitro_. The peptide recognition profile after peptide immunization was unrelated to that obtained after virus immunization. Only a very limited number of these peptide induced responses which recognized the intact virus as well (Table 4).

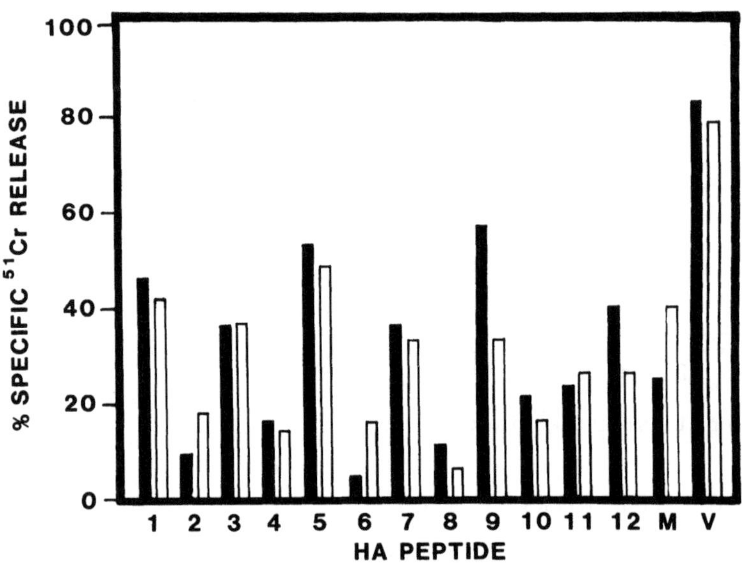

Fig. 6. Complete profile of AX-31 primed CTL recognition of 12 regions at the surface of the hemagglutinin molecule. Closed bars represent the percent of specific lysis by virus-primed CTL of syngeneic target cells exposed to free peptides. C3H/He mice were immunized with live AX-31 virus and their spleen cells used as effectors while L929 cells were used as targets. Open bars represent the specific lysis of virus-infected target cells by virus-primed CTL exposed _in vitro_ to free peptides. C3H/He mice were immunized with virus as before, spleen cells collected and incubated _in vitro_ with peptide. Cytotoxic T cells were harvested and used as effectors in the cytotoxicity assay. Target cells were virus-infected, ^{51}Cr-labeled, L929, as described in Section 3. AX-31 boosted CTL were used as positive control. The AX-31 virus was used as infective alloantoic fluid. Amino acid sequences of the areas studied in the HA molecule and represented above are: peptide 1 (residues HA1 23-36); 2 (residues HA1 124-134); 3 (residues HA1 138-152); 4 (residues HA1 154-167); 5 (residues HA1 175-188); 6 (residues HA1 183-199); 7 (residues HA1 201-218); 8 (residues HA1 272-288); 9 (residues HA1 300-315); 10 (residues HA2 1-11); 11 (residues HA2 56-68), 12 (residues HA2 68-84); M is an equimolar mixture of peptides 1 to 12; V represents AX-31 virus. Figure is from Torres et al. (1988)

CONCLUSIONS

These studies have shown that: (a) Antibody, ThL, CTL and DTH recognition of submolecular regions of the virus hemagglutinin is important in immune responses to the infecting virus; (b) The antibody, ThL, CTL and DTH responses to the individual sites of HA are under separate Ir gene control and (c) T-cells which recognize the virus can be generated only in a few cases by priming with peptide in appropriate mouse strains.

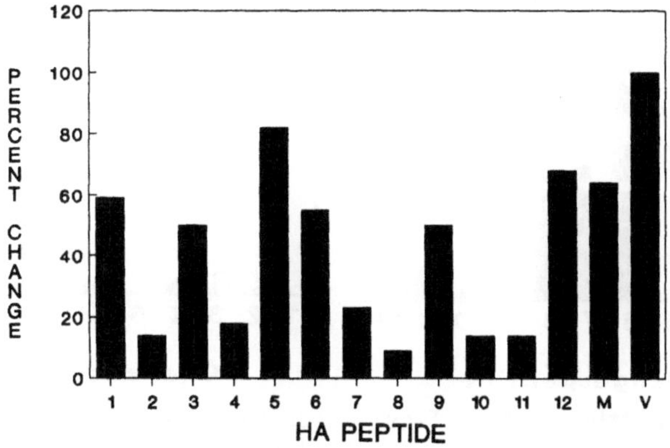

Fig. 7. Delayed type hypersensitivity response of virus-primed C3HμHe mice to hemagglutinin peptides. Mice were immunized i.p. at day 0 and 21 with 100/μl of 1024 HAU infectious alloantoid fluid. On day 28 mice were challenged with 25 μg of appropriate free peptide, equimolar peptide mixture (M) or virus (v) in the right footpad and 25 μg of unrelated peptide in the left footpad as control. Footpad swelling was measured 48 hours after injection with a pressure caliper. Results are reported as percent change based on the response to a live virus challenge.

It is clear that peptides can induce both peptide-specific humoral and cell mediated immune responses. These immune responses could be potentially useful only if they cross-reacted with the virus. Focusing soley on peptides which produce antibodies that will bind to virus overlooks the role of ThL, CTL and DTH responses in defense against the virus. Furthermore, other levels of complexity stem from the finding that the responses to the individual recognition sites in each one of these immune compartments is under separate genetic control. This makes it extremely difficult to design a peptide vaccine which is effecitve in protection and is universal irrespective of the determinant restriction by the MHC of the host.

Classical methods of vaccination and, more recently, subunit and recombinant vaccines are clearly still the methods of choice. However, many of the previous attempts to develop effective peptide vaccines were made on naive assumptions. Knowledge has continued to accumulate, and with it, new ideas and approaches have developed (see above). Moreover, vaccines based on peptide designs have unique advantages which include the ease and relatively low cost of mass production, lack of side reactions and absence of infection by the injected vaccine.

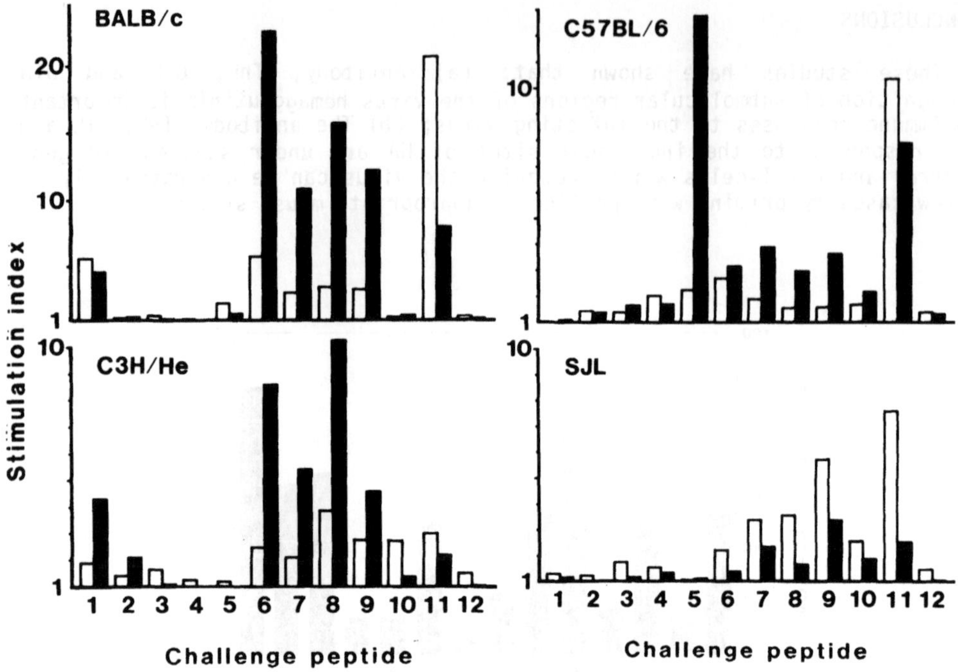

Fig. 8. Comparison of the peptide T-cell recognition profile after immunization of four mouse strains with X31 virus (open bars) or with 100/μg mouse) equimolar mixture (solid bars) of the synthetic peptides shown in Figure 1. Figure is from Atassi and Kurisaki (1986).

ACKNOWLEDGEMENTS

This work was supported by a grant (Q994) from the Welch Foundation and the award to M.Z. Atassi of the Robert A. Welch Chair of Chemistry. The work was also partially supported by an NIAID contract (AI G2517). J.V. Torres is supported by a graduate fellowship from the National Aeronautics and Space Administration (NASA).

REFERENCES

Anderer, F.A. (1963) Preparation and properties of an artificial antigen immunologically related to tobacco mosaic virus. Biochim. Biophys. Acta 71:246-248.

Anderer, F.A. and Schlumberger, H.D. (1965) Properties of different artificial antigens immunologically related to tobacco mosaic virus. Biochim. Biophys. Acta 97:503-509.

Andrew, M.E., Braciale, V.L., Henkel, T.J., Hatch, J.A. and Braciale, T.J. (1985) Stable expression of cytolytic activity by influenza virus-specific cytotoxic T lymphocytes. J. Immunol. 135:3520-3525.

Arnon, R., Ed) (1987) "Synthetic Vaccines", I, pp. 1-155 and Volume II, pp. 1-161, CRC Press, Inc., Boca Raton, Fl.

Atassi, M.Z. (1975) Antigenic structure of myoglobin. The complete immunochemical anatomy of a protein and conclusions relating to antigenic structure of proteins. Immunochem. 12:423-438.

Atassi, M.Z. (1978) Precise determination of the entire antigenic structure of lysozyme: Molecular features of protein antigenic structures and potential of 'surface-stimulation' synthesis: A powerful new concept for protein binding sites. Immunochem. 15:909-936.

Atassi, M.Z. (1981) Investigation by synthesis whether antigenic sites of polymeric cytochrome coincide with locations of sequence differences between the immunizing and host cytochromes. Mol. Immunol. 18:1021-1025.

Atassi, M.Z. (1982) Immune recognition of serum albumin. 15. Localization by synthetic of antigenic site 4 of bovine serum albumin to the region around the disulfide bond 166-175. Biochim. Biophys. Acta 704:552-555.

Atassi, M.Z. (1984) Antigenic structures of proteins. Their determination has revealed important aspects of immune recognition and generated strategies for synthetic mimicking of protein binding sites. Europ. J. Biochem. 145:1-20.

Atassi, M.Z. (1985) in: "High Technology Route to Virus Vaccines", G. Dreesman and R. Kennedy, eds., pp. 17-29, Amer. Soc. Microbiol., Washington, D.C.

Atassi, M.Z. and Bixler, G.S. (1988) in: "Vaccines: New Concepts and Developments" (H. Kohler and P.T. LoVerde, eds., pp. 58-70, Longman Scientific and Technical, Harlow, England.

Atassi, M.Z. and Kazim, A.L. (1978) First Consequences of the determination of the entire antigenic structure of sperm-whale myoglobin. Adv. Exptl. Med. Biol. 98:19-40.

Atassi, M.Z. and Kurisaki, J.I. (1984) A novel approach for localization of the continuous protein antigenic sites by comprehensive synthetic surface scanning: Antibody and T-cell activity to several influenza hemagglutinin synthetic sites. Immunol. Com. 13:539-551.

Atassi, M.Z. and Kurisaki, J.I. (1986) Synthetic localization of T- and B-cell recognition sites of influenza virus hemagglutinin and the antibody of the T-cell responses to the sites. Protides Biol. Fluids 34:157-160.

Atassi, M.Z. and Webster, R.G. (1983) Localization, synthesis and activity of an antigenic site on influenza virus hemagglutinin. Proc. Natl. Acad. Sci. USA 80:840-844.

Atassi, M.Z., Sakata, S. and Kazim, A.L. (1979) Localization and verification by synthesis of five antigenic sites of bovine serum albumin. Biochem. J. 179:327-331.

Benjamini, E. (1977) in: "Immunochemistry of Proteins", M.Z. Atassi, ed., pp. 265-310, Plenum Press, New York.

Benjamini, E., Michaeli, D. and Young, J.D. (1972) Antigenic determinants of proteins of defined sequences. Curr. Top. Microbiol. Immunol. 58:85-134.

Bixler, G.S. and Atassi, M.Z. (1983) Molecular localization of the full profile of the continuous regions recognized by myoglobin primed T-cells using synthetic overlapping peptides encompassing the entire molecule. Immunol. Commun. 12:593-603.

Bixler, G.S. and Atassi, M.Z. (1984) T-cell recognition of myoglobin. Localization of the sites stimulating T-cell proliferative responses by synthetic overlapping peptides encompassing the entire molecule. J. Immunogenet. 11:339-353.

Bixler, G.S., Yoshida, T. and Atassi, M.Z., (1984a) T-cell recognition of lysozyme. I. Localization of regions stimulating T-cell proliferative responses by synthetic overlapping peptides encompassing the entire molecule. Expl. Clin. Immunogen 1:99-111.

Bixler, G.S., Yoshida, T. and Atassi, M.Z. (1984b) Localization and genetic control of the continuous T-cell recognition sites of lysozyme by synthetic overlapping peptides representing the full profile of the protein chain. J. Immunogenet. 11:327-338.

Bosch, F., Orlich, M., Kilenk, H.D. and Rott, R. (1979) The structure of the hemagglutinin, a determinant for the pathogenicity of influenza viruses. Virol. 95:197-201.
Braciale, T.J.J. (1979) Specificity of cytotoxic T cells directed to influenza virus hemagglutinin. J. Exp. Med. 149:856-868.
Caton, A.J., Yewdell, G.G., Brownlee, J.W. and Gerhard, W. (1982) The antigenic structure of the influenza virus hemagglutinin (H1 subtype). Cell 31:417-422.
Fearny, F.J., Leung, C.Y., Young, J.D. and Benjamini, E. (1971) The specificity of antibodies to a peptide determinant of the tobacco mosaic virus protein induced by immunization with the peptide conjugate. Biochim. Biophys. Acta 243:509-514.
Kazim, A.L. and Atassi, M.Z. (1980) A novel and comprehensive synthetic approach for the elucidation of protein antigenic structures: Determination of the full antigenic profile of the alpha chain of human hemoglobin. Biochem. J. 191:261-264.
Kazim, A.L. and Atassi, M.Z. (1982) Structurally inherent antigenic sites. Localization of the antigenic sites of the alpha chain of human hemoglobin in three host species by comprehensive synthetic approach. Biochem. J. 203:201-208.
Klenk, H.D. and Garten, W., Keil, W., Niemann, H., Bosch, F.X., Schwarz, R.T., Scholtissek, C. and Rott, R. (1981) Processing of the Hemagglutinin in: "Genetic Variation Among Influenza Viruses" D.P. Nayak, ed., pp. 193, Academic Press, San Francisco, CA.
Kohler, H. and Loverde, P.T. (1988) "Vaccines, New Concepts and Developments" John Wiley and Sons, New York, N.Y.
Krco, C.J., Abramson, E.J., Atassi, M.Z. and David, C.S. (1984) Genetic control of the immune response to haemoglobin. V. Production and characterization of αβ restricted T cell lines and clones. J. Immunogenet. 11:33-43.
Krco, C.J., Kazim, A.L., Atassi, M.Z. and David, C.S. (1981a) Genetic control of the immune response to hemoglobin. I. Separate genetic control of the α- and β-subunits of human hemoglobin and the influence of the H-2D end by in vitro lymphocyte proliferative response. J. Immunogenet. 8:315-322.
Krco, C.J., Kazim, A.L., Atassi, M.Z. and David, C.S. (1981b) Genetic control of the immune response to haemoglobin. III. Variants but not A_c (D2.GD) Ia polypeptides alter T-cell immune responsiveness towards the α-subunit of human haemoglobin. J. Immunogenet. 8:471-476.
Lazarowitz, S.G. and Choppin, P.W. (1975) Enhancement of the infectivity of influenza A and B viruses by proteolytic cleavage of the hemagglutinin polypeptide. Virol. 68:440-445.
Morrison, L.A., Lukacher, A.E., Braciale, V.L., Fan, D.P. and Braciale, T.J. (1986) Differences in antigen presentation to MHC class I- and class II-restricted influenza virus-specific cytotoxic T lymphocyte clones. J. Exp. Med. 163:903-911.
Okuda, K., Twining, S.S., David, C.S. and Atassi, M.Z. (1979) Genetic control of the immune response to sperm-whale myoglobin in mice. II. T-lymphocyte proliferative response to the synthetic antigenic sites. J. Immunol. 123:182-188.
Potter, C.W. and Oxford, J.S. (1979) Determinants of immunity to influenza infection in man. Br. Med. Bull. 35:69-73.
Rott, R. (1979) Molecular basis of infectivity and pathogenicity of myxovirus. Arch. Virol. 59:285-290.
Skehel, J.J. (1972) Polypeptide synthesis in influenza virus-infected cells. Virol. 49:23-28.
Skehel, J.J. and Schild, G.C. (1971) The polypeptide composition of influenza A viruses. Virol. 44:396-401.
Torres, J.V., Wyde, P.R. and Atassi, M.Z. (1988) Cytotoxic T lymphocytes recognition sites on influenza virus hemagglutinin. Immunol. Letters 19:49-54.

Townsend, A.R.M., McMichael, A.J., Carter, N.P., Huddleston, J.A. and Brownlee, G.G. (1984) Cytotoxic T cell recognition of the influenza nucleoprotein and hemagglutinin expressed in transfected mouse L cells. Cell 39:13-17.
Wabuke-Buenoti, M.A.N., Taku, A., Fan, D.P., Kent, S. and Webster, R.G. (1984) Cytolytic T lymphocytes and antibody responses to synthetic peptides of influenza virus hemagglutinin. J. Immunol. 133:2194-2201.
Ward, C.W. (1981) Structure of influenza virus hemagglutinin. Curr. Top. Immunol. Microbiol. 94:1-9.
Webster, R.G., Laver, W.G., Mair, G. and Schild, G.C. (1982) Molecular mechanisms of variation in influenza viruses. Nature 296:115-118.
White, J., Kartenbeck, J.S. and Helenius, A. (1982) Membrane Fusion Activity of Influenza Virus. EMBO J. 1:217-222.
Wiley, D.C., Skehel, J.J. and Waterfield, M.D. (1977) Evidence from studies with a cross-linking reagent that the haemagglutinin of influenza virus is a trimer. Virol. 79:446-451.
Wiley, D.C., Wilson, I.A. and Skehel, J.J. (1981) Structural identification of the antibody-binding sites of Hong Kong influenza hemagglutinin and their involvement in antigenic variation. Nature 289:373-378.
Yewdell, J.W., Webster, R.G. and Gerhard, W.U. (1979) Antigenic variation in three distinct determinants of an influenza type A haemagglutinin molecule. Nature 279:246-250.
Yoshioka, N. and Atassi, M.Z. (1986) T-cell recognition of human haemoglobin: Localization of the full T-cell recognition profile of the beta chain by a comprehensive synthetic strategy. Biochem. J. 234:457-461.
Young, C.R. and Atassi, M.Z. (1982) Dissection of the molecular parameters for T cell recognition of a myoglobin antigenic site. Adv. Exp. Med. 150:79-93.

HAEMOPHILUS b OLIGOSACCHARIDE-CRM_{197} AND OTHER HAEMOPHILUS b CONJUGATE VACCINES: A STATUS REPORT

David H. Smith*, Dace Viceps Madore*, Ronald J. Eby*, Porter W. Anderson+, Richard A. Insel+, and Cynthia L. Johnson*

*Praxis Biologics, Inc., Rochester, New York
+University of Rochester Medical Center, Rochester, New York

INTRODUCTION

Haemophilus influenzae type b (Haemophilus b) is a major cause of life-threatening and debilitating diseases in infants and young children. In the United States, it is the most common cause of bacterial meningitis, and also causes other invasive diseases, such as epiglottitis, septic arthritis, cellulitis, pneumonia and bacteremia. In 1985, United States Public Health officials estimated that, despite the availability of potentially effective antibiotics and medical care, up to 1 in every 200 children in the USA develops an invasive Haemophilus b disease before the age of five years (Cochi et al., 1985). The 1985 report by the U.S. National Academy of Science on establishing vaccine priorities in the United States (NRC, 1985) estimated the annual vaccine preventable costs of Haemophilus b disease to be approximately 425 million dollars, and ranked Haemophilus b vaccine among the highest priorities for needed new vaccines.

The capsular polysaccharide is the principal virulence determinant of Haemophilus b and a target of antibody (Ab)-mediated immunity. Considerable evidence also indicates that antibody to the capsular polysaccharide is the principal protective host factor.

At birth, maternal IgG confers protection; however, as the level of maternal antibodies in the infant declines, the risk and incidence of Haemophilus b disease rise. As children approach two years of age, antibodies to the capsular polysaccharide begin to appear, stimulated by environmental exposure to Haemophilus b saccharide or cross-reacting antigens (Fothergill and Wright, 1933). In the late 1960's, our group, then at the Children's Hospital in Boston, and Robbin's laboratory at the NIH began an effort to raise the antibody level by active immunization with a vaccine composed of the purified capsular polysaccharide (HbPs) (Schneerson et al. 1971; Smith et al., 1973). The antigen is a linear polymer of ribose, ribitol and phosphate (Crisel et al., 1975), and has been called "PRP".

When injected into mature humans, HbPs triggers a long-lived but non-boostable serum Ab elevation consisting of variable combinations of IgM, IgG and IgA. In children, the antibody response to PRP vaccine develops slowly with age: there is generally little detectable response in infants younger than 18 months of age, after which the rate and level of response increases rapidly through age three years, and then more slowly through age six years. In the youngest responding children, IgM tends to be the predominant class; the participation of IgA and IgG increases with age. While HbPs was effective in protecting healthy children of 18-24 months and older against invasive Haemophilus b disease, it did not protect younger children who have the highest incidence of Haemophilus b disease (Peltola et al., 1977). Furthermore, certain children have a transient immune deficiency to this and certain other bacterial polysaccharides, with or without an associated deficiency in immune globulin isotype or subclass (Granoff and Cates, 1985).

Alternative approaches to increase the immunogenicity of HbPs have been evaluated without success, e.g., a wide range of doses, sequential injections, or aluminum adjuvants (Anderson et al., 1977). Our group found that an outer membrane complex of Haemophilus b, which contained PRP, was immunogenic and gave a boostable response in animals and young infants non-responsive to HbPs (Anderson and Smith, 1977). However, the limited magnitude of the response and the technical difficulties in reproducible production precluded commercial interest. Investigators at the Lederle Laboratories later reported that the immunogenicity of PRP could be increased somewhat and converted to a memory-type (boostable) response by mixing with components of Bordetella pertussis vaccine (King et al., 1981). Initial optimism about this approach, however, has not been borne out.

The Hib capsular polysaccharide vaccine is considered to be a T-independent (T_I) antigen because a high proportion of IgM antibodies are elicited and true secondary immune responses fail to occur. HbPs has little or no immunogenicity in all tested animal systems. Although young infants can respond to protein-like or T-dependent vaccines, they fail to respond to most T_I bacterial polysaccharides, including HbPs, for reasons that are not yet clear. The concept of modifying the polysaccharide to a hapten covalently linked to a protein carrier was suggested more than 50 years ago (Goebel and Avery, 1931) and has now been refined by several research groups.

CONJUGATE VACCINES

Conjugate vaccines differ in their composition with respect to the protein carrier, the size of the hapten saccharide, the type of linkage between the hapten and carrier, and the ratio of saccharide to protein. A comparison of these factors in currently developed Haemophilus b conjugate vaccines is summarized in Table 1.

The first published approach was described by the Robbins laboratory. They coupled PRP to proteins via a 6-carbon adipic spacer by heavily derivatizing the protein and reacting it with CNBr-activated PRP (Schneerson et al., 1980). Connaught Laboratories has adopted this chemistry, using diphtheria toxoid as carrier, yielding a conjugate vaccine called PRP-D (Lepow et al., 1985) (This vaccine is now licensed in the U.S. and Canada as ProHIBiT[R] for use in children of 18 months and older.) A second approach currently favored by the Robbins groups derivatizes PRP and couples it to non-derivatized proteins; they are now focusing on tetanus toxoid as the protein component, creating a conjugate referred to as Hib-TT (Chu et al., 1983). Such a vaccine is being developed commercially by the Merieux Institute. Merck Sharp, and Dohme Laboratories devised a so-called "bigeneric" spacer, whereby saccharide is linked to N-acetylbutylcarbamate, the protein carrier, a complex of a Neisseria meningococcal type b outer membrane protein is linked to N-acetylhomocysteine, and the two are coupled by a thioether linkage to make a conjugate designated Hb-NOMP (Marburg et al., 1986).

Table 1. Properties of Haemophilus b Conjugate Vaccines

	Robbins	Merieux	Connaught	Merck	Anderson	Praxis
Protein	DT and TT	TT	DT	NmOMPs	DT & CRM_{197}	CRM_{197}
Saccharide	Poly	Poly	Poly	Poly & Oligo	Oligo	Oligo
Linker Group	Adipic Dihydrazide	Adipic Dihydrazide	Adipic Dihydrazide	Bigeneric	None	None

Table 2. The Effect of Molecular Size of the Dextran Haptens on Anti-Dextran Responses in Mice to Dextran-Protein Conjugates*

Dextran Molecular Weight	Anti-Dextran Titers** Week 2	Week 6
5-40,000,000	16,000	8,900
10,000	17,000	20,000
1,000	73,000	1,150,000

*Mäkelä, O. et al. (1984), Scand. J. Immunol. 29: 541-550.
**Vaccination at week 0, 4 with 30 μg CHO. Anti-dextran titers expressed as end-point titers by ELISA assay.

The approach in Rochester has been to use oligosaccharide (OS) haptens covalently linked directly to the protein, preferably diphtheria-CRM_{197}. This approach built on earlier observations of Mäkelä et al. (1984) who reported that oligosaccharide-based dextran BSA conjugates are more immunogenic than polysaccharide-based conjugates in mouse models, and the shorter the saccharide, the greater the activity both in initiating a primary response as well as boosting the response (Table 2). Anderson developed a vaccine composed of oligosaccharides of PRP linked covalently (without linkers) to the non-toxic variant of diphtheria toxin, CRM_{197}, developed in Pappenheimer's laboratories (Uchida, 1972) and found it immunogenic in animals and young human infants, producing both primary and secondary antibody responses (Anderson, 1983). These studies provided the basis for the vaccine developed in our laboratories at Praxis.

HAEMOPHILUS b OLIGOSACCHARIDE CRM_{197} CONJUGATE

Composition and Structure

The Praxis vaccine, Haemophilus b Oligosaccharide CRM_{197} Conjugate, abbreviated HbOC (now trademarked HibTITER and licensed in December 1988 for use in children of 18 months of age and older in the U.S.), differs from the other Haemophilus b conjugate vaccine models. Firstly, the carrier protein CRM_{197} is a native protein prepared in a highly purified form. We believe the protein component of a conjugate vaccine is very important in the immunogenicity of the product and is not just a passive "carrier". To guarantee reproducible conjugation of the hapten, and thus a consistent product, as well as the optimal availability of requisite T_H epitopes, we have used CRM_{197} because it can be used without chemical modification.

Most conjugate vaccines are constructed by linking the saccharide hapten to the protein component via amino groups on the lysine residues of the protein. Formaldehyde toxoiding of proteins alters these lysine groups and in an unpredictable manner. Thus, a conjugate vaccine based on a toxoided carrier protein can have undesirable lot-to-lot variations in the critical protein component. Furthermore, toxoiding reduces the number of available conjugation sites, as well as alters the immunogenic nature of the protein. The use of a native protein avoids these problems and permits reproducible access to the lysine groups for covalent coupling. CRM_{197} was selected because of the extensive experience with vaccines containing the diphtheria toxin protein and the considerable definition of the immunochemistry of CRM_{197}. Furthermore, as reported in this volume, Bixler and Pillai have shown in *in vitro* studies using human lymphocytes, that CRM_{197} is superior to diphtheria toxoid in stimulating proliferation.

Table 3. Relative Activity in Binding Polyclonal Human Anti-HbPs Antibody*

Antigen	Binding Activity
HbPs Vaccine	100
Hb Oligosaccharide	1.6
HibTITER^{TM}	7.8
ProHIBiT^{R}	66

*Unpublished results, G. Bixler and S. Pillai

Secondly, our product uses oligosaccharides of defined length rather than polysaccharides. The use of oligosaccharides provides a chemically identifiable constant composition to the vaccine. The shorter saccharide may be also less likely to mask critical T-cell epitopes on the carrier protein, and thus enhance a more protein-like response. In addition, an oligosaccharide-based vaccine binds less anti-polysaccharide antibody in vitro, (Table 3), and we predicted it would have a reduced potential for reducing circulating antibody following vaccination, as has been reported with the HbPs vaccine (Granoff and Cates, 1985).

Thirdly, we directly link the protein and saccharide units by reductive amination. This avoids the use of potentially toxic or immunogenic linker groups and results in the extremely stable amine bond. This linkage creates a novel amino acid, N-hydroxyethyllysine, which can be analyzed to prove and quantitate the covalent bond between the Hib oligosaccharide and CRM_{197} (Table 4) (Seid et al., unpublished observations). The stability of this conjugate has been documented by studies which have demonstrated no significant changes in physiochemical properties or immunogenicity in mice following five cycles at -25°C, then 25°C (each for 24 hours), or five cycles at 50°C, then 25°C (each for 24 hours). Indeed, the vaccine has been autoclaved (20 minutes at 121°C) without loss of mouse immunogenicity. The vaccine is stable in preservatives, such as thimerosal, in which it has retained all properties, including expected immunogenicity for human infants following two years of storage at 2-8°C.

Thus, the HbOC vaccine has a well defined, reproducible, and stable structure.

Table 4. Chemical Analysis of Covalent Linkage Between CRM_{197} and HbO*

$$R\text{-}O\text{-}CH_2\text{-}CH_2\text{-}HN\text{-}(CRM)$$

$$\downarrow \; HCl$$

$$R\text{-}OH + HO\text{-}CH_2\text{-}CH_2\text{-}NH\text{-}(CH_2)_4\text{-}\underset{NH_2}{CH}\text{-}COOH + \text{AMINO ACIDS}$$

(HYDROXYETHYL LYSINE)

R = Haemophilus b Oligosaccharides
CRM = CRM_{197} Protein

*R. Seid et al., unpublished observations

Background and Methods

The primary goals for the Haemophilus b conjugate vaccines are to (1) generate a protective titer of antibody that is long-lived and functional in infants and children of all ages, including those deficient by age or phenotype in response to HbPs, and (2) prime the immune system for "booster" responsiveness, whether induced naturally or by further vaccination. As noted above, it would be very desirable if the vaccine did not decrease pre-existing antibody during the first few days following vaccination.

Earlier studies of the pathogenesis of Haemophilus b and other encapsulated bacterial diseases have taught us that the activity of circulating antibody at the time of infection plays the primary role in preventing invasive disease. This concept was supported and extended by studies conducted at Praxis on the kinetics of post-vaccination antibody activity. Eleven adults and six toddler-aged children who, because of age and pre-existing antibody titers, could be considered "immunologically primed" were vaccinated with HbOC and sera was collected at 1, 3, 7, 14, 21 and 28 days post-vaccination. None of the subjects had antibody detectable increases in antibody (RABA) on day 1 or 3; 15 of 17 had increased titers on day 7 and thereafter, but two adults had no significant antibody increase until 14 days post-vaccination. These data indicate to us that the kinetics of a booster response to this polysaccharide antigen are slower than may have been anticipated, and that the antibody generated following infection cannot be assumed to play a primary role in preventing invasive disease.

What is the protective level of anti-polysaccharide antibody, and the goal for the Haemophilus b conjugate vaccines? Considerable evidence indicates that a titer of $\geq$ 1 µg/ml is protective. This titer correlated best with protection from bacteremia of essentially all infant rats given IgG enriched with anti-PRP antibody and then colonized intranasally with Haemophilus b (Ambrosino, 1983). Secondly, Käyhty et al. found a titer of $\geq$ 1 µg/ml correlated with protection in the large-scale efficacy study of HbPs vaccine in Finland (Käyhty et al., 1983). Finally, although titers $<$ 1 µg/ml have been found prior to or at hospitalization of children with invasive Haemophilus b disease, no child has been reported to date who was hospitalized with an antibody titer $\geq$ 1 µg/ml (Eskola et al., 1987; Santosham et al., 1987).

Immunogenicity of candidate vaccines is assessed by the rise in geometric mean of total anti-PRP antibody activity, as well as antibody isotype and functional activity. As the geometric mean of the antibody response can mask the heterogeneity of response, it is equally important to evaluate the percentage of subjects responding to each vaccine. It can be anticipated that the greater the number of individuals that attain a protective level, the greater the vaccine efficacy.

The classical radioantigen binding assay (RABA) described by Farr (1958) has been adapted for quantitation of antibodies to PRP, and is widely used in a method recently standardized by the FDA.

Safety

The Haemophilus b conjugate vaccines are the first of a novel class of glycoproteins being administered to humans for the first time. Importantly, they are to be given to healthy infants, and a recommendation by professional organizations for universal use in all of the 3.8 million infants born annually in the U.S. can be anticipated. Therefore, studies on the safety of the HbOC vaccine have been extensive and thorough.

To date, approximately 90,000 doses have been administered to about

40,000 children of 1 through 24 months of age. Many of these doses were given at the same time as, but at a different site than, DTP-IPV or DTP vaccines. No unusual or unexpected associated side effects have been observed. A comprehensive analysis of all side effects associated with 1197 doses of HbOC administered without other pediatric vaccines has revealed that local symptoms, e.g. redness, warmth or swelling $\geq$ 2 cm at the injection site, or a temperature > 38.3°C during the first or second post-vaccination day occurred in less than 2% of subjects and that the incidence was similar for dose one, two or three in a vaccine series.

We conclude that HbOC is extremely well tolerated by infants of all ages.

Immunogenicity: Comparison with Haemophilus b Polysaccharide (HbPs) Vaccine

One of the goals of the conjugate vaccine is superiority over the HbPs vaccine. The immunogenicity of HbOC and HbPs vaccines were compared in a study conducted in a single center. Because the HbPs vaccine was licensed for use at $\geq$ 24 months at the time of the study, HbOC had to be administered to younger children. Results of the study involving 42 children of 24 $\pm$ 2 months of age vaccinated with HbPs and 78 children of 18 $\pm$ 2 months vaccinated with HbOC are presented in Table 5.

One month post-vaccination, the HbPs group had: G.M. of total antibody, 2.07 µg/ml; IgG, 2.43 µg/ml; 61.9% with titers $\geq$ 1 µg/ml; and 47.6% with positive bactericidal (BC) activity. The HbOC group had: G.M., 14.8 µg/ml; IgG 11.51 µg/ml; 98.7% with titers $\geq$ 1 µg/ml and 92.2% with positive BC activity. All comparisons were highly statistically significant ($p = 0.00001$) in favor of the HbOC vaccine.

These data not only demonstrate the superiority of HbOC compared to the unconjugated HbPs in this age group, but also document that in essentially all subjects, HbOC generates a very high titer of predominantly IgG and IgG_1 and that the antibody has functional activity (bactericidal activity shown, opsonic activity not shown).

Table 5. Comparison of Anti-HbPs Antibody Responses to a Single Dose of HbOC⁺ in Children of 18 $\pm$ 2 Months of Age or b-CAPSA^R I* in Children 24 $\pm$ 2 Months of Age

	Patients			G.M. Anti-HbPs Antibody Activity**				
Vax	Age (mos.)	No.	Serum	Total	IgG	IgM	IgG_1	IgG_2
b-CAPSA^R I	24$\pm$2	42	Pre	0.13	0.14	0.20	0.15	0.16
			1 mo. Post	2.07	2.43	0.31	1.47	0.45
HbOC	18$\pm$2	78	Pre	0.13	0.15	0.19	0.23	0.18
			1 mo. Post	14.80	11.51	3.69	11.51	0.53

⁺Dose = 10 µg
*Dose = 25 µg
**Total ab in µg/ml assayed by RABA using CBER standard at 70 µg/ml; isotype and subclass Ab assayed by ELISA; values used in calculations of G.M.: lower limit $\leq$ 0.1 = 0.1; upper limit $\geq$ 40 = 40

Note: Data presented includes only subjects with pre-vaccination titer < 0.6 µg/ml

Immunogenicity at Different Ages

The peak incidence of invasive Haemophilus b disease in the U.S. occurs between 6 and 12 months of age. To be effective, therefore, vaccination must generate protective titers by this age. HbPs has essentially no activity in children younger than 18 months, particularly those younger than 12 months. Because routine childhood immunizations are administered at 2, 4, 6, 15, and 18 or 24 months of age, we have analyzed the immunogenicity of the conjugate HbOC given on a three dose schedule with vaccine given at two month intervals to infants first vaccinated younger than 6 months of age and as a one dose regimen for those vaccinated between 15-23 months. A two dose schedule separated by two months was used for children first vaccinated at 7-12 months of age.

Because children are routinely scheduled to receive MMR vaccination at 15 months and DTP at 18-24 months, we analyzed the immune response of children vaccinated with HbOC at 15-17 months and compared it to that of children vaccinated at 18-23 months of age. As evaluated by G.M. of total antibody, IgG, IgM, IgG_1 and IgG_2 titers and percent titers $\geq$ 1 µg/ml, the results are essentially identical: comparing 15-17 vs. 18-23 month specimens respectively, the G.M. of total antibody was 12.55 vs. 14.41; the IgG, 9.88 vs. 8.86; the IgG_1, 10.68 vs. 8.38, and % $\geq$ 1 µg, 98.4 vs. 99. We conclude that the immunogenicity of HbOC is the same when it is administered as a single dose at any time from 15 through 23 months of age.

A single 10 µg (CHO) dose of HbOC was immunogenic in children of all ages. Following one dose of HbOC, there was a direct relationship between the titer of the immune response and the percent of children with titers greater than 1 µg/ml with the age of the subject, up to approximately 12-15 months of age, when the relationship plateaued.

Prior to vaccination, 6% of infants first vaccinated at 2-6 months of age have anti-HbPs greater than 1 µg/ml; following the first, second, and third vaccination, 35%, 94%, and 99% had levels greater than 1 µg/ml. These rises in anti-HbPs antibodies were directly associated with rises in bactericidal activity. Greater than 95% of the sera had bactericidal activity when children of all ages had received two doses of HbOC. As with the older children, HbOC elicits primarily IgG and IgG_1 anti-HbPs antibodies in infants (Table 6).

Immunogenicity: Duration and Boostability of Antibody Response

Table 7 summarizes data from studies directed at the questions of the relation of age to comparative immunogenicity of a complete vaccination schedule and the duration of anti-HbPs generated by that schedule. Three doses of HbOC were administered to 432 subjects of 1-6 months of age; two

Table 6. Isotype of Anti-HbPs Responses of Infants 1-6 Months

Mos. After 1° Vax	Vax Schedule	G.M.* (No. of Subjects) IgG	IgM	IgG_1	IgG_2
0	+	0.18 (70)	0.10 (68)	0.13 (30)	0.20 (30)
2	+	0.66 (70)	0.11 (68)	0.36 (30)	0.17 (30)
4	+	16.00 (65)	0.23 (64)	13.07 (29)	1.80 (29)
5	-	29.27 (24)	0.47 (23)	26.60 (19)	3.93 (19)

*Anti-HbPs in µg/ml assayed by ELISA using OBRR at 70 µg/ml (IgG = 60.9; IgM = 3.5; IgA = 5.6) Analysis range used 0.10-40.0 µg/ml

Table 7. Anti-HbPs Antibody Activity Following Complete HibTITER Vaccination Schedule

Age at First Vax (Mos.)	Total Vax Doses	G.M. (% Subjects ≥ 1 μg/ml) Months Following Last Vaccination 1	4	6	10	15-27
1-8	3	22.3(99.2)	-	6.13(97.0)	-	3.58(87.5)
7-12	2	21.4(97.5)	14.7(99.2)	-	4.22(92.5)	-
15-23	1	13.4(98.5)	-	8.78(99.0)	-	4.12(83.0)

doses to 109 subjects of 7-12 months of age, and one dose to 208 subjects of 15-23 months of age. The results indicate: (1) high geometric mean titers and conversion to ≥ 1 μg/ml were attained at all ages, (2) the multidose regimen in young infants was even more immunogenic than the one dose regimen in toddler-aged children, (3) the antibody activity induced by the vaccine was long-lived with the duration being independent of the subject's age or vaccine schedule, and (4) titers of ≥ 1 μg/ml were observed in ≥ 90% of subjects for 6 to 10 months post-vaccination and in ≥ 80% for 15-27 months. We conclude that the HbOC vaccine meets the goal of generating a long-lived, functional antibody response to PRP.

Although HbOC can stimulate anti-diphtheria toxin antibody activity, use of HbOC and DTP at the same time results in the same total anti-toxin activity as that generated by alone.

Another goal for the conjugate vaccine is the priming of the immune system for a booster response with or without further immunization. Anderson et al. (1987) reported that a prototype of HbOC would prime infants for a true secondary response by subsequent vaccination with HbPs vaccine; we have confirmed those studies. Analysis of the titers of individual subjects has also revealed multiple examples of a spontaneous, "natural" antibody boost many months following vaccination. Selected examples are presented in Table 8.

The data are line listings of antibody titers of individual subjects who received three doses of HbOC given at two months intervals starting at 2-4 months of age, and overall antibody titers of healthy, unvaccinated subjects in our extensive series (which are, of course, titers in specimens taken before vaccination at an older age).

Specimens were available for the unvaccinated subjects at approximately 9-14 months of age (3-6 months post third vaccination) and at 16-24 months of age (12-20 months post last vaccination). The overall geometric means of unvaccinated subjects were 0.13 and 0.22 at those respective ages, and the highest individual titers observed were 3.84 and 12.63, respectively. Although the antibody titers in the vaccinated subjects 3-6 months post-vaccination were lower than their immediate post-vaccination titers, they rose significantly to 23.7-95.5 μg/ml 9 to 12 months later without further vaccination. These titers dramatically exceeded those seen in unvaccinated subjects. Similar observations have been made for children vaccinated with other dosage schedules, including those of 15-23 months who received one dose of HbOC.

We conclude that HbOC primes the immune system for a booster response whether generated by natural exposure or another vaccination.

Table 8. Kinetics of Anti-HbPs Antibody of Individual Infants Vaccinated with 3 Doses of HbOC

Subject ID #	Age at 1° Vax (mos.)	Pre	Anti-HbPs Antibody Post-Vaccination* 1 mo. Post 3° Vax	Post 3rd Vax (Mos.)	
E-10-4-25	2	.34	≥ 40.00	12.05 (3)	95.49 (12)
E-10-4-5	2	2.24	32.61	8.12 (4)	37.04 (12)
E-12-3-37	3	.10	≥ 40.00	11.50 (4)	73.96 (20)
E-12-3-310	3	1.06	28.75	2.66 (4)	24.88 (16)
E-12-3-309	3	.10	≥ 40.00	6.52 (6)	34.75 (16)
E-12-3-65	3	.14	35.71	9.10 (5)	23.70 (16)
E-12-15-9	4	.26	40.00	14.81 (6)	31.46 (14)
E-12-3-345	3	.71	40.00	10.69 (5)	34.92 (17)

Anti-HbPs Antibody in Non-vaccinated Infants

Age Range	9-14 mos.	18-24 mos.
Pre-Vax G.M.	0.13	0.22
Highest Titer Observed	3.84	12.53

*Ab - μg/ml

Immunogenicity: Children with Deficient Anti-PRP Antibody Response

Experiences from extensive clinical use of HbPs revealed that certain children had deficient antibody responses to the vaccine. These included children who had recovered from invasive Haemophilus b disease, those who had transient deficiences in antibody isotype and/or subclass, or those with sickle cell disease (SCD). All children with these conditions vaccinated with HbOC, including 40 with previous invasive Haemophilus b disease (Edwards et al., 1989) and 27 with transient antibody deficiencies (Insel and Anderson, 1986), have responded with a significant antibody response, generally that expected of healthy infants. Table 9 summarizes data from infants with SCD first vaccinated at 1-6 months of age (mean 4 months).

Compatability with Other Childhood Vaccines

In order for a Haemophilus b conjugate vaccine to be introduced into the national childhood vaccination program, it must be compatible with other vaccines. Studies have documented that HbOC given at the same time but at a different site had no effect on the response to the antibody response of diphtheria, tetanus, pertussis, measles, mumps, or rubella vaccines; likewise, those vaccines had no effect on the anti-PRP response generated by HbOC.

Table 9. Anti-HbPs Antibody Following Vaccination of SCD Infants*

Mos. Post 1° Vax	Vax Schedule	G.M.**	% ≥ 1.0 μg/ml	IgG_1/IgG_2
0	+	0.07	0	ND
2	+	0.36	25	0.3
4	+	9.15	90	8.75
6	-	36.10	100	8.98
12	-	4.02	94	ND
18	-	6.92	90	ND

*Gigliotti et al., J. Pediatr., in press
**G.M. μg/ml

COMPARISONS OF THE SEVERAL HAEMOPHILUS b CONJUGATE VACCINES

Composition and General Properties

Table 1 summarizes the significant attributes of the Haemophilus b conjugate vaccines currently under development. Although the total dose of saccharide of the vaccines is in the same general range, the total protein content differs significantly, from 25 μg in HbOC to 150-300 μg in Hb-NOMP. The dose of protein in the latter vaccine would be, by several fold, the largest amount of protein administered in a "purified vaccine" to young infants. Why such a large protein content is needed is not immediately obvious, nor have published reports on this product answered concerns regarding the potential of this large amount of protein to provoke an immunosuppressive effect on repeated doses.

PRP-D and PRP-T utilize toxoids as carriers and thus have the potential for problems outlined above, for use of a chemically treated protein. Hb-NOMP appears to contain a mixture of outer membrane proteins rather than a single protein and a prototype preparation contained a large amount of contaminating lipopolysaccharide (Einhorn et al., 1986). Furthermore, Poolman et al. (1988) have suggested that certain of the Neisseria outer membrane proteins elicit "blocking antibodies" that can prevent bacteriolysis and phagocytosis mediated by antibodies to other bacterial surface polymers. One of the class 4 NOMPs appears to be analogous to a highly conserved protein of Neisseria gonorrhoeae, PIII, which has been shown to elicit antibodies that block the bactericidal activity present in convalescent serum (Blake et al., 1988; Rice et al., 1986). Whether Hb-NOMP elicits such antibodies is unknown but needs to be resolved.

The stability of the proposed conjugate vaccines varies considerably, a factor affecting use and, potentially, immunogenicity. HbOC is quite stable; PRP-D has a more limited shelf life because the saccharide hapten can break off. Since similar linker chemistry is used in PRP-T, the long term stability of that product must be studied carefully. Hb-NOMP is apparently so labile, that the vaccine must be reconstituted at the bedside and used immediately. This instability probably relates to the protein component(s) since the linker groups should be stable. HB-NOMP is the only one of the vaccines that uses an adjuvant; it is adsorbed onto alum (either to enhance immunogenicity or reduce toxicity) (Einhorn et al., 1986; Ahonkhai et al., 1988).

Linker groups are used to couple the hapten to the carrier in three (PRP-D, PRP-T, Hb-NOMP) of the four vaccine models. Presumably these linkers are nontoxic; however, data is not available to answer the important question of whether the linkers are immunogenic. As these products are novel glycoproteins, and it is not desirable to generate antibodies to potential neoantigens, a vaccine without linkers would be preferred. HbOC is the only vaccine that covalently links the Hib capsular polysaccharide directly to the protein carrier.

We believe that HbOC is the only conjugate vaccine that does not significantly reduce pre-existing anti-PRP activity following vaccination and has been proven to have no associated antigenuria. These properties, important to clinicians, most likely result from the structural configuration of the product.

Comparative Immunogenicity

Because of the differences in the chemical composition and structure of the vaccines, it is not surprising that these products will also be found to have differences in immunogenicity. Side-by-side comparisons where vaccines have been administered in the same clinical site and where sera are evaluated in the same laboratory are emerging. It can already be stated definitely, however, that conjugate vaccines are not generic products.

All conjugate vaccines appear to be more immunogenic than HbPs vaccines at all ages (Anderson et al., 1985; Berkowitz et al., 1987; Claesson et al., 1988; Lepow et al., 1985). In adults, the various conjugate vaccines appear to generate similar post-vaccination antibody titers. However, the isotype and subclass of the antibodies generated can differ. Thus, the antibody profile generated by the PRP-D vaccine is quite similar to that generated by HbPs, while HbOC generates a very predominant IgG and IgG_1 response. Käyhty et al. (1988a) hypothesized from these results that the post-vaccination antibody profile of adults may have value in predicting vaccine potency for young infants.

Differences in the immunogenicity of vaccines appear when single doses are given to toddler-aged children. Thus, Käyhty et al. (1988a) reported that at one month post-vaccination, the 10 μg dose of HbOC generated a G.M. of 30.6 μg/ml, while the 25 μg dose of PRP-D generated a G.M. of 11.9 μg/ml. Edwards et al. (personal communication) observed that infants of 12-25 months (mean 17.2 months) vaccinated with HbOC responded with a G.M. seven-fold greater than that of children of 18-24 months vaccinated with PRP-D; 94% of those receiving HbOC attained a titer $\geq$ 1 μg/ml, compared to 65% for those receiving PRP-D. Likewise, the duration that vaccinated subjects continue to maintain a titer $\geq$ 1 μg/ml is significantly longer for those who received HbOC than PRP-D or Hb-NOMP.

However, it is in very young infants that differences in potency of the vaccines become most evident, as predicted by the structure function studies of one conjugate vaccine (Anderson et al., 1986) - and at an age when it is most important medically.

Figure 1 summarizes the data available in the literature on the relative immunogenicity of these vaccines following vaccination of 2 month old infants. The immune response to a single dose of the respective vaccines is shown in the left panel. One dose of HbOC elicits a four-fold rise in anti-HbPs antibodies, while one dose of Hb-NOMP shows a three-fold rise, and PRP-D does not elicit a significant response. Data is not yet available to evaluate the response of very young infants to the tetanus based conjugate.

Response to a second dose of the respective vaccines 1-2 months after the first vaccination (shown in second panel) shows HbOC to be the most effective, boosting the response to the first dose, with a ten-fold greater response than that elicited by Hb-NOMP. Again, PRP-D does not elicit a significant response. When a third boost is given two months following the previous vaccination, or when children are 6-7 months of age, HbOC continues to boost with a further two-fold rise in titer. PRP-D, finally after three doses, elicits a significant antibody response. It appears that the third dose in a sequence with the prototype Hb-NOMP vaccine does not elicit any further antibody rises above that generated by two doses (Einhorn et al., 1986). These results are not those expected for a T_D type vaccine and have led some to raise the question whether the demonstrated activity of this product is due to a polyclonal B cell activation rather than as a true T_D antigen (Anderson and Insel, 1988).

In a parallel comparison of a two dose schedule of HbOC and PRP-D given to infants at 4 and 6 months of age in Finland, Käyhty et al. (1988a, personal communication) reported that at 7 months the G.M. for HbOC was 4.32 μg/ml, while that for PRP-D was 0.63 μg/ml. The authors concluded that the "Praxis vaccine was the most immunogenic".

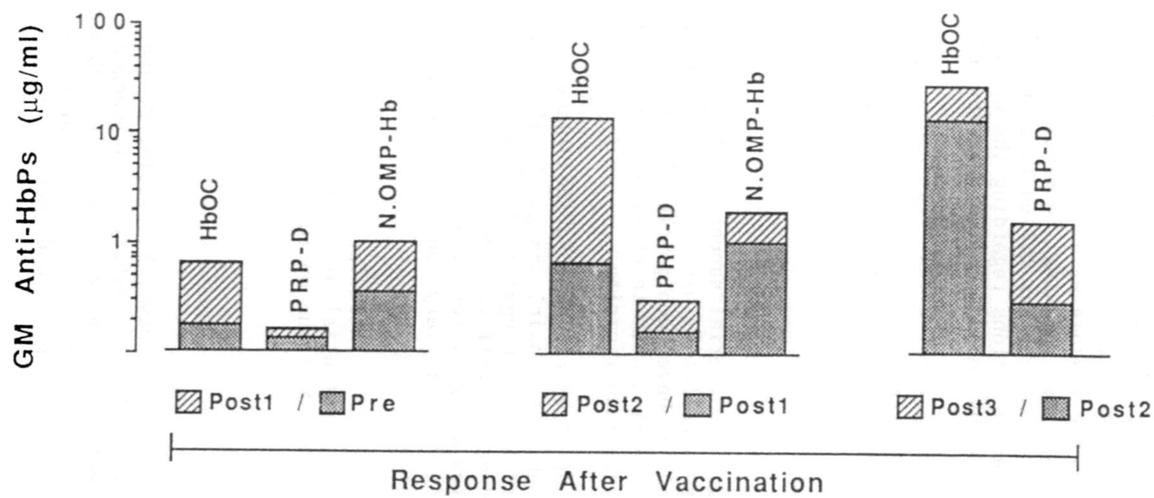

Figure 1. Comparison of Anti-HbPs Response in 2-3 Month Old Infants to Different Conjugate Vaccines.

Einhorn et al. (1986) The Lancet Aug 9:299-302.
Eskola et al. (1985) The Lancet May 25:1184-86.

While PRP-D is not highly immunogenic in younger infants, with only 34-60% of them having antibody levels greater than 1 μg/ml after three vaccinations (Eskola et al., 1985, 1987), ongoing clinical trials in Finland indicate that this vaccine has an efficacy of at least 87% in protecting from disease after three doses and is more efficacious after four doses (Eskola et al., 1987). Thus, it appears that this vaccine is efficacious in Finnish children, in whom the incidence and age-related susceptibility to Haemophilus b disease differs significantly from that in the United States (Granoff and Cates, 1985). Statistics on young infants following one or two doses are beginning to emerge (Eskola et al., 1988); efficacy in this younger age group may not be as promising. A similar trial with three doses of PRP-D given to infants in Alaska has failed to show efficacy (Ward et al., 1988), raising serious questions regarding the use of studies in Finland to project Haemophilus vaccine efficacy in the U.S.

It would seem prudent, therefore, that regulatory agency decisions regarding licensure of any Haemophilus b conjugate vaccine for use in infants in the U.S. should be made on the basis of efficacy trials conducted in U.S. infants.

CONCLUSIONS

The covalent linkage of bacterial saccharide to a protein can lead to a vaccine with greater immunogenicity than the purified polysaccharide. A number of candidate vaccines have been developed and are being evaluated clinically. These vary considerably in chemical composition, structure and, from preliminary results, immunogenicity.

A vaccine composed of sized oligosaccharides linked to a highly purified non-toxic variant of diphtheria toxin, CRM_{197}, by direct reductive amination has the properties of an ideal conjugate: defined structure, well tolerated, does not induce a significant decrease in pre-existing antibodies, and generates a high titered, primarily IgG antibody response which has functional activity, is very long-lived and boostable with or without further vaccination.

Continuing clinical studies, including efficacy studies in young infants, will determine the potential of these agents in preventing a major health problem for infants. Cautious optimism for the future is justified.

REFERENCES

Ahonkhai, V.I., Lukacs, L.J., Matthews, H., Vella, P. and Gerety, R.J. (1988), Continuing evaluation of safety and immunogenicity of *Haemophilus* b conjugate - *N. meningitidis* outer membrane protein vaccine (Hb-NOMP) (Abstract). Programs and abstracts of the Interscience Conference on Antimicrobial Agents and Chemotherapy, 292, Abstract #1027.

Ambrosino, D., Schreiber, J.R., Daum, R.S. and Siber, G.R. (1983), Efficacy of human hyperimmune globulin in prevention of *Haemophilus influenzae* type b disease in infant rats, Infect. Immun. 39, 709-714.

Anderson, P. (1983), Antibody responses to *Haemophilus influenzae* type b and diphtheria toxin induced by conjugates of oligosaccharides of the type b capsule with the nontoxic protein CRM_{197}, Infect. Immun. 39, 233-238.

Anderson, P. and Insel, R.A. (1988), Prospects for overcoming maturational and genetic barriers to the capsular polysaccharide of Haemophilus influenzae type b, Vaccine, 6, 188-191.

Anderson, P., Pichichero, M., Edwards, K., Porch, C.R. and Insel, R.A. (1987), Priming and induction of Haemophilus influenzae type b capsular antibodies in early infancy by Dpo20, an oligosaccharide-protein conjugate vaccine, J. Pediatr. 111, 644-650.

Anderson, P., Pichichero, M.E., Insel, R.A., Betts, R., Eby, R. and Smith, D.H. (1986), Vaccines consisting of periodate-cleaved oligosaccharides from the capsule of Haemophilus influenzae type b coupled to a protein carrier: structural and temporal requirements for priming in the human infant, J. Immunol. 137, 1181-1186.

Anderson, P., Pichichero, M.E. and Insel, R.A. (1985), Immunogens consisting of oligosaccharides from the capsule of Haemophilus influenzae type b, coupled with diphtheria toxoid or the toxin protein CRM_{197}, J. Clin. Invest. 76, 52-59.

Anderson, P., Smith, D.H., Ingram, D.L., Wilkins, J., Wehrle, P.F. and Howie, V.M. (1977), Antibody to polyribophosphate of Haemophilus influenzae type b in infants and children: Effect of immunization of polyribophosphate, J. Infect. Dis. 136, S557-S562.

Anderson, P. and Smith, D.H. (1977), Immunogenicity in weanling rabbits of a polyribophosphate complex from Haemophilus influenzae type b, J. Infect. Dis. 136, S563-S570.

Berkowitz, C.D., Ward, J.I., Meier, K., Hendley, J.O., Brunell, P.A., Barkin, R.A., Zahradnik, J.M., Samuelson, J. and Gordon, L. (1987), Safety and immunogenicity of Haemophilus influenzae type b polysaccharide and polysaccharide diphtheria toxoid conjugate vaccines in children 15 to 24 months of age, J. Pediatr. 110, 509-514.

Blake, M.S., Lytton, E.J., Seiff, M.E. and Gotschlich, E.C. (1988), Studies on gonococcal Protein III, In: Bacterial-Host Cell Interaction (Alan R. Liss, Inc.) 85-97.

Chu, C., Schneerson, R., Robbins, J.B. and Rastogi, S.C. (1983), Further studies on the immunogenicity of Haemophilus influenzae type b and pneumococcal type 6A polysaccharide protein conjugates, Infect. Immun. 40, 245-256.

Claesson, B.A., Trollfors, B., Lagergard, T., Taranger, J., Bryla, D., Otterman, G., Cramton, T., Yang, Y., Reimer, C.B., Robbins, J.B. and Schneerson, R. (1988), Clinical and immunologic responses to the capsular polysaccharide of Haemophilus influenzae type b alone or conjugated to tetanus toxoid in 18 to 23 month old children, J. Pediatr. 112, 695-702.

Cochi, S.L., Broome, C.V. and Hightower, A.W. (1985), Immunization of U.S. children with Haemophilus influenzae type b polysaccharide vaccine, JAMA 253, 521-529.

Crisel, R.M., Baker, R.S. and Dorman, D.E. (1975), Capsular polymer of Haemophilus influenzae type b, J. Biol. Chem. 250, 4926.

Edwards, K.M., Decker, M.D., Porch, C.R., Palmer, P. and Bradley, R. (1989), Immunization after invasive Haemophilus influenzae type b disease. Serologic response to a conjugate vaccine, Am. J. Dis. Child. 143, 31-33.

Einhorn, M.S., Weinberg, G.A., Anderson, E.L., Granoff, P.D. and Granoff, D.M. (1986), Immunogenicity in infants of Haemophilus influenzae type b polysaccharide in a conjugate vaccine with Neisseria meningitidis outer-membrane protein, Lancet 2, 299-302.

Eskola, J., Käyhty, H., Peltola, H., Karanko, V., Mäkelä, P.H., Samuelson, J. and Gordon, L.K. (1985), Antibody levels achieved in infants by course of Haemophilus influenzae type b polysaccharide/diphtheria toxoid conjugate vaccine, Lancet 1, 1184-1186.

Eskola, J., Peltola, H., Takala, A.K., Käyhty, H., Hakulinen, M., Karanko, V., Kela, E., Rekola, P., Rönnberg, P., Samuelson, J.S., Gordon, L.K. and Mäkelä, P.H. (1987), Efficacy of Haemophilus influenzae type b polysaccharide-diphtheria toxoid conjugate vaccine in infancy, N. Engl. J. Med. 317, 717-722.

Eskola, J., Takala, A., Peltola, H. and Mäkelä, P.H. (1988), Clinical efficacy of the Haemophilus influenzae type b (Hib) capsular polysaccharide-diphtheria toxoid conjugate vaccine (PRP-D) in Finland, (Abstract), Program and Abstracts of the Interscience Conference on Antimicrobial Agents and Chemotherapy, 309, Abstract #1126.

Farr, R.S. (1958), A quantitative immunochemical measure of the primary interaction between I* BSA and antibody, J. Infect. Dis. 103, 239-262.

Fothergill, L.D. and Wright, J. (1933), Influenzal meningitis: The relation of age incidence to bactericidal power of blood against causal organism, J. Immunol. 24, 273-284.

Gigliotti, F., Feldman, S., Wang, W.C., Day, S.W., Brunson, G. Immunization of young infants with sickle cell disease with a Haemophilus influenzae type b saccharide-diphtheria CRM_{197} protein conjugate vaccine, Journal of Pediatrics, (in press).

Goebel, W.F. and Avery, O.T. (1931), Chemo-immunological studies on conjugated carbohydrate-proteins, V. The immunological specificity of an antigen prepared by combining one capsular polysaccharide of type 3 pneumococcus with foreign protein, J. Exp. Med. 54, 437-447.

Granoff, D.M. and Cates, K.L. (1985), Haemophilus influenzae type b polysaccharide vaccines, J. Pediatr. 107, 330-336.

Insel, R.A. and Anderson, P.W. (1986), Response to oligosaccharide-protein conjugate vaccine against Haemophilus influenzae b in two patients with IgG_2 deficiency unresponsive to capsular polysaccharide vaccine, N. Engl. J. Med. 315, 499-503.

Käyhty, H., Mäkelä, O., Eskola, J., Peltola, H. and Saarinen, L. (1988a), Quantitative and qualitative characterization of antibody response in infants, children and adults to three Haemophilus influenzae type b (Hib) conjugate vaccines. (Abstract). In: Program and abstracts of the Interscience Conference on Antimicrobial Agents and Chemotherapy, 292, Abstract #1026.

Käyhty, H., Mäkelä, O. and Eskola, J. (1988b), Isotype distribution and bactericidal activity of antibodies after immunization with Haemophilus influenzae type b vaccines at 18 to 24 months of age, J. Infect. Dis. (in press).

Käyhty, H., Peltola, H., Karanko, V. and Mäkelä, P.H. (1983), The protective level of serum antibodies to the capsular polysaccharide of Haemophilus influenzae type b, J. Infect. Dis. 147, 1100.

King, S.D., Ramlal, A., Wynter, H., Moodie, K., Castle, D., Kuo, J.S.C., Barnes, L. and Williams, C.L. (1981), Safety and immunogenicity of a new Haemophilus influenzae type b vaccine in infants under one year of age, Lancet 2, 705-709.

Lepow, M.L., Samuelson, J.S. and Gordon, L.K. (1985), Safety and immunogenicity of Haemophilus influenzae type b polysaccharide diphtheria toxoid conjugate vaccine in infants 9 to 15 months of age, J. Pediatr. 106, 185-189.

Mäkelä, O., Peterfy, F., Outschoorn, I.G., Ritcher, A.W. and Seppälä, I. (1984), Immunogenic properties of α-(1-6)-dextran, its protein conjugates and conjugates of its breakdown products in mice, Scand. J. Immunol. 19, 541-550.

Marburg, S., Jorn, D., Tolman, R.L., Arison, B., McCauley, J., Kniskern, P.J., Hagopian, A. and Vella, P.P. (1986), Bimolecular chemistry of macromolecules - synthesis of bacterial polysaccharide conjugates with Neisseria meningitidis membrane protein, J. Am. Chem. Soc. 108, 5282-5287.

National Research Council, New Vaccine Development Establishing Priorities, I Diseases of Importance in the United States, National Academy Press, Washington, D.C., (1985).

Peltola, H., Käyhty, H., Sivonen, A. and Mäkelä, H. (1977), Haemophilus influenzae type b capsular polysaccharide vaccine in children 3 months to 5 years of age in Finland, Pediatrics 60, 730-737.

Poolman, J.T., Timmermans, H.A.M., Hopman, C.T.P., Teerlink, T., VanVaught, P.A.M., Witvliet, M.H. and Beuvery, E.C. (1988), Comparison of meningococcal outer membrane protein vaccines solubilized with detergent or C polysaccharide, In: Gonococci and Meningococci, Poolman, J.T., Zanen, H.C., Meyer, T.F., Heckles, J.E., Mäkelä, P.R.H., Smith, H. and Beuvery, E.C., eds. Kluwer Academic Publishers, Norwell, MA, 159-165.

Rice, P.A., Vayo, H.E., Tam, M.R. and Blake, M.S. (1986), Immunoglobulin G antibodies directed against protein III block killing of serum resistant Neisseria gonorrhoeae by immune sera, J. Exp. Med. 164, 1735-1748.

Santosham, M., Reid, R., Ambrosino, D.M., Wolff, M.C., Hill, J.A., Priehs, CC., Aspery, K.M., Garrett, S., Croll, L., Foster, S., Burge, G., Page, P., Zacher, B., Moxon, R., Chir, B. and Siber, G.R. (1987), Prevention of Haemophilus influenzae type b infections in high-risk infants treated with bacterial polysaccharide immune globulin, N. Engl. J. Med. 317, 923-929.

Schneerson, R., Barrera, O., Sutton, A. and Robbins, J.B. (1980), Preparation, characterization, and immunogenicity of Haemophilus influenzae type b polysaccharide-protein conjugates, J. Exp. Med. 152, 361-376.

Schneerson, R. Rodrigues, L.P., Parke, J.C. Jr., and Robbins, J.B. (1971), Immunity to disease caused by Haemophilus influenzae type b II, Specificity and some biological characteristics of "natural" infection-acquired and immunization-induced antibodies to the capsular polysaccharide of Haemophilus influenzae type b, J. Immunol. 107, 1081-1089.

Smith, D., Peter, G., Ingram, D.L., Harding, A.L. and Anderson, P. (1973), Responses of children immunized with the capsular polysaccharide of Haemophilus influenzae type b, Pediatrics 52, 637-645.

Uchida, T., Pappenheimer, A.M. Jr. and Harper, A.A. (1972), Reconstitution of diphtheria toxin from two nontoxic cross-reacting mutant proteins, Science 175, 901-903.

Ward, J.I., Brenneman, G., Letson, G., Heyward, W. and Alaska Vaccine Efficacy Trial Study Group. (1988a), Limited protective efficacy of an H. influenzae type b conjugate vaccine (PRP-D) in native Alaskan infants immunized at 2, 4 and 6 months of age, (Abstract), Program and abstract of the Interscience Conference on Antimicrobial Agents and Chemotherapy, 309, Abstract #1127.

THE APPLICATION OF MOLECULAR BIOLOGY TO THE DEVELOPMENT OF NOVEL VACCINES

P.J. Kniskern[1], A. Hagopian[1], P. Burke[1], N. Dunn[1], D.L. Montgomery[1], L.D. Schultz[1], C.A. Schulman[1], C.E. Carty[1], R.Z. Maigetter[1], D.E. Wampler[2], E.D. Lehman[1], S. Yamazaki[2], D.J. Kubek[2], E.A. Emini[1], W.J. Miller[1], W.M. Hurni[1], and R.W. Ellis[1]

Departments of Virus and Cell Biology[1] and
Biochemical Process Research and Development[2]
Merck Sharp & Dohme Research Laboratories
West Point, Pennsylvania 19486 (USA)

INTRODUCTION

Before the dawn of molecular biology and recombinant DNA technology, vaccines were essentially natural products. Such vaccines comprise live attenuated disease agents, killed organisms, or biochemically isolated subcomponents or subfractions of the disease causing agent. The fact that these classical vaccines are natural products ensured that, if sufficient care were taken (during attenuation, inactivation, or subunit isolation), the important protective epitopes represented on the disease agent could be preserved on the vaccine candidate in proper immunogenic form.

The development of such vaccines was limited, however, to diseases for which a source of the etiologic agent either existed naturally in large quantities or to diseases caused by agents which could be propagated easily in laboratory culture. In addition, the use of inactivated whole organism vaccines or subunit vaccines, isolated biochemically from the intact agents, often introduces the potential for undesired adverse reactions in vaccinees which are associated with non-vaccine molecules co-administered with the desired immunogens.

The application of recombinant DNA technology to vaccine development theoretically has removed the limitation imposed by the lack of commercially exploitable natural sources of potential vaccines. In addition, the perceived safety profile has been improved greatly for recombinant-derived vaccines due to the fact that such vaccines can be defined at the genetic level to contain only the chosen protective immunogen(s). However, recombinant vaccines have the potential disadvantage that they comprise gene products synthesized in foreign host cells. As such, they do not possess *a priori* the immunologic conformation of the native epitopes displayed on the disease agent. Therefore, care must be taken in the design of recombinant expression systems and purification techniques for such vaccines; in addition, the development of suitable methodologies for demonstration of immunological equivalence becomes paramount.

Because the human vaccines for the prevention of diseases caused by hepatitis B virus (HBV) have evolved from a successful first-generation "natural product" vaccine (isolated from the plasma of chronic carriers) to an equally efficacious second generation recombinant vaccine which is expressed in bakers' yeast (*Saccharomyces cerevisiae*), these vaccines establish a paradigm for such a development effort. In addition, work in our and other laboratories now is being extended to the evaluation of potential third-generation recombinant vaccines for the prevention of HBV disease which may confer even greater breadth or longer duration of protection than that afforded by the existing highly effective vaccines, even though no such vaccine has been identified to date.

This monograph will address the development and evaluation of recombinant vaccines. Relevant points will be highlighted by tracking the ongoing research effort which has been directed at establishing and evaluating the successive generations of hepatitis B vaccine.

HEPATITIS B VIRUS

HBV is a member of the hepadnaviridae group and is the major etiologic agent for a variety of infectious human liver diseases. Other family members are the related, but species-specific, viruses which cause hepatic diseases in woodchucks (WHV), ground squirrels (GSHV), Pekin ducks (DHBV), tree squirrels, herons and snakes (Robinson & Marion, 1988). HBV-associated diseases are endemic in all parts of the world, with primary modes of transmission being via bodily fluids or through perinatal infection from chronically infected mothers to their newborns. Many individuals who are infected by HBV suffer through an acute disease which is followed by recovery. However, a percentage of individuals (including most perinatally infected newborns) fail to clear their infection and become chronic carriers of the virus. The number of chronic carriers worldwide is estimated at over two hundred fifty million; hundreds of thousands from this cohort die annually from the long-term consequences of hepatitis B, *viz.*, cirrhosis and hepatocellular carcinoma (Szmuness, 1978; Szmuness et *al.*, 1978; Beasley & Hwang, 1984).

The envelope of HBV is composed of three related proteins which are the translational products of a large open reading frame (ORF). This ORF is demarcated into three domains by in-frame ATG codons, each of which is capable of functioning as a translational initiation site *in vivo* (see Figure 1). These domains are referred to in their 5' to 3' order as PreS1 (108 amino acids), PreS2 (55 amino acids) and S (226 amino acids) and the resultant translational products are referred to as PreS1+PreS2+S (large), PreS2+S (middle) and S (major or HBsAg) (Tiollais et al., 1985). In addition to virions (Dane particles), the plasma of chronic carriers contains noninfectious 20 nm particles, each of which is composed of *ca.* 100 individual envelope polypeptides (primarily S with minor amounts of PreS2+S and traces of PreS1+PreS2+S) (Heermann et al., 1984; Tiollais et al., 1985).

PLASMA-DERIVED HBsAg: THE FIRST-GENERATION HEPATITIS B VACCINE (HEPTAVAX-B®)

The search for a vaccine for prevention of hepatitis B initially led to the development of a vaccine comprised of the 20 nm particles (HBsAg) which were purified from the plasma of chronic carriers. This effort, which involved numerous laboratories, culminated with the commercial availability of a vaccine in 1981 (Heptavax-B®, H-B-Vax®, Merck Sharp & Dohme) which has been demonstrated to be highly effective in preventing diseases caused by HBV (Szmuness et al., 1981). In addition, this vaccine is well-tolerated and demonstrably free of risk from AIDS (Francis et al.,

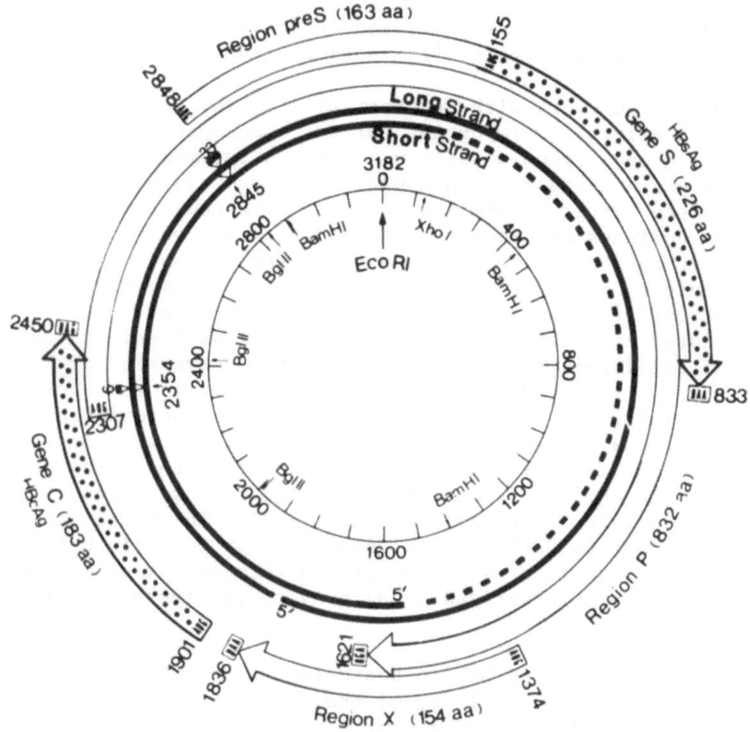

Fig. 1. The hepatitis B virus genome

1986); the process developed for isolation and purification of the vaccine particle employs three independent inactivation steps [urea treatment, pepsin digestion at pH 2.0 and formalin treatment], each of which is independently capable of total inactivation of human immunodeficiency virus [HIV] (Tabor et al., 1983; Francis et al., 1986). The development of Heptavax-B® has been reviewed extensively elsewhere (Hilleman et al., 1983), and the details will be discussed herein only in terms of biochemical and immunological comparisons to the subsequent recombinant vaccines for which it serves as the "bench-mark".

RECOMBINANT-DERIVED HBsAg (RECOMBIVAX HB®): THE SECOND-GENERATION HEPATITIS B VACCINE

The commercial production of Heptavax B® requires the availability of a large supply of plasma from human carriers. For such production to be commercially feasible, this plasma must contain high titers of the HBsAg vaccine antigen. In areas of the world where HBV disease is highly endemic, transmission generally occurs at birth from chronically infected mothers to their newborn infants who frequently progress to become chronic carriers. However, neonatally infected individuals seldom develop into carriers whose plasma contains sufficient titers of HBsAg to be suitable for vaccine isolation. In other geographic regions with lower incidence of hepatitis B disease, individuals generally are infected later in life via bodily fluids; the plasma of such individuals who become chronic carriers generally contains a sufficiently high titer of HBsAg to be an acceptable source material for vaccine isolation. In the late 1970's concern arose regarding the continuing availability of such high-titered plasma from chronic carriers. Therefore, even during the latter phases of the development of the plasma-derived vaccine, we initiated an active research program to seek an alternative starting material for vaccine production.

We knew from our work with Heptavax-B® that the conformation of the 20 nm particle was critically important for the immunogenicity and display of the protective group-common a epitope. Thus, a recombinant expression system should express HBsAg which would: (1) form visible (electron microscopy) HBsAg particles similar to those obtained from natural sources; (2) possess a epitopes as present on the plasma-derived particle, (3) induce specific antibodies to these epitopes and (4) be shown to protect against HBV infection in experimentally inoculated chimpanzees and humans in clinical trials (Hilleman et al., 1978; McAleer et al., 1984). To this end, we initiated a collaborative research effort with scientists at the University of California (San Francisco) and the University of Washington. The HBV genome was isolated, made double stranded (from the nicked and gapped form in which it resides in the virion) with the endogenous polymerase reaction (Hruska et al., 1977) and cloned into suitable bacterial plasmids (Valenzuela et al., 1979). The DNA sequence of the viral genome was determined and the ORF's, as demarcated in Figure 1, were identified (Galibert et al., 1979; Valenzuela et al., 1980). The ORF in the viral genome corresponding to HBsAg(s) was identified by comparison of the imputed amino acid sequence with a partial N-terminal amino acid sequence determined for the plasma-derived HBsAg polypeptide (Peterson et al., 1977).

Efforts to express this ORF in bacterial systems failed due to a combination of an apparent instability of the S polypeptide in E. coli and the inability of the polypeptides to assemble into 20 nm particles [an attribute previously demonstrated to be necessary for the plasma-derived vaccine to elicit a protective immune response (Szmuness et al., 1982)].

We next turned our efforts to developing and exploiting the expression systems for S. cerevisiae which, at that time, were beginning to be studied in detail. The advantages and safety of yeast systems for development of recombinant vaccines have been reviewed (Hilleman & Ellis, 1986; Ellis, 1989) and the utility and versatility of such yeast expression systems also will be highlighted below. These efforts culminated in the construction of the expression vector shown diagramatically in Figure 2A (Valenzuela et al., 1982). This vector, when transformed into a suitable yeast host strain, resulted in expression of the S polypeptide which assembled into 20 nm particles which were biochemically and physically highly similar to plasma-derived particles (Valenzuela et al., 1982; McAleer et al., 1984; Wampler et al., 1985). An isolation and purification procedure was developed (McAleer et al., 1984; Wampler et al., 1985), and a vaccine was formulated which was subjected to extensive preclinical and clinical testing.

The recombinant vaccine (Recombivax HB®, H-B-VaxII®, Merck Sharp & Dohme) was shown to have equivalent immunogenicity to the plasma-derived vaccine in mice as well as in grivit monkeys (McAleer et al., 1984). Active protection in chimpanzees (McAleer et al., 1984) also was shown to be equivalent for the two vaccines. A further important and relevant in vitro correlate predictive of protective efficacy is the ability to elicit a strong immune response to the a epitope on the S polypeptide (Bhatnagar et al., 1982; Prince et al., 1982). A monoclonal antibody (MAB) to this determinant has been prepared (Brown et al., 1984) which can neutralize the virus in vitro and prevent subsequent infection of chimpanzees (Iwarson et al., 1985). The availability of such a reagent allowed us to evaluate the sera from humans immunized with the vaccine for the presence of antibodies capable of binding to this protective epitope and displacing the specific MAB in vitro; such a result would give evidence of display of the a epitope

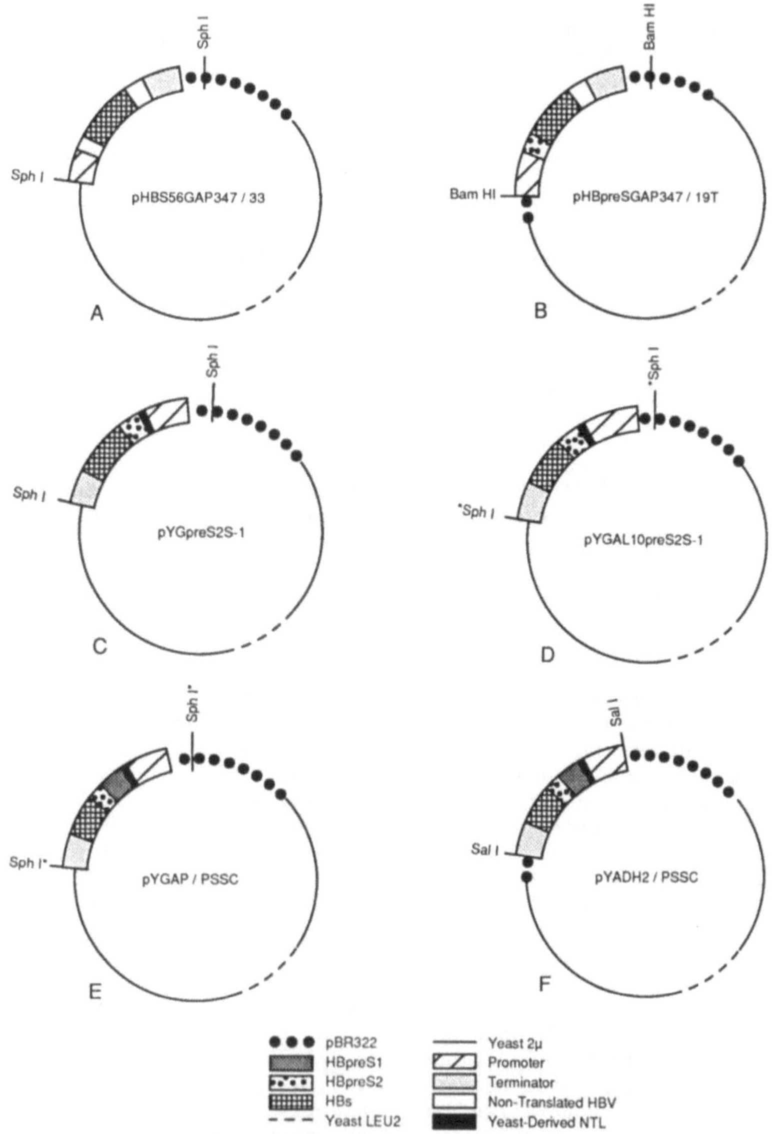

Fig. 2. Plasmid constructions for expression, in yeast, of: (A) The S protein (Recombivax HB®); (B) PreS2+S from the GAP promoter (original construction); (C) PreS2+S from the GAP promoter (improved construction); (D) PreS2+S from the GAL10 promoter; (E) PreS1+PreS2+S from the GAP promoter; (F) PreS1+PreS2+S from the ADHII promoter.

on the vaccine candidate in immunogenic form. Representative sera from recipients of either the recom- binant-derived or plasma-derived vaccine have comparable potency in elicit- ing an anti-a response (Emini et al., 1986). In addition, numerous clinical trials have demonstrated the high efficacy of both vaccines in preventing infections in high-risk groups (Szmuness et al., 1981; Francis et al., 1982; Stevens et al., 1984; Zajac et al., 1986; Stevens et al., 1987).

IMPROVEMENTS IN RECOMBIVAX HB®

Since the introduction of the vaccine, a second production master seed has been prepared by transformation of the original yeast host strain with the plasmid shown in Figure 2A. An improved method for establishment of the master seed bank has resulted in a cloned recombinant stock which contains a significantly higher number of plasmids per cell as well as a higher stability over time. This improved production seed produces higher levels of HBsAg as a percentage of yeast protein during fermentation. This increased productivity, when combined with an improved purification process, gives rise to a product with an improved purity profile ($\geq$99% as compared to the previous $\geq$96%), as well as enhanced immunogenicity (West et al., 1988). Three injections of this improved vaccine seroconverted 99% of recipients to an anti-HBs response of $\geq$10 mIU/ml and gave a geometric mean titer of ca. 1800 mIU/ml (West et al., 1988), which is ca. 2.5-fold higher than that elicited by the original master seed and production process.

EVIDENCE IMPLICATING THE PreS REGIONS IN A POTENTIAL PROTECTIVE IMMUNE RESPONSE

The plasma-derived and yeast-derived vaccines consist of particles containing the translational products of the S domain of the envelope ORF. However, in some target groups such as the elderly and renal-dialysis patients (and other immune compromised individuals), these vaccines elicit lower antibody responses (Stevens et al., 1984; Jilg et al., 1986), suggesting a diminished efficacy in these populations for vaccines containing only the S polypeptide. However, the envelope ORF encodes two additional polypeptides (PreS2+S and PreS1+PreS2+S) which, along with the S polypeptide, comprise the envelope of the 42 nm virion. Several recent observations suggest that it would be useful to test the immunogenicity in man of a PreS+S-containing vaccine: (1) the PreS domain is probably essential to the infectious cycle of HBV in that all hepadnaviruses contain PreS as well as S domains (Robinson & Marion, 1988), and PreS peptides block binding of the virion to hepatocytes *in vitro* (Neurath et al., 1986; Neurath *et al*., 1987); (2) some inbred mouse strains show poor immune response to the S domain alone, and immunization with PreS+S overcomes this block (Milich et al., 1985; Neurath et al., 1985; Milich et al., 1986); (3) anti-PreS antibodies neutralize viral infectivity *in vitro* (Neurath et al., 1986), and immunization with PreS antigens protects chimpanzees against challenge with HBV (Itoh et al., 1986).

PROTECTION OF CHIMPANZEES WITH THE PreS2 PEPTIDE

To further examine the potential benefit which might be derived from inclusion of PreS sequences in a protective immune response, we immunized chimpanzees with a synthetic peptide comprising the 55 amino acids of the PreS2 domain (serotype *adw*) (Emini et al., 1989). Such immunization resulted in significant anti-PreS2 titers in the absence of an anti-S response. Such immunized chimpanzees were completely prevented from developing all signs of HBV disease following inoculation with infectious virus of either serotype *adw* or *ayw*, thereby demonstrating a protective benefit of an anti-PreS2 response in the absence of an anti-HBs response (Emini et al., 1989).

RECOMBINANT PreS2+S: THE POTENTIAL THIRD-GENERATION HBV VACCINE

In light of the above observations regarding PreS and because of the proven utility of recombinant yeast in producing hepatitis B vaccines, we have turned our attention to the development of yeast-derived hepatitis B vaccine candidates containing not only the S domain but also all or part of the PreS domain (Kniskern et al., 1988; Ellis et al., 1988). The first of these experimental vaccines (comprising PreS2+S) will be discussed in this section, and the second (comprising PreS1+PreS2+S) will be covered later.

The yeast-derived hepatitis B vaccines successfully developed to date depend on the strong constitutively active promoter for the yeast enzyme glyceraldehyde-3-phosphate dehydrogenase (pGAP491); and for our initial work with PreS2+S, we utilized this promoter. An initial construction (Figure 2B) was employed for these studies (Valenzuela et al., 1985). Based on our previous report describing the importance of flanking sequences on the expression of hepatitis B core antigen (Kniskern et al., 1986) we assembled a second construction (Figure 2C) with pGAP which exploited these potential improvements (Ellis et al., 1988). This second construction in fact did improve expression levels by ca. 5-fold over those obtained from the construction initially employed (Ellis et al., 1988).

However, a problem arose when we attempted to subculture either of these productive clones in order to establish master seed and production working stock seed banks. As shown in Table 1, the productivity of the working stock seeds was 3-10% of that obtained for the corresponding earlier seeds. In addition, plasmid levels (as measured by retention of

Table 1. Production of HBV PreS2+S by Subcultures of GAP Clones

	Relative Antigen Productivity(c)	
Passage	Original Clone(a)	Improved Clone(b)
Premaster	1.0	5.0
Master Seed	1.0(d)	7.0
Working Stock Seed	0.1	0.2

(a) Host strain: MATa ade1 Leu2-04 cir°
Plasmid: pHB preSGAP 347/19T

(b) Host strain: Mat a/α ade1 Leu2-04 ura3 cir°
Plasmid: pYGpreS2S-1

(c) Cultures were grown and assayed for PreS2+S antigen by radioimmune assay and for protein content as described previously (Ellis et al., 1988; Carty et al., 1989; Kniskern et al., 1989).

(d) The relative antigen productivities in μg antigen/mg protein are standardized to a value of 1.0 for this transformant.

Table 2. Cultivation of the Improved pGAP Culture[a] in Fermenters

Fermenter Size (L)	Cellular Growth (A660)	Relative Antigen Productivity[c]	% Retention of Leu^+ Phenotype
16	12.0	5.0[b]	33-45
250	10.8	5.0	77

(a) Host strain: MATa ade1 Leu2-04 cir°
Plasmid: pHB preSGAP 347/19T

(b) Antigen productivities in μg antigen/mg protein are expressed relative to a culture of the original master seed which was standardized to a value of 1.0 (see Table 1).

(c) Cultures were grown and assayed for PreS2+S antigen by radioimmune assay and for protein content as described previously (Ellis et al., 1988; Carty et al., 1989; Kniskern et al., 1989).

the leu^+ phenotype during fermentation) of the master seed cultures dropped to 40-70% of the initial values by the end of fermentation (Table 2). [This observation was also confirmed by employing a plasmid copy number assay (Nasmyth & Reed, 1979).] Although the antigen productivity remained high during fermenta- tion (Table 2), we felt that our ability to control and reproducibly perform production-scale fermentations might be compromised by clones which were unstable upon subculture and which had significant losses of plasmid during growth.

We sought to overcome this problem with the development of physiologically regulated expression systems. We reasoned that because the expression described above probably was achieved in the face of toxicity of the foreign polypeptides for the yeast host cells which manifested itself in decreased growth rate and plasmid copy number loss, we could prevent such deleterious effects by employing promoters which could be activated late in the growth cycle. One such physiological promoter which offers this advantage is pGAL10, which is repressed in yeast cells which are grown in the presence of glucose and induced by the addition of galactose to the culture medium in the absence of glucose. Because the pGAL10 previously had been employed successfully for expression of the Epstein- Barr virus membrane antigen [EBMA; another protein with severe toxicity for yeast] (Schultz et al., 1987a) we sought to exploit this promoter system for expression of the PreS2+S antigen.

A plasmid was constructed (Figure 2D) which is identical to that shown in Figure 2C except for the promoter sequence which is pGAL10. This vector was transformed into a yeast host which was wild-type for the galactose regulatory genes, and a clone was established as a pre-master seed. As expected, this clone easily could be established as both a master seed and working stock seed without decrease in antigen production (Table 3). These seeds could be grown in fermenters with virtual complete retention of the leu^+ phenotype. However, the levels of PreS2+S antigen productivity obtained were only 10-20% of those obtained for the improved pGAP construction (Table 2 vs. Table 4).

The inducible pGAL10 has an additional potential advantage in that it may be up-regulated in a host strain containing a chromosomally integrated expression cassette comprising the pGAL10 fused to the GAL4 positive

Table 3. Subculturing the GAL10 Construct[a]: Production of HBV PreS2+S

Passage	Relative Antigen Productivity[b]
Premaster	1.0
Master Seed	1.0
Working Stock Seed	1.0

(a) Plasmid: pYGAL10preS2S-1
Host: MATa ade1 Leu2-04 ura3 cir°

(b) Antigen productivities in μg antigen/mg protein are expressed relative to a culture of the original master seed which was standardized to a value of 1.0 (see Table 1).

regulatory gene in addition to the normal GAL4 gene (Schultz et al., 1987b). It has been shown that the GAL4 gene produces a DNA binding protein that activates transcription from the GAL promoters (Perlman & Hopper, 1979; Hashimoto et al., 1983). However, the GAL4 protein is present in limiting amounts in non-integrant yeast cells. Upon expression of a foreign gene such as EBMA (Schultz et al., 1987a) or PreS2+S from pGAL10 on a high copy number plasmid (≥100 copies/cell), there is not enough GAL4 protein to activate transcription from all the copies of the foreign gene (Johnston et al, 1982, 1986; Laughon et al., 1984). In yeast cells carrying the integrated extra gene copy (pGAL10/GAL4), galactose induction provides enough GAL4 protein to get high-level transcriptional activation from such plasmids. When we employed this expression system for PreS2+S, there was a 3-5-fold increase in PreS2+S production over that seen for the non-integrant culture to similar levels of expression of PreS2+S as those observed for the improved pGAP culture (Table 5).

Table 4. Production of HBV PreS2+S by a Wild-Type Galactose Strain[a]

Seed Medium	Production Medium	Cellular Growth (A600)	Relative Antigen Productivity[b] uninduced	Relative Antigen Productivity[b] induced	% Retention of Leu^+ Phenotype
selective	complex	14.2	<0.01	1.0	98
complex	complex	15.5	<0.01	1.0	96

(a) Plasmid: pYGAL10preS2S-1
Host: Mata ade1 Leu2-04 ura3 cir°

(b) Antigen productivities in μg antigen/mg protein are expressed relative to a culture of the original master seed which was standardized to a value of 1.0 (see Table 1).

Table 5. Production of HBV PreS2+S in pGAL Cultures as Compared to pGAP Cultures

Culture	Relative Antigen Productivity
GAL10 (nonintegrant)	1.0
GAL10pGAL4 (integrant)	5.0
Improved GAP	5.0
Original GAP	1.0[a]

[a] The relative antigen productivities in μg antigen/mg protein are standardized to a value of 1.0 for this transformant.

The PreS2+S antigen which is expressed by either of the systems described above is comprised of p30 peptides and its gp34 glycosylated derivative (Figure 3, lane 2) and assembles into 20 nm particles which are similar to those composed of the S polypeptide and isolated from either carrier plasma or recombinant yeast. However, these particles have one important difference in that they display epitopes on their surface specific for PreS2 sequences as evidenced by their ability to be bound by a PreS2-specific antibody (Ellis et al., 1988). Such PreS2+S particles are highly immunogenic in mice and in grivit monkeys, in which they elicit not only anti-S but also anti-PreS2 immune responses (Ellis et al., 1988). Thus, desired preclinical potency clearly has been demonstrated; but the utility of this vaccine candidate for improving the efficacy of the current vaccines in humans must await the results of clinical trials which are being initiated in patient populations which give weaker responses to the S vaccines.

RECOMBINANT PreS1+PreS2+S

The inclusion of the PreS1 sequence, in addition to PreS2 and S sequences, may provide a means of overcoming nonresponsiveness to S and PreS2 epitopes in inbred mice (Milich et al., 1986). In addition, it has been suggested that the viral binding site for hepatocytes exists in the PreS1 portion of the envelope protein (Neurath et al., 1987). For these reasons, we have begun developing an additional vaccine candidate which contains the complete envelope polypeptide (PreS1+PreS2+S) and which theoretically might expand the breadth of immunogenicity and protection of an hepatitis B vaccine.

To this end, we first sought to exploit the high-level expression generally obtained in yeast with pGAP by constructing a plasmid expression vector depicted in Figure 2E (Kniskern et al., 1988). However, yeast transformed with this plasmid failed to recover from the transformation event, suggesting an even higher level of toxicity of PreS1 sequences for yeast than that previously observed for PreS2 sequences. This toxicity for yeast was verified by making a frame-shift mutation in the PreS1 domain which aborted the synthesis of the polypeptide and allowed normal yeast growth.

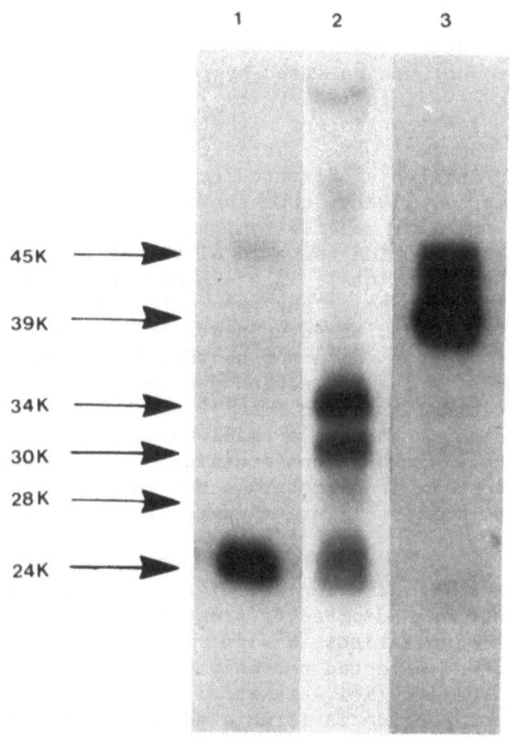

Fig. 3. SDS-PAGE analysis in 12.5% gels. Immunoblotting with rabbit anti-S (P24) antibodies. Lane 1 = purified S; Lane 2 = purified PreS2+S; Lane 3 = purified PreS1+PreS2+S. Molecular weights in kilodaltons of polypeptides reactive in immunoblots are shown, as calculated relative to the following marker proteins: phosphorylase b (93 kD); bovine serum albumin (68 kD); ovalbumin (46 kD); carbonic anhydrase (30 kD) and lactoglobulin A (18 kD).

In order to circumvent this toxicity, we next examined the use of another regulatable promoter which could be kept repressed during selection and scale-up in order to prevent accumulation of the PreS1+PreS2+S polypeptide and its deleterious effect on yeast growth. The promoter for the yeast gene encoding the enzyme alcohol dehydrogenase II (pADH2), which is responsible for the conversion of ethanol to acetaldehyde, is repressed during the fermen- tative growth of yeast in glucose-containing medium and derepressed within a few hours following the depletion of glucose from the medium; at this time, the yeast shift to aerobic growth and metabolize ethanol as a carbon source (Ciriacy, 1975). Therefore, a plasmid was constructed (Figure 2F), in which pADH2 would direct the expression of PreS1+PreS2+S in yeast. Yeast transformed with this vector recovered successfully and could be established as stable master seeds (Kniskern et al., 1988).

Following purification (Yamazaki et al., manuscript in preparation), the PreS1+PreS2+S product was analyzed by polyacrylamide gel electrophoresis in sodium dodecyl sulfate (SDS-PAGE) [Figure 3, lane 3]. The 39-kD species (molecular weight consistent with that predicted for the translation product of the complete ORF) reacted strongly with antibodies directed to S as well as with antibodies directed to the amino-terminal 22 residues of PreS1 (Kniskern et al., 1988), thus confirming the presence of the full-length polypeptide. Similar reactivity is detected with antibodies directed to a synthetic peptide composed of the 55 amino acids of the PreS2 region (unpublished observations). A 45-kD species (Figure 3, lane 3) shared these antigenic properties and apparently is the glycosylated version of the 39-kD species, based upon its ability to react with the periodic acid-Schiff reagent (data not shown); both polypeptides have been reported to be present in the HBV virion (Heermann et al., 1984). Electron microscopic examination of purified PreS1+PreS2+S preparations revealed few typical 20 nm particles. Although a small number of apparently 20 nm particles were seen, these preparations contained predominantly a mixture of small particles (2 to 3 nm in diameter) and a polydisperse population of loosely organized assemblies of larger aggregates (15 to 50 nm in diameter) (data not shown).

In order to evaluate the potential of the complete envelope protein as a vaccine, the product was adsorbed to preformed alum and used to immunize rabbits. After two immunizations, a significant immune response occurred in all rabbits which was directed not only to the S domain but also to both the PreS2 and PreS1 domains (Kniskern et al., 1988). Therefore, the purified PreS1+PreS2+S polypeptide is formulated as an immunogenic structure capable of eliciting antibodies to each domain. However, the response to the S domain was lower than that elicited in rabbits by recombinant- or plasmid-derived antigens which contain the S polypeptide alone.

SUMMARY

In summary, we have shown that yeast is the preferred host for the expression of recombinant-derived hepatitis B vaccines, and that a yeast expression system which is productive, stable and scaleable can be developed for each of the three HBV envelope proteins. The versatility of regulated and integrated yeast expression systems in the production of foreign polypeptides with biomedical utility also has been highlighted.

We also have shown that careful attention to the development of recombinant clones helps to optimize the entire production process leading to highly purified products which share many biochemical properties with the plasma-derived vaccine. Furthermore, immunization with PreS2 sequences is capable of protecting chimpanzees from HBV infection.

The availability of PreS2+S and PreS1+PreS2+S proteins expressed in yeast now provides the opportunity for establishing the relevance of such candidate vaccines in preventing human disease, thereby highlighting the utility of molecular biology in modern vaccine development.

ACKNOWLEDGMENTS

We thank J. Broach, K. Hofmann, J. Hopper, J. Tomassini and P. Valenzuela for some of the yeast strains, plasmids and DNA used in these studies; A. Bertland, M. Johnston, B. Rich, E. Wasmuth and A. Wolfe for analytical testing; R. Charbonneau, N. Grason, J. Harder, F. Leu, M. Mudri T. Schaefer and Amanda Sha for isolation of antigens; F. Banker, P. Conard, V. Larson and W. Schleif for animal testing; J. Bailey for fermentation and cell harvesting; E. Arcuri, J. Broach, A. Conley, S. Drew, R. Gerety, R. Greasham, W. Herber, J. Hopper and R. Johnson, M. King, D. Omstead, R. Sitrin for helpful discussions; R. Zeigler for artwork and D. Sagel for careful preparation of the manuscript.

REFERENCES

Beasley, R.P. and Hwang, L.-Y. (1984), Epidemiology of hepatocellular carcinoma, Viral Hepatitis and Liver Disease. (G.N. Vyas, J.H. Dienstag and J.H. Hoofnagle eds.), Grune & Stratton, Orlando, pp. 209-224.

Bhatnagar, P.K., Papas, E., Blum, H.E., Milich, D.R., Nitecki, D., Karels, M.J. and Vyas, G.H. (1982), Immune Response to Synthetic Peptide Analogues of Hepatitis B Surface Antigen Specific for the a Determinant. Proc. Natl. Acad. Sci. (U.S.A.) 79, 4400-4404.

Brown, S.E., Howard, C.R., Zuckerman, A.J. and Steward, M.W. (1984), Determination of the Affinity of Antibodies to Hepatitis B Surface Antigen in Human Sera. J. Immunol. Meth. 72, 41-48.

Carty, C.E., Montgomery, D.L., Hagopian, A. Burke, P., Dunn, N., Schultz, L.D., Kniskern, P.J., Maigetter, R.Z., and Ellis, R.W. (1989), Regulated Expression of Hepatitis B Surface Antigen Containing the PreS2 Domain in Recombinant Yeast. Submitted to J. Industrial Micro.

Ciriacy, M. (1975), Genetics of Alcohol Dehydrogenase in Saccharomyces cerevisiae. II. Two Loci Controlling Synthesis of the Glucose-repressible ADHII. Mol. Gen. Genet. 138, 157-164.

Ellis, R.W., Kinskern, P.J., Hagopian, A., Schultz, L.D., Montgomery, D.L., Maigetter, R.Z., Wampler, D.E., Emini, E.A., Wolanski, B., McAleer, W.J., Hurni, W.M., Miller, W.J. (1988), Preparation and Testing of a Recombinant-derived Hepatitis B Vaccine Consisting of PreS2+S Polypepties, "Viral Hepatitis and Liver Disease", A.J. Zuckerman (ed.), Alan R. Liss, New York, pp. 1079-1086.

Ellis, R.W. (1989), Recombinant Yeast-derived Hepatitis B Vaccine: The Prototype for Biotechnologically-derived Old Vaccines. For submission to: "Biotechnology Handbook", Chiu and Gueriguian, Editors, New York, Marcel Dekker.

Emini, E.A., Ellis, R.W., Miller, W.J., McAleer, W.J., Scolnick, E.M. and Gerety, R.J. (1986), Production and Immunological Analysis of Recombinant Hepatitis B Vaccine. J. Inf. 13(A), 3-9.

Emini, E.A., Larson, V., Eichberg, J., Conard, P., Garsky, V.M., Lee, D.R., Ellis, R.W., Miller, W.J., Anderson, C.A. and Gerety, R.J. (1989), A Synthetic Peptide Comprising the Complete PreS2 Region of the Hepatitis B Virus (HBV) Surface Protein Confers Protection in Chimpanzees Against HBV Challenge. Submitted to J. Med. Virology.

Francis, D.P., Hadler, S.C., Thompson, S.E, Maynard, J.E., Ostrow, D.G., Altman, N., Braff, E.H., O'Malley, P., Hawkins, D., Judson, F.N., Penley, K., Nylund, T., Christie, G., Mayers, F., Moore, J.N., Gardner, A., Doto, I.L., Miller, J.H., Reynolds, G.H., Murphy, B.L., Schable, C.A., Clark, B.T., Curran, J.W. and Redeker, A.G. (1982), The Prevention of Hepatitis B with Vaccine. Ann. Int. Med. 97, 362-366.

Francis, D.P., Feorino, P.M., McDougal, S., Warfield, D., Getchell, J., Cabradilla, C., Tong, M., Miller, W.J., Schultz, L.D., Bailey, F.J., McAleer, W.J., Scolnick, E.M. and Ellis, R.W. (1986), The Safety of the Hepatitis B Vaccine. J. Am. Med. Assn. 256, 869-872.

Galibert, F., E. Mandart, F. Fitoussi, P. Tiollais and P. Charnay (1979), Nucleotide Sequence of the Hepatitis B Genome (subtype ayw) Cloned in E. coli. Nature (London) 281, 646-650.

Hashimoto, H., Kikuchi, Y., Nogi, Y. and Fukasawa, T. (1983), Regulation of Expression of the Galactose Gene Cluster in Saccharomyces cerevisiae - Isolation and Characterization of the Regulatory Gene GAL4. Mol. Gen. Genet. 191, 31-38.

Heermann, K.H., Goldmann, U., Schwartz, W., Seyffarth, T., Baumgarten, H., Gerlich, W.H. (1984), Large Surface Proteins of Hepatitis B Virus Containing the Pre-S Sequence. Journal of Virology 52, 396-402.

Hilleman, M.R., Bertland, A.V., Buynak, E.B., Lampson, G.P., McAleer, W.J., Roehm, R.R. and Tytell, A.A. (1978), Clinical Laboratory Studies of HBsAg Vaccine, Viral Hepatitis (G.N. Vyas, S.N. Cohen and R. Schmid, eds.). Franklin Institute Press, Philadelphia, pp. 525-537.

Hilleman, M.R., McAleer, W.J., Buynak, E.B. and McLean, A.A. (1983), The Preparation and Safety of Hepatitis B Vaccine. J. Infect. 7, 3-8.

Hilleman, M.R. and Ellis, R.W. (1986), Vaccines Made from Recombinant Yeast Cells. Vaccine 4, 75-76.

Hruska, J.F., Clayton, D.A., Rubenstein, J.L.R. and Robinson, W.S. (1977), Structure of Hepatitis B Dane Particle DNA Before and After the Dane Particle DNA Polymerase Reaction. J. Virology 21, 666-672.

Itoh, Y., Takai, E., Ohnuma, H., Kitajima, H., Tsuda, F., Machida, A., Mishiro, S., Nakamura, T., Miyakaway, Y., Mayumi, M. (1986), A Synthetic Peptide Vaccine Involving the Product of the Pre-S(2) Region of Hepatitis B Virus DNA: Protective Efficacy in Chimpanzees. Proc. Natl. Acad. Sci. USA 83, 9174-9178.

Iwarson, S., Tabor, E., Thomas, H.C., Godall, A., Waters, J., Snoy, P., Shih, J.W.-K. and Gerety, R.J. (1985), Neutralization of Hepatitis B Virus Infectivity by a Murine Monoclonal Antibody: An Experimental Study in the Chimpanzee. J. Med. Virol. 16, 89-96.

Jilg, W., Schmidt, M., Weinel, B., Kuttler, Th., Brass, H., Bommer, J., Muller, R., Schulte, B., Schwarzbeck, A. and Deinhardt, F. (1986), Immunogenicity of Recombinant Hepatitis B Vaccine in Dialysis Patients. J. Hepatology 3, 190-195.

Johnston, S.A. and Hopper, J.E. (1982), Isolation of the Yeast Regulatory Gene GAL4 and Analysis of its Dosage Effects on the Galactose/ Melibiose Regulon. Proc. Natl. Acad. Sci. USA 79, 6971-6975.

Johnston, S.A., Zavortink, M.J., Debouck, C. and Hopper, J.E. (1986), Functional Domains of the Yeast Regulatory Protein, GAL4. Proc. Natl. Acad. Sci. USA 83, 6553-6557.

Kniskern, P.J., Hagopian A., Montgomery, D.L., Burke, P., Dunn, N.R., Hofmann, K.J., Miller, W.J., Ellis, R.W. (1986), Unusually High-Level Expression of a Foreign Gene (Hepatitis B Virus Core Antigen) in Saccharomyces cerevisiae. Gene 46, 135-141.

Kniskern, P.J., Hagopian, A., Burke, P., Dunn, N.R., Emini, E.A., Miller, W.J., Yamazaki, S. and Ellis, R.W. (1988), A Candidate Vaccine for Hepatitis B Containing the Complete Viral Surface Protein. Hepatology 8, 82-87.

Kniskern, P.J., Montgomery, D.L., Carty, C.E., Burke, P., Schulman, C.A., Hofmann, K.J., Bailey, F.J., Hagopian, A., Dunn, N.R., Maigetter, R.Z., Schultz, L.D., and Ellis R.W. (1989), Constitutive and Regulated Expression of Hepatitis B Virus (HBV) PreS2+S Protein in Yeast. For publication in: The ACS Symposium on Microbiological and Biochemical Technology.

Laughon, A., Driscoll, R., Wills, N. and Gesteland, R.F. (1984), Identification of Two Proteins Encoded by the Saccharomyces cerevisiae GAL4 gene. Mol. Cell. Biol. 4, 268-275.

McAleer, W.J., Buynak, E.B., Maigetter, R.Z., Wampler, D.E., Miller, W.J., Hilleman, M.R. (1984), Human Hepatitis B Vaccine from Recombinant Yeast. Nature 307, 178-180.

Milich, D.R., McNamara, M.K., McLachlan, A., Thornton, G.B., Chisari, F.V. (1985), Distinct H-2-linked Regulation of T-cell Responses to the Pre-S and S Regions of the Same Hepatitis B Surface Antigen Polypeptide Allows Circumvention of Nonresponsiveness to the S Region. Proc. Natl. Acad. Sci. USA 82, 8168-172.

Milich, D.R., McLachlan, A., Chisari, F.V., Kent, S.B., Thorton, G.B.

(1986), Immune Response to the Pre-S(1) Region of the Hepatitis B Surface Antigen (HBsAg): A Pre-S(1)-specific T Cell Response can Bypass Nonresponsiveness to the Pre-S(2) and S Regions of HBsAg. J. Immunol. 137, 315-322.

Nasmyth, K.A. and Reed, S.I. (1980), Isolation of Genes by Complementation in Yeast: Molecular Cloning of a Cell-Cycle Gene. Proc. Natl. Acad. Sci. USA 77, 2119-2123.

Neurath, A.R., Kent, S.B., Strick, N., Stark, D., Sproul, P. (1985), Genetic Restriction of Immune Responsiveness to Synthetic Peptides Corresponding to Sequences in the Pre-S Region of the Hepatitis B Virus (HBV) Envelope Gene. J. Med. Virol. 17, 119-25.

Neurath, A.R., Kent, S.B., Parker, K., Prince, A.M., Strick, N., Brotman, B., Sproul, P. (1986), Antibodies to a Synthetic Peptide from the Pre-S 120-145 Region of the Hepatitis B Virus Envelope are Virus-Neutralizing. Vaccine 4, 35-37.

Neurath, A.R., Kent, S.B.H., Strick, N., Parker, K., Courouce, A.-M., Riottot, M.M., Petit, M.A., Budkowska, A., Girard, M., Pillot, J. (1987), Antibodies to Synthetic Peptides from the Pre-S1 and Pre-S2 Regions of One Subtype of the Hepatitis B Virus (HBV) Envelope Protein Recognize all HBV Subtypes. Molecular Immunology 24, 975-980.

Perlman, D. and Hopper, J.E. (1979), Constitutive Synthesis of the GAL4 Protein, a Galactose Pathway Regulator in Saccharomyces cerevisiae. Cell 16, 89-95.

Peterson, D.L., Roberts, I.M. and Vyas, G.N. (1977), Partial Amino Acid Sequence of Two Major Component Polypeptides of Hepatitis B Surface Antigen. Proc. Natl. Acad. Sci. U.S.A. 74, 1530-1534.

Prince, A.M., Ikram, H. and Hopp, T.P. (1982), Hepatitis B Virus Vaccine: Identification of HBsAg/a and HBsAg/d but not HBsAg/y Subtype Antigenic Determinants on A Synthetic Immunogenic Peptide. Proc. Natl. Acad. Sci. (U.S.A.) 79, 579-582.

Robinson, W.S. and Marion, P.L. (1988), Biological Features of Hepadna Viruses, Viral Hepatitis and Liver Disease (A.J. Zuckerman, ed.), Alan R. Liss, Inc., N.Y., pp 449-458.

Schultz, L.D., Tanner, J., Hofmann, K.J., Emini, E.A., Condra, J.H., Jones, R.E., Kieff, E., Ellis, R.W. (1987a), Expression in Yeast of the Gene Encoding the Major Envelope Glycoproteins (gp350/gp220) of Epstein-Barr Virus. Gene 54, 113-123.

Schultz, L.D., Hofmann, K.J., Mylin, L.M., Montgomery, D.L., Ellis, R.W. and Hopper, J.E. (1987b), Regulated Overproduction of the GAL4 Gene Product Greatly Increases Expression from Galactose-Inducible Promoters on Multi-Copy Expression Vectors in Yeast. Gene 61, 121-133.

Stevens, C.E., Alter, H.J., Taylor, P.E., Zaug, E.A., Harley, E.J. and Szmuness, W. (1984), Hepatitis B Vaccine in Patients Receiving Hemodialysis: immunogenicity and efficacy. N. Engl. J. Med. 311, 496-501.

Stevens, C.E., Taylor, P.E., Tong, M.J., Toy, P.T., Vyas, G.N. and Krugman, S. (1987), Yeast-Recombinant Hepatitis B Vaccine: Efficacy with Hepatitis B Immune Globulin in Prevention of Perinatal Hepatitis B Virus Transmission. J. Am. Med. Assoc. 257, 2612-2616.

Szmuness, W. (1978), Hepatocellular Carcinoma and the Hepatitis B Virus: Evidence for a Causal Association. Prog. Med. Vir. 24, 40-69.

Szmuness, W., Harley, E.J., Ikram, H. and Stevens, C.E. (1978), Sociodemographic Aspects of the Epidemiology of HBV, Viral Hepatitis (G.N. Vyas, S.N. Cohen and R. Schmid, eds.), Franklin Institute Press, Philadelphia, pp. 297-320.

Szmuness, W., Stevens, C.E., Zang, E.A., Harley, E.J., Kellner (1981), A Controlled Clinical Trial of the Efficacy of the Hepatitis B Vaccine (Heptavax-B®): A Final Report. Hepatology 1, 377-385.

Tabor, E., Buynak, E.B., Smallwood, L.A., Snoy, P., Hilleman, M.R. and Gerety, R.J. (1983), Inactivation of Hepatitis B Virus by Three Methods: treatment with pepsin, urea, or formalin. J. Med. Virol 11, 1-10.

Tiollais, P., Pourcel, C., Dejean, A. (1985), The Hepatitis B Virus. Nature (London) 317, 489-495.

Valenzuela, P., Gray, P., Quiroga, M., Zaldivar, J., Goodman, H.M., Rutter, W.J. (1979), Nucleotide Sequence of the Gene Coding for the Major Protein of Hepatitis B Virus Surface Antigen. Nature 280, 815-819.

Valenzuela, P., M. Quiroga, J. Zaldivar, P. Gray and W.J. Rutter (1981), The Nucleotide Sequence of the Hepatitis B Viral Genome and the Identification of the Major Viral Genes, In B. Fields, R. Jaenisch and C.F. Fox (ed.), Animal virus genetics. Academic Press, Inc., New York, pp.57-70.

Valenzuela, P., Medina, A., Rutter, W.J., Ammerer, G., Hall, B.D. (1982), Synthesis and Assembly of Hepatitis B Virus Surface Antigen Particles in Yeast. Nature 298, 347-352.

Valenzuela, P., Coit, D., Kuo, C.H. (1985), Synthesis and Assembly in Yeast of Hepatitis B Virus Surface Antigen Particles Containing the Polyalbumin Receptor. Biotechnology 3, 317-320.

Wampler, D.E., Lehman, E.D., Boger, J., McAleer, W.J. and Scolnick, E.M. (1985), Multiple Chemical Forms of Hepatitis B Surface Antigen Produced in Yeast. Proc. Natl. Acad. Sci. U.S.A. 82, 6830-6834.

West, D.J., Zajac, B.A., Ellis, R.W. and Gerety, R.J. (1988), Improved Immunogenicity of a Yeast-Derived Hepatitis B Vaccine. Infektionsklinik 1, 1-7.

Zajac, B.A., West, D.J., McAleer, W.J. and Scolnick, E.M. (1986), Overview of Clinical Studies with Hepatitis B Vaccine Made by Recombinant DNA. J. Infect. 13(A), 39-45.

STUDIES ON THE CLONALITY OF THE ANTIBODY RESPONSE TO A PROTEIN EPITOPE

Eli Benjamini[1], Matthew L. Andria[1], Soshana Levy[2] and Cherry Y. Leung[1]

[1]Department of Medical Microbiology and Immunology, School of Medicine, University of California, Davis, CA 95616

[2]Department of Medicine, Stanford University, School of Medicine, Stanford, CA 94305

INTRODUCTION

During the past decade, great advances have been made in the synthesis and preparation of various formulations of compounds to be used as vaccines. Indeed, the field of "vaccinology" has experienced a dramatic surge. Yet, while many preparations, such as hepatis B immunogens, constitute effective vaccines, attempts to prepare effective vaccines against other diseases, such as malaria or AIDS, have encountered formidable difficulties.

Amongst the many attributes of an effective vaccine, whether it consists of killed or attenuated microorganisms or portions thereof, obtained synthetically or by recombinant DNA technology, is the capacity to induce an immune response against the relevant epitopes. This is well exemplified by cases where immunization against a certain virus leads to the production of anti-virus antibodies but those which lack the capacity to neutralize the cytopathic effect of the virus. In such cases, the antibodies may be directed against various epitopes on the virus but not against that viral area which is involved in the attachment of the virus to its target cell. Thus, the lack of virus-neutralizing activity following immunization with a certain viral immunogen may be due to the absence of an antibody population directed to a single critical epitope. It, therefore, becomes of more than just an academic interest to elucidate the various mechanisms which control the antibody response and the clonality of the response to a single epitope.

The sections which follow describe some of our studies on the immune response to one epitope of a model protein antigen, the tobacco mosaic virus protein (TMVP). These studies include investigations on the development of the response and on the clonality of the response to a single epitope consisting of a decapeptide representing residues 103-112 of TMVP. In addition, the amino acid sequences of several monoclonal antibodies directed to this epitope have been determined and the genetic origin of the antibodies, as well as the relationship between these sequences and the fine specificity of the monoclonal antibodies will be discussed.

THE RESPONSE TO THE DECAPEPTIDE EPITOPE

The tobacco mosaic virus protein, TMVP, contains several epitopes. A region of the protein consisting of tryptic peptide number 8, representing amino acid residues, 93-112 of the protein contains epitopes recognized by B cells and T cells (reviewed by Benjamini, 1987). In many strains of mice a T cell response is directed against the N-terminal portion while the antibody response is directed against the C-terminal portion of peptide 8. The humoral response to the C-terminal portion of peptide 8 is not confined to mice only. In fact, immunization of several different animal species which include rabbits, mice, guinea pigs and sharks with TMVP induces antibodies which exhibit binding with the C-terminal decapeptide portion of peptide 8 (henceforth referred to as the "decapeptide") representing residues 103-112 of TMVP and having the amino acid sequence Thr-Thr-Ala-Glu-Thr-Leu-Asp-Ala-Thr-Arg.

Further studies have revealed that while all the mouse strains tested responded to TMVP and to peptide 8, the response to the decapeptide was genetically linked to the immunoglobulin allotype (Morrow et al., 1983). Moreover, during many experiments, using hundreds of mice, it became evident that even in some "decapeptide responder" strains, although all the individuals make anti-TMVP antibodies, not all of the individuals make anti-decapeptide antibodies (Morrow et al., 1983). For example, whereas over 90% of A/J mice immunized with TMVP produced anti-decapeptide antibodies, only approximately 50% of BALB/c mice exhibited such a response, and in C3H.SW (CSW) mice a little over 50% of the individuals immunized with TMVP produced anti-decapeptide antibodies.

The fine specificity of the anti-decapeptide antibodies may be discerned by the use of synthetic analogues of the decapeptide that contain amino acid substitutions at different loci on the decapeptide. Thus, it was observed that different strains of mice immunized with TMVP produce anti-decapeptide antibodies with fine specificities characteristic for a given strain and that even within a strain of highly inbred mice the fine specificities of the anti-decapeptide antibodies differ among different individuals (Morrow et al., 1983).

Four synthetic analogues have been found to be very useful in discerning the fine specificity of C3H.SW mice. These analogues are: Ala-Ala-Ala-Ala-Ala-Leu-Asp-Ala-Thr-Arg (referred to as the $(Ala)_5$ analogue); an analogue in which the Glu_{106} was replaced Tyr (referred to as the Tyr_{106} analogue); an analogue in which the Thr_{107} was replaced by Met (referred to as the Met_{107} analogue), and an analogue in which the Arg_{112} was replaced by Cys (referred to as the Cys_{112} analogue). An example of the fine specificity frequency of anti-decapeptide antibodies following immunization of C3H.SW mice with TMVP is given in Table 1.

In view of the observations that only a certain proportion of TMVP immunized C3H.SW mice make anti-decapeptide antibodies and that not all the mice exhibit the same fine specificity for the anti-decapeptide antibodies it was of importance to ascertain the possible mechanisms that regulate the clonality of the response to the decapeptide epitope. If the clonality is indeed restricted, as is indicated by the above observations and supported by other observations (Rennick et al., 1982), then the question must be asked whether this restriction is due to a limited repertoire of decapeptide recognizing clones or whether it is due to operating mechanisms which select clones for activation from a potentially large repertoire of decapeptide recognizing clones.

TABLE 1

The fine specificity of decapeptide-binding antibodies from TMVP immunized C3H.SW mice as assessed by their binding with synthetic analogues of the decapeptide.

Number of Animals	Binding with Analogue $(Ala)_5$	Tyr_{106}	Met_{107}	Cys_{112}
8	+	+	+	+
15	+	+	+	-
11	-	+	+	+
7	-	+	+	-
9	+	-	+	+
2	+	-	+	-
20	-	-	+	+
11	-	-	+	-
Total 83=100%	41%	49%	100%	47%

One or more of several mechanisms may be responsible for the observed clonality of the anti-decapeptide response following immunization with TMVP. One such mechanism may consist of the idiotypic network as originally proposed by Jerne (1974). It appears that in a variety of systems T helper cells or T suppressor cells can select for particular major idiotypes of B cells from a larger repertoire of potential idiotypes (Slaohi et al., 1984). Other observations suggest that T helper cells may restrict the B cell clones that are selected for activation (Ceska et al., 1976; Katz et al., 1983). It has also been proposed (Berzofsky, 1983) that the spacing between those epitopes recognized by T helper cells and those recognized by B cells may result in the selective activation of only certain B cell populations and thus be reflected in the presence or absence of an antibody response to a particular determinant or in the fine specificity of the response to that determinant. Other regulatory mechanisms may consist of factors inherent in the B cell clones themselves which may be competing for antigen: B cells with higher affinity for the epitope would successfully compete with lower affinity B cells to become activated and expanded. Various B cell clones may also compete when antigen is reintroduced at a later time. This is evident in the phenomenon of the "original antigenic sin" described by Fazakas de St. Groth (1967) in which B cell clones activated during the initial exposure to an antigen become preferentially activated during subsequent exposure to a variant form of the antigen if that latter form contains one or more epitopes in common with the original antigen.

The experiments described below were designed to ascertain which, if any, of the above regulatory mechanisms is responsible for the selection for activation of anti-decapeptide recognizing clones and for the selection of the fine specificity of the anti-decapeptide antibodies in individuals of the inbred C3H.SW mouse strain immunized with TMVP. Some of the results of experiments recently reported (Benjamini et al., 1988) are highlighted below.

THE RANDOM SELECTION OF NAIVE EPITOPE RECOGNIZING CLONES BY ANTIGEN

An experiment was performed in which spleen cells were pooled from a large number (forty-eight) naive mice. The cells were divided into two equal portions. One portion was treated with anti-Thy 1.2 monoclonal antibodies in the presence of complement to remove the T cells; the other portion was not. Cells (1×10^8), untreated or treated with anti-Thy 1.2 and complement were transferred into 21 and 22 recipients respectively which have been irradiated with 400 rads on the same day, previous to transfer. This dose of irradiation (as well as the dose of 600 rads used in other experiments described herein) kills B cells and T suppressor cells but leaves the T helper cells unaffected (Katz et al., 1973; Huchings et al., 1985). One day after transfer all the recipients were immunized with 100 μg TMVP in Complete Freund's Adjuvant (CFA) followed 3 weeks later by an aqueous boost with 100 μg TMVP. The animals were bled 10 days after the boost and the presence of anti-TMVP and anti-decapeptide antibodies was determined by solid phase radioimmunoassay. The results in Table 2 demonstrates that every immunized animal made anti-TMVP antibodies, demonstrating that the transferred cells had competent B cells and that, following irradiation, the recipients had competent T helper cells. In contrast, results in Table 2 clearly show that only approximately half of the immunized recipients made anti-decapeptide antibodies and, moreover, that this frequency was the same whether or not the recipients were transplanted with whole spleen cells or with spleen cells depleted of T cells. The frequency of decapeptide positive sera of the recipients is very similar to that seen in the general population of mice of this strain (C3H.SW) immunized with TMVP.

TABLE 2

The frequency of antibodies, capable of binding with TMVP or with the decapeptide, obtained from naive recipients of pooled naive spleen cells which were subsequently immunized with TMVP.

	RECIPIENTS	
Test Antigen	Received Naive Spleen Cells	Received Naive Spleen Cells Treated with Anti-Thy 1.2 Antibody and Complement
TMVP	21/21 (100%)	22/22 (100%)
Decapeptide	10/21 (48%)	11/22 (50%)

Seven decapeptide positive sera from mice that had received naive spleen cells in the absence of T cells were tested by the Farr assay for binding with the four synthetic analogues of the decapeptide which, as mentioned earlier, have been found to be very useful in discerning the fine specificity of anti-decapeptide antibodies of this strain. Results in Table 3 show that the frequency of the responses to the $(Ala)_5$, Tyr_{106}, Met_{107}, and Cys_{112} analogues is approximately the same as that seen decapeptide positive sera obtained from normal naive mice of this strain immunized with TMVP (Table 1).

TABLE 3

The fine specificity of decapeptide binding antibodies obtained from naive recipients of pooled naive B cells which were subsequently immunized with TMVP.

Test Antigent	Frequency of Response
$(Ala)_5$ Analogue	57%
Tyr_{106} Analogue	48%
Met_{107} Analogue	100%
Cys_{112} Analogue	29%

The results presented so far, strongly suggest that 1) T cells have no discernible effect on the expressed clonality of the B cells and 2) the clonality of the activated B cells is selected, at random, by the antigen from a larger repertoire of B cells. The roughly 50% frequency of anti-decapeptide responsiveness suggests that the B cell clones which recognize the decapeptide are restricted in distribution and that immunization with TMVP which always results in the induction of anti-TMVP antibodies, may or may not select decapeptide recognizing B cell clones. Moreover, those decapeptide recognizing clones which are selected for activation at random by the antigen may or may not include clones with a certain fine specificity.

THE DEVELOPMENT OF MEMORY AND THE "LOCKED IN" CLONALITY OF THE RESPONSE

Having shown that the clonality of the response to the decapeptide is selected at random by the antigen it was of interest to investigate the fate of this clonality and to assess whether or not it changes without or with subsequent immunizations with the antigen.

For this purpose we have immunized eleven mice with TMVP. Of these, all responded with anti-TMVP while only seven responded with anti-decapeptide antibodies (Table 4). The fine specificity of the anti-decapeptide antibodies produced by each of these mice is also shown in Table 4. All of the mice were bled periodically during the ensuing seven months and their antisera were assayed for TMVP, for anti-decapeptide antibodies, and for the capapcity to bind the synthetic analogues. Results of these tests showed that none of the mice which initially did not respond with anti-decapeptide antibodies, converted to responders. Moreover, the fine specificity of the anti-decapeptide antibodies of the responding mice did not change. After seven months the anti-decapeptide titers of the responding mice fell below detectable levels. At this time the mice were given a second set of immunizations consisting of 100 μg TMVP in FCA followed three weeks later by two weekly aqueous boost with 100 μg TMVP. The sera were collected one week after the second boost and were tested for anti-TMVP, for anti-decapeptide and for the fine specificity of the anti-decapeptide antibodies. Again, all the mice made anti-TMVP antibodies. The results in Table 4 clearly demonstrate that the anti-decapeptide responsiveness and the fine specificity of this response were virtually identical after the first and second set of immunizations:

those mice which did not respond with anti-decapeptide antibodies after the first set of immunizations did not respond after the second set of immunizations; those mice which responded after the first set responded also after the second set. Moreover, the clonality of the response, as judged by the fine specificity of the antibodies, also remained the same after both sets of immunizations: there was no discernible indications that clones have disappeared or that new clones have appeared.

To exclude the possibility that the clonality expressed during the second set of immunizations is regulated by the antigen specific T helper cells which have participated in the initial response to the antigen, an experiment was performed in which fifteen TMVP-immunized mice exhibiting anti-decapeptide titers of known fine specificity were rested for seven months at the end of which time their anti-decapeptide titers fell below detectable levels. The mice then received 600 rads, a dose of radiation which eliminates B cells and T suppressor cells but does not affect T helper cells. On the same day, immediately after irradiation, each mouse received 1 x 10^8 naive spleen cells from a pool derived from a large number of syngeneic donors. One day later the recipients were immunized with 100 μg TMVP in FCA followed three weeks later by two weekly aqueous boost with 100 μg TMVP. The mice were bled one week after the last boost.

TABLE 4

The frequency and fine specificity[1)] of anti-TMVP antibodies capable of binding with the decapeptide and its synthetic analogues after two sets of immunizations with TMVP seven months apart.

	After First Set of Immunizations				After Second Set of Immunizations			
	Test Antigens				Test Antigens			
Animal Number	Deca-Peptide	$(Ala)_5$ Analogue	Tyr_{106} Analogue	Cys_{112} Analogue	Deca-Peptide	$(Ala)_5$ Analogue	Tyr_{106} Analogue	Cys_{112} Analogue
1	+	-	-	++++	+	-	-	++++
2	+	+++	++	-	+	++++	++	-
3	+	++	++	++	+	+++	++	+
4	+	++++	++++	-	+	++++	++++	+/-
5	+	-	+	+	+	-	+	+
6	+	++++	+	++++	+	++++	+++	++++
7	+	++	++	+++	+	++++	++++	+
8	-	ND[2)]	ND	ND	-	ND	ND	ND
9	-	ND	ND	ND	-	ND	ND	ND
10	-	ND	ND	ND	-	ND	ND	ND
11	-	ND	ND	ND	-	ND	ND	ND

1) Binding with the decapeptide is represented by + or - signs denoting binding or no binding respectively; binding with analogues is normalized for binding with the decapeptide as described by Morrow et al., (1983).

2) Not done.

The results in Table 5 show that only nine of the fifteen mice (60%) made anti-decapeptide antibodies, a proportion similar to that seen in naive mice of these strains following immunization with TMVP. Moreover, the results show that 88%, 100% and 33% made anti-decapeptide which bind with the $(Ala)_5$, Tyr_{106} and Cys_{112} analogues. Thus, there appears to be a change in the frequencies of binding with the analogues following the second immunization: the frequencies of binding with the $(Ala)_5$ and Tyr_{106} analogues are higher than those seen after the first set of immunizations, while the frequency of binding with the Cys_{112} analogue is lower. These results demonstrate that the primed T helper cells of the recipients had no effect on the clonality of the response expressed by the transferred naive B cells and indicate that the expressed clonality is a function of the transferred naive B cells.

CLONAL DOMINANCE

So far the results which show less than 100% responsiveness to the decapeptide and to the synthetic analogues, indicate that in each individual of this strain the frequency of clones recognizing the decapeptide is rather limited. Even if an individual is transplanted with an aliquot from a pool of naive B cells representing the entire repertoire of epitope recognizing clones of this strain, only a limited number of clones encounter the epitope, at random, become activated and establish the clonality. These clones expand; some of these cells will become

TABLE 5

The frequency of binding with the decapeptide and its synthetic analogues of antibodies produced by immunization with TMVP and by a reimmunization with TMVP 7 months later following transfer of a pool of naive spleen cells.

Test Antigen	After the First Set of Immunizations	After Second Set of Immunizations Seven Months Later Following Transfer of Naive Spleen Cells
Decapeptide	70%	60%
$(Ala)_5$ Analogue	45%	88%
Tyr_{106} Analogue	55%	100%
Cys_{112} Analogue	45%	33%

antibody secreting plasma cells, others will become memory B cells which now are present in higher frequencies and become activated by antigen easier than naive cells (Klinman, 1972; Klinman et al., 1973; Gray et al., 1986), thus favorably competing with naive cells for antigen and thereby maintaining their clonal dominance.

The process of the establishment of clonal dominance appears to be inherent in the B cells themselves and is independent of antigen specific T cells. Such independence is reflected not only in the expression of the response of naive B cells (Table 2, 4) but also in the expression of memory B cells. This is evident from data in Table 6 which were obtained from an experiment in which mice were first immunized with TMVP and their anti-decapeptide titers and fine specificity of the anti-decapeptide

TABLE 6

The fine specificity of anti-decapeptide antibodies produced by mice which have been immunized with TMVP and boosted seven months later after receiving a pool of spleen cells from TMVP immunized syngeneic donor mice with anti-decapeptide antibodies of determined fine specificity.

Test Antigen	Donors Reactivity	Recipients Reactivity: First Set of Immunization	Recipients Reactivity: Booster Immunization
$(Ala)_5$ Analogue	27%	53%	95%
Tyr_{106} Analogue	45%	68%	95%
Cys_{112} Analogue	59%	58%	95%

antibodies were established. The mice were then rested for 7 months until their anti-decapeptide antibodies were undetectable. Each mouse was then irradiated with 600 rads and, on the same day, transplanted with 1 x 10^8 spleen cells obtained from a large pool of mice which were previously immunized to TMVP and which had anti-decapeptide titers of various fine specificities. The recipient mice were boosted immediately after transfer with 100 μg aqueous TMVP followed one week later with another such aqueous boost. The animals were bled one week later and their anti-decapeptide titers and the fine specificity of the anti-decapeptide antibodies were determined. Results shown in Table 6, clearly indicate that upon the initial set of immunizations the recipients and the donors exhibited frequencies of anti-decapeptide titers with fine specificity distributions characteristic to this strain of mice. In contrast, after the transfer of the pooled memory cells, virtually all of the recipient made anti-decapeptide antibodies expressing all the fine specificities. The results clearly indicate that the primed T cells of the recipients, which as expected, remained functional after irradiation, or the primed T cells of the donors had no effect upon the expressed fine specificity of the transferred memory B cells. the results indicate that the transferred memory cells contained expanded populations of decapeptide specific memory B cells expressing the various fine specificities. Their presence in relatively high frequency in the pool of transferred spleen cells and the ease at which memory cells are activated by antigen compared to virgin cells resulted in the activation, in virtually every recipient, of enough memory cells to result in the expression of anti-decapeptide antibodies exhibiting all the fine specificities.

In summary, the above results demonstrate that the clonality of the response to a given epitope is selected at random by antigens from a larger repertoire of epitope recognizing clones. Once activated, these selected clones expand. Part of them become antibody secreting plasma cells; others become memory cells which are now present in higher frequencies and which, upon subsequent exposure to antigen, are activated easier than virgin cells resulting in the preferential activation of these memory cells thus expressing their clonal dominance.

Once established, the clonality of the response becomes "locked in" for the life of the individual. This process appears to be independent of antigen specific helper T cells. Moreover, the experiments presented herein suggest that the effect of the idiotype network on the established clonality is minimal, if any. In addition, the results indicate that although antigen driven somatic mutations may play a significant role in the development of the clonality upon initial exposure to antigen, such antigen driven mutations do not play a significant role once clonal dominance has been established. This conclusion is supported by data from sequence analysis of anti-decapeptide antibodies which will be given and discussed later.

The above results may help explain a common observation, namely, that often, immunization of an individual with an immunogen results in a certain response which, for example, does not include the capacity of the antibodies to exhibit a certain desired function (e.g. viral neutralization); repeated re-immunizations of this individual often do not result in a change of the response such as to bring about the production of antibody populations which do exhibit the particular desired function: once the initial clonality is established, certain epitope specific clones become clonally dominant and usually successfully compete with naive B cell clones of different fine specificities or which recognize other epitopes on the antigen.

The question must be asked whether or not, once established, the clonality of the response can be changed either naturally or artificially. Results presented herein clearly show that the elimination of memory B cells by irradiation results in the change of the clonality (Table 5). This of course does not constitute a practical way of changing clonalities and overcoming clonal dominance. However, if a response of a certain clonality to a certain epitope is desired but is not achieved after a set of immunizations with a multi-epitope antigen, due to the establishment of dominance by other clones of different specificities recognizing other epitopes on the antigen, it may be possible to conjugate the epitope in question to another non-related antigen carrier and reimmunize with this conjugate. It is conceivable that upon reimmunization with this new antigen new naive clones which recognize the desired epitope will be activated and will become part of a new set of dominant clones. It remains for future experiments to asses this approach.

COMPARISON OF THE COMPLEMENTARITY DETERMINING REGIONS (CDR'S) OF MONOCLONAL ANTI-DECAPEPTIDE ANTIBODIES

The "locked in" nature of the clonality of the anti-decapeptide response in which no discernable new clones appear and no clones disappear suggests that, following the establishment of the clonality, antigen driven somatic mutations do not play a major role in the subsequent clonality of the response to the antigen. It was therefore of interest to investigate on a molecular level, the relationship between various monoclonal antibodies (MAbs) with different fine specificity but all of which recognize the decapeptide. To this end we have determined the nucleic acid sequences of cDNA clones of the H and L chains as described (Levy et al., 1987) and used these sequences to deduce the amino acid sequences of the chain. Table 7 depicts the binding patterns of the various monoclonal antibodies; Figure 1 depicts the nucleotide and amino acid sequences the CDRs of their L and H chains. Although the sequences of some H chains are incomplete results shown in Figure 1 give rise to a general picture indicating that antibodies with extremely different amino acid sequences in the CDR regions of the L and H chains, can still bind with the same ligand, namely, the decapeptide. Even antibodies with similar fine specificities, namely CM2, AF1 and AF2 have very different L chains (the sequences of some of the H chains are currently being determined). Interestingly, strong homology is seen in the L chains of antibodies with very different binding patterns. Thus, monoclonals C10

TABLE 7

The binding of monoclonal antibodies with the decapeptide and its synthetic analogues.

Monoclonal Antibody	Deca-peptide	$(Ala)_5$ Analogue	Tyr_{106} Analogue	Met_{107} Analogue	Leu_{110} Analogue	Cys_{112} Analogue
C10	+	+	+	+	+	-
AF2	+	-	+	-	-	+
CM2	+	-	+/-	-	+	+
CM4	-	-	-	-	-	-
CR1	+	+/-	-	-	-	-
AF1	+	-	+	-	-	+

and AF2 have similar L chains despite differences in binding with the analogues, and CM4 and CM2 have nearly identical L chains although CM4 does not bind the decapeptide (but it does bind with an area consisting of the N-terminal portion of the decapeptide and residues present on tryptic peptide 8 located in the N terminal half of peptide 8).

```
Light Chains:
                       CDR1               CDR2       CDR3
Antibody   2  2       2     3      5      5    8        9
           4  7abcdef8      4      0      6    8        6
 C10       KSSQSLLNSSSQKNYLA       FASTRES     QQHYSTPYT
 AF2       ----------GN-N---T      W------     -ND-RY-L-
 CM2       R---NIVHT.NGNT--E       KV-N-F-     F-GSHV-F-
 CM4       R----IVHT.NGNT--E       RV-N-F-     F-GSHV-W-
 CR1       R--K---H-.NGNT--Y       RM-NLA-     M--LEY-L-
 AF1       RA--......DIS---N       YT-RLH-     --GNTL-W-

Heavy Chains:

           CDR1          CDR2                      CDR3
                                                   1      1
Antibody   3   3  5  5  5                 6  9     0      0
           1   5  0  3ab4                 5  5     0abcde 2
 C10       DSYMD  YIYF..NNGGTNYNQKFKG     ...........WT
 CM2       -YWIH  E-N-..S----S--E---T     FPLSYGNFAW.F-
 CM4       TYW-H  E-D-..SDTY-YH--N---     RSMGHYG....LA
 CR1                                      DGSYGNYHVPYFD
```

Figure 1. Amino acid sequences of complementarity determining regents of L and H chains of monoclonal anti-decapeptide antibodies. Amino acids are designated by single letters.

The numerous differences in nucleic acid sequences between the CDRs of the anti-decapeptide monoclonal antibodies indicate that these antibodies are coded by very different germ line genes. Diversity in CDR sequences has been seen in antibodies to some haptens, e.g. poly-dextran (Akolar et al., 1987), digoxin (Mudgett-Hunter et al., 1985), alprenolol (Nahmias et al., 1988), and in antibodies to the T cell independent antigen nitrophenol on ficoll (Maizels and Bothwell, 1985). However, most studies have shown strong homology between MAbs that bind a given epitope, e.g. phosphorylcholine (Crews et al., 1981), T cell dependent responses to nitrophenol (Bothwell, 1984), epitopes on lysozyme (Darsley and Rees, 1985), the random polymer GAT (Schiff et al., 1983), and many others.

These restricted responses may indicate that regulatory mechanisms select the quality of the antibody response. In the present communication we demonstrate that antibody binding sites capable of binding the same protein epitope can be composed of very different sequences. Such diverse antibody sequences illustrate the flexibility of the immune system in generating binding sites to a particular epitope. In their initial description of hypervariable regions, Wu and Kabat (1970) noted that "...it should be borne in mind that the contour of an antibody site having a given specificity or binding affinity for a given determinant could conceivably be formed by several kinds of patterns of amino acid sequences". Our findings provide direct support for this conjecture. Moreover, the differences in the fine specificities of the antibodies reported herein appear not to be due to somatic mutations of a "prototype" anti-decapeptide V region sequence. Rather, the differences result from the use of different constellations of amino acids, each constellation forming a site capable of binding with the decapeptide epitope.

The reverse aspect of the concept that different sequences may bind to the same ligand and perform the same function, is exemplified by the finding that the same amino acid sequence can code for portions of antibodies that bind different antigens (Akolar et al., 1987). Indeed, several kappa chains of the anti-decapeptide MAbs reported herein are each highly homologous to previously reported kappa sequences for antibodies specific to small haptens. For example, the entire L chain of C10 has 94% homology with the L chain of an antibody to DNP (TF2-76) reported by Riley et al., (1986). Similarly the L chain of CM2 has 90% homology with the L chain of an antibody to 2-phenyloxazolone HQ45.1 (Kaartinen et al., 1983) and the L chain of AF1 has 93% homology with the L chain of the antibody to p-azophenylarsonate 36-35 (Wysocki et al., 1988). These homologies indicate that each antibody chain has the potential to individually participate in the binding with several different antigens, depending upon the interaction with its partner H or L chain, thus illustrating the importance of combinatorial associations of L and H V regions in forming the antibody combining site.

The phenomenon, that protein sites composed of totally different amino acid sequences may bind with the same ligand, appears not to be unique to antibodies. For example, Ma and Ptashne (1987) showed that the RNA polymerase binding domain within a given yeast transcription activator could be composed of many different amino acid sequences, and Kaiser et al., (1987) demonstrated that a very large number of amino acid sequences can function as secretion signal sequences for yeast invertase. Also, Ogata et al., (1987) found that two proteins which have entirely different amino acid sequences and crystal structures, have affinity for the same sweet receptor.

The examples for the different proteins and for the antibodies described herein appear to illustrate that the amino acid sequences within proteins are considerably degenerate.

CONCLUSIONS

1. It appears that the clonality of the antibody response to a single epitope is restricted: a certain number of B cell clones is selected for activation, at random, by the immunogen from a larger repertoire of epitope recognizing clones.

2. Once selected, the epitope recognizing clones expand and differentiate, some into antibody secreting plasma cells, others into long lived memory B cells. Because of their higher numbers and ease of activation compared to virgin B cells, subsequent exposure to antigen preferentially activates these memory cells which thus establishes themselves as dominant clones.

3. These mechanisms explain the observed phenomena that, once established, the clonality of the response appears to be "locked in" for the life of the individual.

4. Clonal dominance may be "broken" by means which selectively remove the dominant memory cells, thus allowing for the activation of naive epitope recognizing cells and the establishment of new dominant clones.

5. Once established, the clonality of the response is not significantly altered by exposure to antigen thus minimizing the importance of antigen driven somatic mutations in changing the clonality of the response.

6. Comparisons of cDNA and amino acid sequences of several monoclonal antibodies which bind with the same epitope indicates that different V regions and hence different constellations of amino acids in the CDRs of the L and H chains may constitute binding sites for the same epitope;

7. Similar V regions of the H or L chains may participate in forming binding sites to different epitopes, depending upon the interaction with the partner H or L chain.

ACKNOWLEDGEMENT

The work reported herein was supported in part by Grant No. 8411235 from the National Science Foundation and Grants No. IM401 and IN32-38 from the American Cancer Society.

REFERENCES

Akolar, P. N. Sikder, S. K., Bhattacharya, S. B., Liao, J., Gruezo, F., Morrison, S. L. and Kabat, E. A, 1987, Different V_2 and V_H germ line genes and used to produce similar confining sites with specificity for α(1-->6) dextran, J. Immunol., 138:4472.

Benjamini, E, 1987, Expression of B- and T-cell responses to a defined epitope of a protein antigen in: Immunogenicity of Protein Antigens: Repertoire and Regulation, Vol. I, E. E. Sercarz and J. A. Berzofsky, Eds. CRC Press Inc. Boca Raton, Florida.

Benjamini, E., Andria, M. L. , Estin, C. D., Norton, F. L. and Leung, C. Y, 1988, Studies on the clonality of the response to an epitope of a protein antigen: Randomness of activation of epitope recognizing clones and the development of clonal dominance, J. Immunol., 141:55.

Berzofsky, J. A., 1983, T-B reciprocity. An Ia restricted epitope-specific circuit regulating T-cell - B-cell interaction and antibody specificity, Surv. Immunol. Res, 2:223.

Bothwell, A. L. M., 1984, The genes encloding anti-NP antibodies in inbred strains of mice, in: The Biology of Idiotypes, Greene, M. I., and Nisonoff, A., Eds., Plenum Press, N. Y.

Ceska, J. M., Stratton, J. A., Miller, A. and Sercarz, E., 1976, Structural aspects of immune recognition and its consequences for antibody specificity, Eur. J. Immunol., 6:639.

Crews, S., Griffin, J., Huang, H., Calame, K., and Hood, L., 1981, A single V_H gene segment encodes the immune response to phosphorylcholine: somatic mutation is correlated with the class of antibody, Cell, 25:59.

Darsley, M. J., and Rees, A. R., 1985, Nucleotide sequences of fine anti-lysozyme monoclonal antibodies, EMBO J, 4:393.

Fazakas de St. Groth, S, 1967, Cross recognition and cross reactivity. Cold Spring Harbor Symposium Quant. Biol., 32:525.

Gray, D. I., MacLennan, C. M. and Lane, P. J., 1986, Virgin B cell recruitment and the life span of memory clones during antibody response to 2,4-dimitrophenyl hemocyanin, Eur. J. Immunol., 16:641.

Jerne, N. K., 1974, Towards a network theory of the immune system, Ann. Immunol. (Paris), 125C:373.

Kaartinen, M., Griffiths, G. M., Markham, A. F. and Milstein, C., 1983, MRNA sequences define an unusually restricted IgG response to 2-phenyloxazolone and its early diversification, Nature, 304:320.

Kaiser, C. A., Pruess, D., Grisafi, P. and Botslein, D., 1987, Many random sequences functionally replace the secretion signal sequence of yeast invertase, Science, 235:312.

Katz, M. E., Miller, A., Krzych, V., Wicker, L., Maizels, R., Clark, J., Shastri, N., Oki, A. and Sercarz, E., 1983, Hierarchial relationships among epitopes; only certain potentially utilizable epitopes on protein antigens are actually presented in a particular haplotype, in: In Gene: Past, Present, and Future. Pierce, C. W., Cohen, S. E., Kapp, J. A., Schwartz, B. D. and Schreffler, D. C., eds. Humana Press, Clifton N. J.

Klinman, N. R., 1972, The mechanism of antigenic stimulation of primary and secondary clonal precursor cells, J. Exp. Med., 136:241.

Klinman, N. R., Press, J. L. and Segal, G. P., 1973, Overlap stimulation of primary and secondary B cells by cross reacting determinants, J. Exp. Med., 138:1276.

Ma, J. and Ptashne, M., 1987, A new class of yeast transcriptional activators, Cell, 51:113.

Maizels, N. and Bothwell, A., 1985, The T cell independent immune response to the hapten NP uses a large repertoire of heavy chain genes, Cell, 43:715.

Morrow, P. R., Renick, D. M. and Benjamini, E., 1983, The antibody response to a single antigenic determinant of the tobacco mosaic virus protein (TMVP): Effect of allotype-linked genes and restricted heterogeneity of the response, J. Immunol., 131:2875.

Mudgett-Hunter, M., Anderson, W., Haber, E. and Margolies, M. N., 1985, Bending and structural diversity among high affinity monoclonal anti-digoxin antibodies, Mol. Immunol., 22:477.

Nahmias, C., Strassberg, A. D. and Emorine, L. J., 1988, The immune response towards β adrenergic ligands and their receptors, VIII, extensive diversity of V_H and V_L genes encoding anti-alprenolol antibodies, J. Immunol., 148:1304.

Ogata, C., Hatada, M., Tomlinson, G., Shin, W. and Kim, S., 1987, Crystal structure of the intensely sweet protein monellin, Nature, 328:739.

Rennick, D. M., Morrow, P. R. and Benjamini, E., 1983, Immunological studies with tobacco mosaic virus protein: differential activation of B cell subpopulations, Adv. Exp. Med. and Biol., 150:201.

Riley, S. C., Connors, S. J., Klinman, N. K. and Ogata, R. T., 1986, Preferential expression of variable region heavy chain gene segments by predominant 2,4-dimitrophenyl-specific BALB/c neonatal antibody clonotype, Proc. Natl. Acad. Sci. (USA), 83:2589.

Schiff, C., Milili, M. and Fougereau, M., 1983, Immunoglobulin diversity: analysis of the germ live V_H gene repertoire of the murine anti-GAT response, Nucleic Acid Res., 11:4007.

Slaohi, M., Leo, O., Marvel, J., Maser, M., Hierioux, J. and Urbain, J., 1984, Idiotypic analysis of potential and available repertoire in the arsonate system, J. Exp. Med., 160:1.

Wu, T. T. and Kabat, E. A., 1970, An analysis of the sequences of the variable regions of Bence Jones proteins and myeloma light chains and their implications for antibody complementarity, J. Exp. Med., 132:211.

Wysocki, L., Manser, T. and Gefter, M. L., 1986, Somatic evolution of variable region structures during an immune response, Proc. Natl. Acad. Sci. (USA), 83:1847.

ANTI-IDIOTYPE TUMOR VACCINES

Malaya Bhattacharya-Chatterjee* and Heinz Kohler+

*Roswell Park Memorial Institute, Buffalo, NY 14263 and +University of California at San Diego, Dept. of Pathology, San Diego, CA 92161

INTRODUCTION

The immunotherapy of human cancer using tumor cells or tumor-derived vaccines has been disappointing for several reasons. It has been consistently difficult to obtain large quantities of pure tumor antigens which are often chemically ill-defined and difficult to purify. In addition, there is the question whether there exists truly tumor-specific antigens and finally, there remains the problem of the immunobiological response potential against tumor antigens, or in other words, the question whether a cancer patient can mount effectively an immune response against his tumor. Tumor-associated antigens are often a part of "self" and usually evoke a very poor immune response in a tumor-bearing host due to a state of T cell-mediated suppression (Haubeck and Kolsch, 1982; Howie and McBride, 1982) and also because of tolerance to the antigen (Greene, 1980; McBride and Howie, 1986). Immunobiologists have learned that a poor antigen can be turned into a strong antigen by simply changing the molecular environment of the haptenic structure. Such changes of the carrier for haptens allow T cell help to become active, making the overall immune response stronger. In addition, changing the carrier can also turn a tolerogenic antigen into an effective antigen. Furthermore, often the immunological status of a cancer patient is suppressed and only able to respond to certain T-dependent antigens and not to other antigen forms. From these considerations, it would make sense to introduce molecular changes into the tumor antigens before using these as vaccines. Unfortunately, this is impossible to accomplish for most tumor antigens, because they are not well-defined and very hard to purify.

The network hypothesis of Niels Jerne (1974) offers an elegant approach to transform epitope structures into idiotypic determinants (Id) expressed on the surface of antibodies. According to this theory, certain anti-idiotypic antibodies express three dimensional shapes which resemble the structures of external antigens. The existence of such internal antigenic idiotopes comes as a statistical necessity of the diverse and "complete" network of idiotopes (Bona and Kohler, 1984; Kohler et al, 1985). The internal antigens are stereochemical copies of nominal antigens and can induce specific immune responses similar to responses induced by nominal antigens and also can compete with nominal antigens in binding assays.

Assuming one is aiming for a tumor specific immunotherapy, the anti-idiotype hybridoma route of making surrogate antigens could alleviate the problem of preparing a large amount of purified antigens associated with a given tumor. Furthermore idiotype-based tumor antigens are free of the danger of transmitting tumor viruses which may contaminate cell-based tumor vaccines. Parenthetically, it makes no difference in this approach whether a de-novo tumor antigen is the target for immunotherapy or whether the target is a "normal" but highly tumor-associated antigen. If a tumor specific hybridoma is available or can be made, then a second anti-anti-hybridoma, (an anti-idiotypic hybridoma) can be made and selected for the structure which represents the tumor antigen. Figure 1 below should help clarify this approach. An anti-TAA antibody (Ab1) is generated and is used to make a second antibody or an anti-idiotype antibody (Ab2).

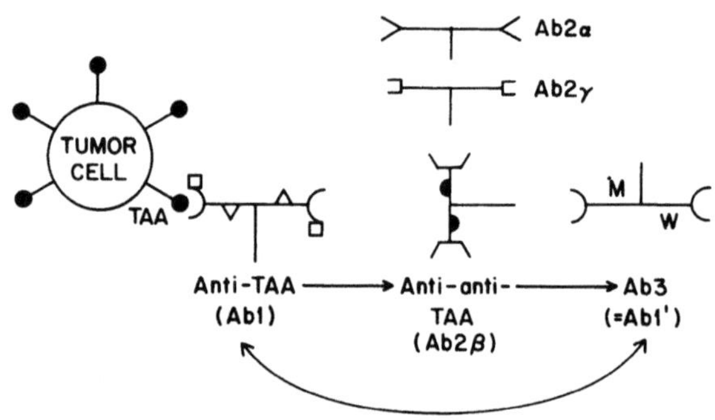

Figure 1. Idiotype Tumor Network

At this point one must distinguish between three different kinds of Ab2's: if the target for Ab2 is an idiotope which is not too close to the binding site of Ab1, then this has been called by Jerne (1974) the α type; if the target idiotope is close to the binding site so it can interfere with the antigen binding, Bona and Kohler (1984) have called this the γ type; if however Ab1 binds to an idiotypic determinant of Ab2, then this becomes the β type (Jerne et al, 1982; Jerne, 1984).

The β type of Ab2 is the internal image of the antigen for Ab2 (see Fig. 1). If this β type is used like an antigen, an Ab3 antibody is induced which has a similar binding site (paratope) as Abl. Because of this paratope similarity, the Ab3 is also called Abl' to indicate that it might differ in its other idiotopes from Abl.

This cyclic nature of complementary binding sites and idiotopes is the basis for the approach to make idiotope vaccines. The experimental evidence that idiotypes or anti-idiotypes mimic antigens is strong (Kohler, 1985; Lindeman, 1973; Nisonoff and Lamoyi, 1981). Stimulation via internal antigen is operationally different from anti-idiotype stimulation. While with anti-idiotype stimulation, the genetic predisposition to make an idiotype must exist, stimulation with internal antigen is not idiotypically restricted. For inducing a response with internal antigen, only the general ability to mount a response must be there, regardless of how restricted or polyclonal that response might be.

This alternative and safe approach of immunotherapy with the use of internal image antigens as vaccines has generated a great deal of interest in recent years in the treatment of cancer and infectious diseases. Studies have indicated that anti-idiotype vaccine can be effective in inducing protective immunity against a variety of infectious agents (reviewed by UydeHaag et al, 1986). Idiotype vaccines may also find potential application in cancer immunotherapy.

ANTI-IDIOTYPE ANTIBODY AS POSSIBLE VACCINE FOR HUMAN T CELL LEUKEMIA AND LYMPHOMA

To explore the feasibility of the anti-idiotype vaccine approach for human cancer, we have initiated our studies with human T cell leukemias and lymphomas. T cell acute lymphoblastic leukemia (T-ALL), which represents a common form of childhood leukemia, is associated with poor prognosis and a high rate of relapse (Sallan et al, 1980; Greaves, 1981; Bowman et al, 1981). Cutaneous T cell lymphoma (CTCL) is an indolent but incurable disease. In this regard, exploitation of new approaches to curing T-ALL and CTCL appear to be important. A new approach in the proposed study is to use anti-idiotype antibodies as an antigen substitute for the induction of therapeutic immunity. A number of murine monoclonal antibodies against various unique human T-cell leukemia-associated antigens have been previously described. These monoclonals were made by Seon and associates (Seon et al, 1983; 1984; 1986). The monoclonal antibody SN2 was selected for production of anti-idiotypes because it defines the protein moiety of a relatively unique human T-cell leukemia/lymphoma-associated cell-surface glycoprotein, gp37 (MW 37,000), which has a high density on the surface of human T-acute lymphoblastic leukemia cells and CTCL cells and has been partially characterized (Seon et al, 1986). The gp37 antigen is expressed by the majority of T-ALL and CTCL cells and has a limited heterogeneity among cells removed from patients at different stages of therapy or phases of the cell cycle. SN2 has a high specificity for T-leukemia/ lymphoma cells and does not react with any normal tissues including thymocytes, lymphocytes, bone marrow cells, liver, kidney, etc. Furthermore, gp37 is barely detectable on phytohemagglutinin- and concanavalin A-activated T-cells. SN2 (isotype IgG1,k) is highly cytotoxic to T-leukemia cells when conjugated with the A chain of ricin (Seon, 1984), but it was not cytotoxic to normal haematopoietic progenitors, i.e. CFU-GM, CFU-E and CFU-GEMM (Tebbi et al, 1985). In a recent study, complete suppression of in vivo growth of human T-leukemia cells in nude mouse models was achieved by using SN2-immunotoxins (Hara and Seon, 1987). The absence of antigenic modulation of gp37 by SN2 further indicated its usefulness as an appropriate target for immunotherapy of T-cell leukemia/lymphoma patients.

Generation and Characterization of Anti-Idiotype Hybridomas

$F(ab')_2$ fragment of SN2 coupled to KLH was used to generate Ab2 in syngeneic BALB/c mice. Culture supernatants of primary fusion cells were initially screened on the basis of their binding to the $F(ab')_2$ fragment of SN2 and their failure to bind to $F(ab')_2$ of pooled normal BALB/c mice sera. Fifteen Ab2 hybridomas were obtained from two fusions which were specific for the SN2-idiotope and did not react with isotype or allotype determinants. The Ab2 hybridomas were cloned twice by limited dilution. To find out whether these anti-idiotypes were directed against the paratope of SN2, the binding of radiolabeled SN2 to human leukemic MOLT-4 cells and JM-cells which contain gp37 as a surface constituent, was studied in the presence of different amounts of these Ab2. Ab2 clone 4EA2 inhibited the binding 100 percent at a concentration of 50 ng and 4DC6 inhibited 90 percent at a concentration of 250 ng. A third clone 4DD6 gave about 50 percent inhibition. Similar was the inhibition of SN2 binding to insolubilized MOLT-4 antigen or cell membrane preparation. The binding of SN2 (Ab1) to 4EA2 and 4DC6 was also inhibited by semipurified preparation of gp37 antigen. These results demonstrate that at least two of the anti-idiotope antibodies are binding either at or near the binding site idiotope of SN2. Next, we investigated whether any of these two purified anti-idiotope antibodies 4EA2 and 4DC6, could induce an anti-tumor antibody response in syngeneic BALB/c mice. Sera from mice immunized with 4EA2 and 4DC6 coupled to KLH contained antibodies binding specifically to semi-purified gp37 (Figure 2). Antibodies to gp37 were not detectable in the sera of animals immunized with unrelated Ab2-KLH (2F10, 3A4) or pre-immune mice. When MOLT-4 cells were used as target, we detected significant binding with Ab2 immune sera when compared to pre-immune control. Immune sera also inhibited the binding of iodinated Ab2 to Ab1, indicating that Ab3 in mice shares idiotopes with Ab1 (SN2). Also, the binding of iodinated Ab2 to Ab1 was inhibited by rabbit antisera specific for gp37. The ability to interact with an anti-gp37 antibody generated in a different species was additional evidence that the anti-Id may be an Ab2β (Bhattacharya-Chatterjee et al, 1987).

We have obtained two other anti-Ids 1E11 (IgG2a, k) and 4G10 (IgG1, k) from a third fusion experiment which also seem to be directed against the paratope of SN2. These two anti-Ids will be characterized further.

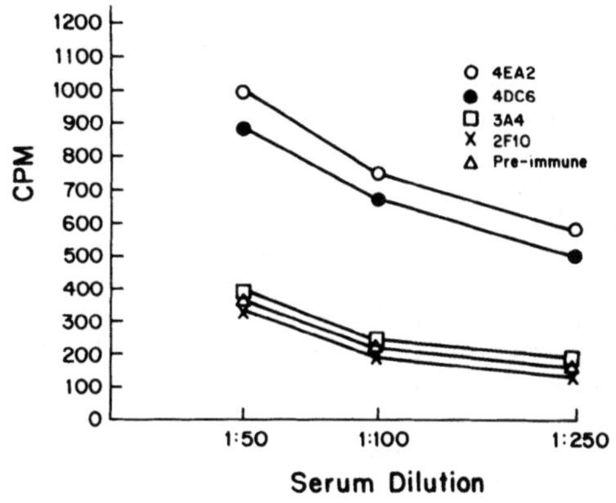

Figure 2. Assay of anti-gp37 antibodies induced by anti-idiotopes. BALB/c mice were immunized with 100 μg of Ab2 coupled to KLH on day 0,14, and 28. Sera were obtained on day 35 and assayed for antibodies binding to semipurified gp37. Plates were coated with gp37 antigen extract. Sera from mice immunized with Ab2 were added at different dilutions and developed with sheep anti-mouse IgG F(ab')2-^{125}I (~20,000 cpm). The results are presented as the mean cpm (n = 3). The SD of the data was <10% for each assay.

Generation of Rabbit Ab2β Against Murine Monoclonal Antibody SN2

We also made polyclonal anti-idiotype antibody in rabbits against SN2. Rabbits were immunized by s.c. injection of 0.5 mg of SN-IgG1 F(ab')$_2$ fragment mixed with Freund's complete adjuvant followed by two booster injections with the same amount of antigen in IFA during the next 3 months. Anti-idiotope antiserum was made specific by sequential adsorption through normal BALB/c Ig and 1E3 (IgG1,k, BALB/c) coupled to Sepharose 4B followed by absorption and elution from SN2-Sepharose 4B column. The anti-Id activity of the purified antibody was assayed by two methods: 1) by measuring their capacity to bind ^{125}I-labeled SN2 but not unrelated Ab1 of the same isotype and 2) by measuring the inhibition in a radioimmunoassay in which the binding of ^{125}I-labeled SN2 to MOLT-4 cells or gp37 antigen was determined in presence of rabbit anti-Id. The purified rabbit-anti-idiotypes will be useful for analysis of idiotype profiles in the sera of animals immunized with murine monoclonal Ab2.

Generation of Monoclonal and Polyclonal Anti-anti-idiotype Antibodies (Ab3)

If the Ab2s are true internal images, then they should induce the production of antigen-specific Ab3 in the absence of exposure to antigen in a genetically unrestricted way and across the species barriers (Nisonoff and Lamoyi, 1981; Kennedy et al, 1985). To confirm this, we immunized BALB/c mice and rabbits with Ab2 4DC6 and 4EA2 for the production of Ab3 that might share idiotopes with Ab1 and exhibit identical binding specificity.

Monoclonal anti-anti-idiotypic hybridomas were generated after fusion of spleen cells from immunized mice with non-secretor P3-653 mouse myeloma cells. Culture supernatants of primary fusion cells were initially screened on the basis of their binding to the IgG1 fraction of 4DC6 or 4EA2 and their failure to bind to the IgG1 fraction of an unrelated syngeneic monoclonal Ab1, 1E3. The idiotype positive fusion wells were then checked for their ability to bind to cultured human tumor cells which are SN2-defined antigen gp37 positive (MOLT-4 and JM). The Id-reactive and Ag-reactive fusion wells were then cloned by limiting dilution. Several monoclonal Ab3, mostly of IgM, k isotype and few IgG1-k, were obtained from the fusion experiment. The monoclonal Ab3 1A3 (IgG1-k) and 2C3 (IgM-k) were used in our study.

Anti-isotypic and anti-allotypic antibodies present in the rabbit Ab3 sera were absorbed out on mouse Ig coupled Sepharose 4B column.

Identification of Tumor Antigen-Specific Ab3 (Ab1')

Next, the monoclonal and polyclonal Ab3 were tested for the antibody binding to gp37 antigen preparation. Microtiter plates were coated with semi-purified preparation of gp37 antigen and reacted with monoclonal Ab3 culture supernatant of 1A3 and 2C3 or sera from rabbits immunized with 4DC6 and 4EA2. The bound antibody was detected by ELISA assay. Monoclonal antibody 2C3 showed binding to gp37 up to 1:100 dilution of supernatant. Somewhat less binding was observed with 1A3. No binding was obtained with the negative control, 1E3. SN2 (Ab1) was used as a positive control in this assay. Immunization of rabbits with 4DC6 and 4EA2 induced antibodies that also bind to insolubilized gp37 antigen. The highest titer was present in the serum of rabbit immunized with 4DC6. No binding was obtained when an unrelated ovarian tumor glycoprotein (Bhattacharya et al, 1982) was used as plate coat. When MOLT-4 cells were used as targets, sera from rabbits immunized four times with 4DC6 or 4EA2 contained antibodies which showed binding to MOLT-4 cells as compared to pre-immune sera.

Comparison of the Binding Specificity of Monoclonal Ab1 and Ab3

The immunoreactivity of murine monoclonal Ab3 and Ab1 was compared against a panel of human leukemia/lymphoma cell lines and normal EBV virus transformed B cell lines by cell binding radioimmunoassay. The binding specificity of monoclonal Ab3 was identical with SN2. We also compared the immunoreactivity of monoclonal Ab3 and Ab1 (SN2) against many normal human adult tissues by a sensitive immunoperoxidase (IP) staining method previously described by us (Gangopadhyay et al, 1985). The IP assay was performed on 5µm sections of frozen tissue blocks with Ab2 used as control antibody. None of the normal adult tissues tested showed any significant staining with monoclonal Ab1 (SN2), or Ab3 indicating the absence of gp37 in those organs. Frozen section of MOLT-4 cell blocks showed intense staining with both Ab1 and Ab3. We tested CTCL cells from 6 patients and 5 out of 6 were strongly positive for gp37 antigen with intense surface membrane staining (manuscript under preparation).

Competition of Abl and Ab3 for Binding to MOLT-4 Cells

If Ab3 has a similar binding site as Abl, it should compete with Abl for binding to the antigen gp37. A fixed amount of the radiolabeled SN2 was co-incubated with different dilutions of Ab3 or Abl preparations and MOLT-4 cells. Fifty percent inhibition of binding was obtained by using 10 ng of purified SN2 IgGl (Abl). Monoclonal Ab3 2C3 (ascites) and polyclonal adsorbed Ab3, rabbit No. 627 sera, produced 50 percent inhibition at a dilution of about 1:100 indicating that the monoclonal Ab3 in mice and polyclonal Ab3 in rabbits bind to the same antigen as Abl. Control monoclonal 1E3 ascites and preimmune rabbit sera did not produce any inhibition.

Inhibition of Abl Binding to Ab2 by Ab3

The question whether the anti-anti-idiotype antibody (Ab3) induced in syngeneic mice and rabbits by immunization with Ab2 shares SN2 idiotopes with SN2 was also addressed. Adsorbed sera from rabbit immunized with 4DC6, at 1:10 dilution, inhibited binding of iodinated Abl to Ab2 by 100 percent whereas sera from rabbit immunized with 4EA2 inhibited the binding by 68 percent. No significant inhibition was observed with preimmune sera. About 80 percent inhibition was obtained with monoclonal Ab3 2C3 ascites at a dilution of 1:100 and 54 percent inhibition was obtained with Ab3 1A3. No significant inhibition was obtained with control monoclonal ascites 1E3 at the same dilution. Although steric hindrance by Ab3 cannot be excluded in these assays, the results suggest the presence of Ab3 antibodies in mice and rabbits which share idiotopes with Abl (SN2).

Immunoprecipitation of gp37 Antigen by Abl and Ab3

Since the gp37 antigen has been partially purified and characterized (Seon et al, 1983; 1986), it was of interest to determine whether the Ab3 antibodies obtained from different species would precipitate the same glycoprotein as Abl. Such experiments have been successfully carried out by other investigators (Herlyn et al, 1986). For this, the iodinated gp37 antigen was immunoprecipitated by different Ab3 preparations as well as Abl and analyzed by SDS-PAGE (Figure 3). The results indicate that both murine monoclonal Ab3 1A3 and 2C3 precipitated a 37K major band which had the same mobility as the band precipitated by SN2 (Abl). Similar was the result obtained with rabbit polyclonal Ab3 No. 627 and No. 626 sera. When the iodinated gp37 antigen, pretreated with either monoclonal Ab3 or polyclonal Ab3, was reacted with SN2, there was no significant immunoprecipitation suggesting that the iodinated preparation was depleted of the gp37 antigen (Bhattacharya-Chatterjee et al, 1988).

In these preliminary studies we have characterized two monoclonal syngeneic anti-idiotypic antibodies which induce TAA specific humoral responses in an idiotype cascade. These responses are induced in the mouse which is the origin of the Abl and in a different species (rabbit). By this criteria, such anti-idiotypic antibodies are referred to as internal images or Ab2β (Bona and Kohler, 1984; Kohler et al, 1985). A

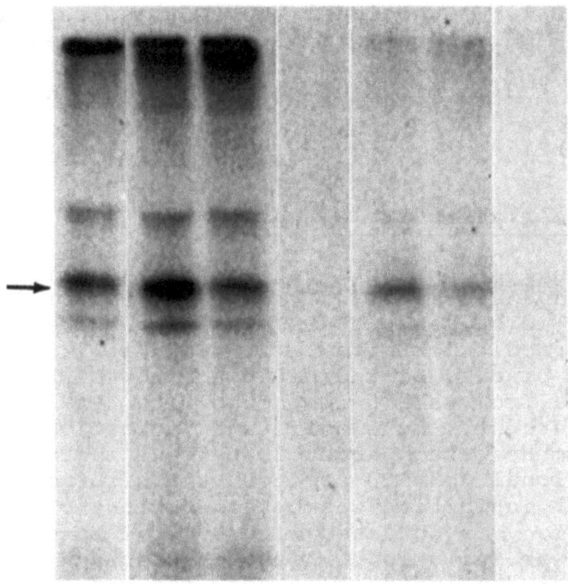

Figure 3. SDS-PAGE pattern of ^{125}I-labeled gp37 Ag after immunoprecipitation with SN2 (lane 1), Ab3 2C3 (lane 2), Ab3 1A3 (lane 3), unrelated Ab1 1E3 (lane 4), rabbit Ab3 No. 627 sera (lane 5), rabbit Ab3 No. 626 sera (lane 6), and pre-immune rabbit sera (lane 7). Marker sizes are shown in kDa. The major band at 37 kDa (arrow) represents the ^{125}I-labeled gp37 Ag. Amount of antibodies used per gel was either 10 µl of ascites or 10 µl of absorbed rabbit sera. gp37 Ag was labeled by the chloramine-T method and pre-cleaned by treatment with an unrelated murine mAb 1E3 twice. Approximately 5% of the radioactivity was non-specifically bound to unrelated monoclonal mouse IgG. This treatment also removed non-specific binding to normal rabbit IgG. The pre-cleaned radiolabeled Ag was then incubated with different Ab1 and Ab3 preparations, and the precipitated molecules were analyzed by SDS-PAGE and autoradiography. The film was exposed for 3 days.

molecular interpretation of the biological effects induced by anti-idiotypic antibodies is the concept that the anti-idiotype mimics the molecular contacts which exist between antigen and antibody (Ab1). In our system 4DC6 and 4EA2 can establish inter-molecular contacts with the anti-TAA and Ab1 like Ig receptors on B cells in a similar manner as the gp37 TAA. This conclusion is derived from the following observations: (1) Ab2 inhibited the binding of Ab1 to gp37 antigen and target cells expressing gp37 as a surface constituent, (2) the binding of Ab2 to Ab1 was antigen inhibitable, (3) Ab2 reacted with rabbit antisera made operationally specific for gp37, and (4) Ab2 induced the production of antigen specific Ab3 (Ab1') in mice and rabbits in the absence of exposure to gp37 antigen. Collectively these data indicate that these Ab2 may be useful as idiotype vaccine for human T cell leukemia/lymphoma.

STUDIES RELATED TO TUMOR-SPECIFIC ANTI-ID VACCINES

There are several reports for the production of network antigens (anti-idiotopes) which induce responses against human tumor-associated antigens. Nepom et al. (1984) described polyclonal anti-idiotypic antibodies raised in rabbits against murine monoclonal antibody 8.2, an antibody specific for a human melanoma-associated marker called p97. Mice immunized with anti-idiotype demonstrated delayed-type hypersensitivity reaction when challenged with p97-positive melanoma cells. Herlyn et al. (1986) produced anti-idiotypic antibodies in goat against a mouse monoclonal antibody to human gastric carcinoma antigen GA733. When the anti-idiotype was injected into mice and rabbits, it elicited the production of Ab3 which reacted with the same tumor antigen as the monoclonal antibody Ab1, that had induced the anti-idiotype. This demonstrated that the anti-idiotype carried the internal image of the tumor antigen. Recently, Kusama et al. (1987) have developed a monoclonal anti-idiotype antibody against a syngeneic monoclonal antibody defining a high molecular weight human melanoma-associated antigen which was binding site specific. However, they did not produce Ab3 or Ab1' antibody to characterize these anti-Ids further. Viale et al. (1987) have reported on the characterization of two Ab2β anti-idiotypic antibodies which induced anti-human tumor antibodies in mice and rabbits.

Recently, Herlyn et al. (1987) has started clinical trials in human colorectal patients with a polyclonal anti-Id raised against the MoAb 17-1A that recognizes a colon carcinoma antigen. Thirty patients with advanced CRC were immunized with goat Ab2 to Ab1 CO17-1A. Patients were treated with alum-precipitated Ab2 in doses between 0.5 and 4 mg per injection. All patients developed Ab3 with binding specificities on the surfaces of cultured tumor cells similar to the specificity of Ab1. Furthermore, the Ab3 competed with Ab1 for binding to CRC cells. Fractions of Ab3-containing sera, obtained after elution of the serum immunoglobulins from CRC cells bound to purified tumor antigen and inhibited binding of Ab2 to Ab1. Although the aim of this trial was to evaluate Ab3 responses in patients with advanced CRC, objective clinical improvement was observed in 13 of the 30 treated patients. In another clinical trial 15 patients with disseminated melanoma were immunized with monoclonal anti-idiotypic antibodies to anti-melanoma monoclonal antibody. Seven patients developed Ab3 antibodies which inhibited the binding of Ab2 to Ab1 and 3 patients showed tumor reduction. Immunization of patients with Ab2 effectively elicited anti-tumor immune responses that were highly specific (Ziai et al, 1988).

These results showing the effects of anti-Id treatment in human colorectal carcinoma and melanoma are quite encouraging. In addition, active immunization with tumor-specific idiotype vaccines has been shown to inhibit the growth of tumors in animal models (Sugai et al, 1974; Stevenson and Gordon, 1983; Kaminski et al, 1987). Raychaudhuri et al. (1986, 1987) have undertaken an extensive study on the effect of anti-idiotype therapy in a controlled mouse leukemia model L1210 in DBA/2 mice, which has provided us with the basic information on B and T cell induced responses using the anti-idiotype approach. They generated a number of anti-idiotypic hybridomas against monoclonal antibody to the L1210 tumor, which have been shown to induce tumor-specific DTH, inhibition of tumor growth, CTL, antibodies and T helper cells in this system. These findings are very promising since they demonstrate a crossreaction of nominal antigen and internal image antigen for a tumor-associated antigen system at the T- and B-cell level. In a recent study, 100 percent cure of established tumors was achieved in DBA/2 mice by combining anti-Id vaccines with cyclophosphamide whereas 50 percent cure rate was obtained with anti-Id therapy alone. Similar findings have been also obtained by Levy and co-workers who observed that cyclophosphamide (100 mg/kg), administered in combination with idiotype vaccines to mice bearing ten day old, 1-2 cm diameter subcutaneous B cell lymphoma (38C13), resulted in a dramatic survival benefit. As compared to no therapy, immunotherapy with idiotype vaccine alone also prolonged survival ($p<.001$) (Campbell et al, 1988). Kennedy et al. (1985) used an anti-Id antibody to induce immunity to SV-40 transformed cells. Mice vaccinated with this anti-Id demonstrated prolonged survival after tumor transfer. Recently Nelson et al. (1987) have examined the role of idiotype interactions in regulating the immune response of mice to chemically induced, syngeneic sarcomas. They showed that treatment of mice with established sarcomas MCA-1490 and MCA-1511, with the use of i.p. injection of the appropriate anti-idiotype monoclonal antibody, had significant antitumor activity.

CONCLUSIONS

One other major advantage of idiotype vaccines is that by their virtue of being proteins, they can be easily manipulated; they can be coupled to potent immunogenic carriers to become T cell dependent antigens which can receive full T help. This further alteration of the molecular environment for expression of the idiotope antigens was indeed required in the work on the generation of an idiotope vaccine for Pneumococcal infection (McNamara et al, 1984). Without coupling the idiotope antigen to a carrier, the internal image was ineffective in producing a serum titer or in providing immunity against infection. Studies from Dr. Levy's laboratory have demonstrated the immunoprophylactic effects of idiotype vaccination against a murine B cell lymphoma (Kaminski et al, 1987; Campbell et al, 1988). These studies also revealed that conjugation of the idiotype protein to an immunogenic carrier protein, such as KLH, and administration with an adjuvant, such as the muramyl dipeptide analog (MDP-derivative) were necessary for optimal anti-tumor responses. Thus, we might have to include in our procedure of making a tumor vaccine the step of modifying the internal antigen idiotope. On the other hand, if mouse anti-Id is used in humans, the mouse-specific Fc portion may function as a carrier determinant and no further carrier may be needed.

One other point which must be considered is the compatability of mouse hybridoma internal antigen in patients. Previous experience with murine hybridomas in humans, however have not shown major anaphylactic type reactions (Oldham et al, 1984; Schroff et al, 1985). Many allergic reactions including low-grade fever, nausea, diarrhea and urticaria are not uncommon. These reactions have not led to discontinuation of therapy, however. We may find even less toxicity using fragments of antibody such as Fab or $F(ab')_2$ and increase the immunogenicity of the fragments by incorporation of an adjuvant and/or carrier such as tetanus toxoid or KLH and MDP-derivative. Nevertheless, multiple immunizations may decrease the effectiveness of the anti-Id vaccination by producing immune complex formation.

Alternatively, potential heterologous and homologous anti-Id vaccines may be replaced by synthetic peptide antigens. In brief, it is possible to obtain the three dimensional shape of idiotope internal antigens using computer modeling on the sequence of internal antigen hybridomas (Kieber-Emmons et al, 1986). The molecular modeling studies will indicate whether the internal idiotope antigen is expressed as a linear sequence or a complex determinant involving non-contiguous sequence contributions to the molecular architecture. After obtaining sufficient information as to the most likely peptide regions which depict the antigenic structure, it should be possible to synthesize the relevant peptides which can mimic the structure and couple them to new synthetic adjuvants to increase the relative level of immunogenicity.

Heterogeneous expression and modulation of target antigens represent other potential problems to immunotherapy of tumors induced by anti-Id vaccine preparations. However, these problems are not insurmountable and they may be minimized by several complementary approaches. Heterogeneity of TAA expression may be addressed by utilizing cocktails of anti-Id vaccine preparations directed against multiple target antigens collectively expressed by the vast majority of tumor cells. With respect to antigenic modulation by Ab3 induced in response to Ab2β, it is known that modulation rates for different antigens may vary considerably. Hence, TAAs with slow modulation characteristics should be chosen as targets for anti-Id active immunotherapy. Furthermore, recent studies (Liao et al, 1982; Carrel et al, 1985) have indicated that some biological response modifiers such as IFN, are capable of inducing the expression of TAAs in several tumor systems. Hence, in addition to providing general stimulation to the immune system, the administration of biological response modifiers known to increase the expression of relevant TAAs may further enhance the efficacy of anti-Id tumor vaccine application.

The anti-idiotype approach needs to be compared to other tumor therapies; both established and experimental. A realistic assessment of the anti-Id therapy predicts that complete remission cannot be expected at the present state. However, evidence exists already that partial remission and responses can be achieved with anti-idiotypes. It is apparent that successful therapy of established tumors will depend on a multi-modality approach. Since the immune response has a limited capacity, immunotherapeutic manipulations to destroy a tumor will probably be effective only against small tumor masses. Consequently, immunotherapy with idiotype vaccine should prove most curative in the adjuvant setting where the disease has been controlled or stabilized with conventional therapies. Compared to chemotherapy or lymphokine therapy, the idiotype approach is a safer and less toxic form of treatment and may improve the quality of life in cancer patients which is sometimes compromised by more aggressive chemo- and/or radiotherapy.

ACKNOWLEDGEMENTS

This work was supported by United States Public Health Service Grant CA 41636.

REFERENCES

Bhattacharya, M., Chatterjee, S. K., Barlow, J. J., and Fuji, H., 1982, Monoclonal antibodies recoginzing tumor-associated antigen of human ovarian mucinous cystadenocarcinomas, Cancer Res., 42:1650.

Bhattacharya-Chatterjee, M., Pride, M. W., Seon, B. K., and Kohler, H., 1987, Idiotype vaccines against human T cell acute lymphoblastic leukemia. I. Generation and characterization of biologically active monoclonal anti-idiotopes, J. Immunol., 139:1354.

Bhattacharya-Chatterjee, M., Chatterjee, S. K., Vasile, S., Seon, B. K., and Kohler, H., 1988, Idiotype vaccines against human T cell leukemia. II. Generation and characterization of a monoclonal idiotype cascade (Ab1, Ab2 and Ab3), J. Immunol., 141:1398.

Bona, C. A., and Kohler, H., 1984, Anti-idiotypic antibodies and internal images, in: "Probes for Receptor Structure and Function", J. C. Venter, C. M. Fraser, and J. Lindstrom, eds., Alan R. Liss, Inc., New York, pp. 141-150.

Bowman, W. P., Melvin, S. L., Aur, R. J. A., and Mauer, A. M., 1981, A clinical perspective on cell markers in acute lymphocytic leukemia, Cancer Res., 41:4794.

Campbell, M. J., Carroll, W., Kon, S., Thielemans, K., Rothbard, J. B., Levy, S., and Levy, R., 1988, Idiotype vaccination against murine B cell lymphoma. Humoral and cellular responses elicited by tumor-derived IgM and its molecular subunits, J. Immunol., in press.

Campbell, M. J., Esserman, L., and Levy, R., 1988, Immunotherapy of established murine B cell lymphoma. Combination of idiotype vaccine and cyclophosphamide, J. Immunol., 141:3227.

Carrel, S., Schmidt-Kessen, A., and Giuffre, L., 1985, Recombinant interferon-gamma can induce the expression of HLA-DR and -DC on DR negative melanoma . cells and enhance the expression of HLA-ABC and tumor-associated antigens, Eur. J. Immunol., 15:118.

Gangopadhyay, A., Bhattacharya, M., Chatterjee, S. K., Barlow, J. J., and Tsukada, Y., 1985, Immunoperoxidase localization of a high molecular weight mucin recognized by monoclonal antibody 1D3, Cancer Res., 45:1650.

Greaves, M. F., 1981, Analysis of the clinical and biological significance of lymphoid phenotypes in acute leukemia, Cancer Res., 41:4752.

Greene, M. I., 1980, Cellular and genetic basis of immune reactivity to tumor cells, Contemp. Top. Mol. Immunol., 11:81.

Hara, H., and Seon, B. K., 1987, Complete suppression of in vivo growth of human leukemia cells by specific immunotoxins: Nude mouse models, Proc. Natl. Acad. Sci. USA, 84:3390.

Haubeck, H. D., and Kolsch, E., 1982, Tumor-specific T suppressor cells induced at early stage of tumorigenesis act on the induction phase of cytotoxic T cells, Immunology, 47:503.

Herlyn, D., Ross, A. H., and Koprowski, H., 1986, Anti-idiotypic antibodies bear the interal image of a human tumor antigen, Science, 232:100.

Herlyn, D., Wettendorff, M., Schmoll, E., Iliopoulos, D., Schedel, I., Drejkhausen, U., Raab, R., Ross, A. H., and Koprowski, H., 1987, Anti-idiotype immunization of cancer patients: Modulation of the immune response, Proc. Natl. Acad. Sci. USA, 84:8055.

Howie, S. M., and McBride, W. H., 1982, Tumor specific T helper activity can be abrogated by two distinct suppressor cell mechanisms, Eur. J. Immunol., 12:671.

Jerne, N. K., 1974, Towards a network theory of the immune system, Ann. Immunol., (Paris), 124C:171.

Jerne, N. K., 1984, Idiotypic networks and other preconceived ideas, Immunol. Rev., 79:5.

Jerne, N. K., Roland, J., and Cazenave, P. A., 1982, Recurrent idiotypes and internal images, EMBO J., 1:243.

Kaminski, M. S., Kitamura, K., Maloney, D. G., and Levy, R., 1987, Idiotype vaccination against murine B cell lymphoma. Inhibition of tumor immunity by free idiotype protein, J. Immunol., 138:1289.

Kennedy, R. C., Dreesman, G. R., and Kohler, H., 1985, Vaccines utilizing internal image anti-idiotypic antibodies that mimic antigens of infectious organism, Biotechniques, 3:404.

Kennedy, R. C., Dreesman, G. R., Butel, J. S., and Lanford, R. E., 1985, Suppression of in vivo tumor formation induced by simian virus 40 transformed cells in mice receiving anti-idiotypic antibodies, J. Exp. Med., 161:1432.

Kieber-Emmons, T., Ward, R. E., Raychaudhuri, S., Rein, R., and Kohler, H., 1986, Rational design and application of idiotope vaccines, Intern. Rev. Immun., 1:1.

Kohler, H., 1985, The immune network revisited, in: "Idiotypes in Biology and Medicine", H. Kohler, J. Urbain, and P. Cazenave, eds., Academic Press, New York, pp. 3-28.

Kohler, H., Muller, S., and Bona, C., 1985, Internal antigen and the immune network, Proc. Soc. Exp. Biol. Med., 178:189.

Kusama, M., Kageshita, T., Kim, K. T., and Ferrone, S., 1987, Antiidiotypic monoclonal antibodies to antihuman high molecular weight melanoma-associated antigen monoclonal antibodies, Proc. Am. Assoc. Cancer Res., 28:361.

Liao, S. K., Kwong, P. C., Khosravi, M., and Dent, P. B., 1982, Enhanced expression of melanoma-associated antigens and β_2-microglobulin on cultured melanoma cells by interferon, J. Natl. Cancer Inst., 68:19.

Lindemann, J., 1973, Speculation of idiotypes and homobodies, Ann. Immunol., 124C:171.

McBride, W. H., and Howie, S. M., 1986, Induction of tolerance to a murine fibrosarcoma in two zones of dosage-the involvement of suppressor cells, Br. J. Cancer, 53:707.

McNamara, M. K., Ward, R. E., and Kohler, H., 1984, Monoclonal idiotype vaccine against Streptococcus pneumoniae infection, Science, 226:1325.

Nelson, K. A., George, E., Swenson, C., Forstrom, J. W., and Hellstrom, K. E., 1987, Immunotherapy of murine sarcomas and auto-anti-idiotypic monoclonal antibodies which bind to tumor-specific T cells, J. Immunol., 139:2110.

Nepom, G. T., Nelson, K. A., Holbeck, S. L., Hellstrom, I., and Hellstrom, K. E., 1984, Induction of immunity to a human tumor marker by in vivo administration of anti-idiotypic antibodies in mice, Proc. Natl. Acad. Sci. USA, 81:2864.

Nisonoff, A., and Lamoyi, E., 1981, Implications of the presence of an internal image of the antigen in anti-idiotypic antibodies: possible application to vaccine production, Clin. Immunol. Immunopathol., 21:397.

Oldham, R. K., Morgan, A. C., Woodhouse, C., Schroff, R. W., Abrams, P., and Foon, K., 1984, Monoclonal antibodies in the treatment of cancer: preliminary observations and future prospects, Med. Oncol. and Tumor Pharmacother., 1:51.

Raychaudhuri, S., Saeki, Y., Fuji, H., and Kohler, H., 1986, Tumor-specific idiotype vaccines. I. Generation and characterization of internal image tumor antigen, J. Immunol., 137:1743.

Raychaudhuri, S., Saeki, Y., Chen, J. J., Iribe, H., Fuji, H., and Kohler, H., 1987, Tumor-specific idiotype vaccines. II. Analysis of the tumor-related network response induced by the tumor and by internal image antigens (Ab2), J. Immunol., 139:271.

Raychaudhuri, S., Saeki, Y., Chen, J. J., and Kohler, H., 1987, Tumor-specific idiotype vaccines. III. Induction of T helper cells by anti-idiotype and tumor cells, J. Immunol., 139:2096.

Sallan, S. E., Ritz, J., Pesando, J., Gelber, R., O'Brien, C., Hitoncock, S., Coral, F., and Schlossman, S. F., 1980, Cell surface antigens: prognostic implications in childhood acute lymphoblastic leukemia, Blood, 55:395.

Schroff, R. W., Foon, K. A., Wilburn, S. B., Oldham, R. K., and Morgan, A. C., 1985, Human anti-murine immunoglobulin responses in patients receiving monoclonal antibody therapy, Cancer Res., 45:879.

Seon, B. K., 1984, Specific killing of human T-leukemia cells by immunotoxins prepared with ricin A chain and monoclonal anti-human T-cell leukemia antibodies, Cancer Res., 44:259.

Seon, B. K., Negoro, S., and Barcos, M., 1983, Monoclonal antibody that defines a unique human T-cell leukemia antigen, Proc. Natl. Acad. Sci. USA, 80:845.

Seon, B. K., Negoro, S., Barcos, M. P., Tebbi, C. K., Chervinsky, D., and Fukukawa, T., 1984, Monoclonal antibody SN2 defining a human T cell leukemia associated cell surface glycoprotein, J. Immunol., 132:2089.

Seon, B. K., Fukukawa, T., Jackson, A. L., Chervinsky, D., Tebbi, C. K., Freeman, A. L., and Matsuzaki, H., 1986, Human T-cell leukemia-associated cell surface glycoprotein GP37: studies with three monoclonal antibodies and a rabbit antiserum, Mol. Immunol., 23:569.

Stevenson, F. K., and Gordon, J., 1983, Immunization of idiotypic immunoglobulin protects against development of B lymphocytic leukemia, but emerging tumor cells can evade antibody attack by modulation, J. Immunol., 130:970.

Sugai, S., Palmer, D. W., Talal, N., and Witz, I. P., 1974, Protective and cellular immune responses to idiotypic determinants on cells from a spontaneous lymphoma of NZB/NZWF mice, J. Exp. Med., 140:1547.

Tebbi, C. K., Chervinsky, D., and Seon, B. K., 1985, Effect of specific anti-human leukaemia immunotoxins on the colony formation by human haematopoietic progenitors and by human leukaemia cells, Clin. Exp. Immunol., 60:496.

UydeHaag, F. G. C. H., Bunschoten, H., Weijer, K., and Osterhaus, A. D. M. E., 1986, From Jenner to Jerne: Towards idiotype vaccines, Immunol. Rev., 90:93.

Viale, G., Grassi, F., Pelagi, M., Alzani, R., Menard, S., Miotti, S., Buffa, R., Gini, A., and Siccardi, A. G., 1987, Anti-human tumor antibodies induced in mice and rabbits by "internal image" anti-idiotypic monoclonal immunoglobulins, J. Immunol., 139:4250.

Ziai, M. R., Armandola, E., Hyashibe, K., Giordano, A., Kageshita, T., Mittelman, A., and Ferrone, S., 1988, Triggering of the idiotype network in a patient with melanoma injected with murine anti-idiotypic monoclonal antibodies, Proc. Am. Assoc. Cancer Res., 29:436.

A VH REGION SYNTHETIC PEPTIDE INDUCES ANTIBODIES WHICH BIND NATIVE IMMUNOGLOBULINS AND AUGMENT AN IMMUNE RESPONSE TO ANTIGEN

F. A. Bonilla, R. S. Anderson*, M. Z. Atassi*, and C. A. Bona

Department of Microbiology, The Mount Sinai School of Medicine of The City University of New York New York, NY 10029, USA

*Department of Biochemistry, Baylor College of Medicine, Houston, TX 77030, USA

ABSTRACT

The BALB/c myeloma protein ABPC48 (A48) binds the polysaccharide bacterial levan; its Vh is encoded by a gene derived from the VhX24 family. This antibody has been shown to cross-react idiotypically with the phosphorylcholine-binding BALB/c myeloma protein MOPC167 whose Vh shares homology with A48 from residues 32-44. We have synthesized a peptide corresponding to residues 32-44 of the Vh encoded by a germline gene of the VhX24 family. Anti-peptide antisera from rabbits were purified by affinity chromatography with peptide or intact antibody. Several myeloma proteins and monoclonal antibodies with varying degrees of homology to the peptide have been analyzed for reactivity with purified rabbit antibodies in solid-phase RIA. We observed that the specificities within rabbit antisera are heterogeneous, and that purification with antibody versus peptide yields preparations containing different specificities, albeit demonstrably peptide-related. We also show that injection of mice at birth with small amounts of purified rabbit antibodies can affect the magnitude of the response to bacterial levan and the expression of A48 idiotopes in that response.

INTRODUCTION

The BALB/c plasmacytoma ABPC48 (A48) secretes antibody binding fructans containing beta(2-6) linkages such as bacterial levan and grass levan (Lieberman et al., 1975); and expresses an idiotype designated the A48Id. Well-defined treatments with polyclonal or monoclonal anti-A48Id antibodies can significantly increase expression of the A48Id among antibodies induced by levan immunization (Hiernaux et al., 1981; Rubinstein et al., 1982, 1983). This phenomenon is of considerable interest within models of immunoregulation which include a role for processes mediated by idiotype recognition (Jerne, 1974).

Another beta(2-6) fructan-binding BALB/c myeloma protein, UPC10 (U10), exhibits one or more A48 idiotopes (Legrain et al., 1981; Goldberg et al., 1983). As part of ongoing investigations of the occurrence of the A48Id among antibodies specific for levans or other antigens, we previously reported the screening of 168 murine and human myeloma proteins for presence of the A48Id (Goldberg et al., 1983). Only the phosphorylcholine-binding BALB/c myeloma protein MOPC167 (M167) was found to inhibit the reaction of syngeneic polyclonal anti-A48Id antibodies with A48.

U10 and A48 express genes of the VhX24 and Vk10 families (Auffray et al., 1981; Potter et al., 1982; Brodeur and Riblet, 1984; Legrain and Buttin, 1985). M167 expresses genes of the VhS107 and Vk24 families (Rudikoff and Potter, 1976; Potter et al., 1982; Brodeur and Riblet, 1984). Significant homology between these three proteins occurs only in a region containing part of CDR1, from residues 32 through 44 (see Figure 2B).

Synthetic peptides representing fragments of larger molecules have found wide application in studies of structure-function relationships and antigenic properties of the parent proteins. Idiotypic relationships of antibodies have also been probed in this manner (McMillan et al., 1983; Chen et al., 1984a, 1984b, 1985a, 1985b; Goldfien et al., 1985; Go i et al., 1985; Seiden et al., 1986). In order to further characterize the contribution of Vh residues 32-44 to the A48Id, we have synthesized a peptide corresponding to residues 32-44 of the protein encoded by the germline Vh gene expressed by both A48 and U10 (Auffray et al., 1981; Ollo et al., 1981) hereafter designated ID 32-44.

ID 32-44-specific antisera prepared in rabbits have been used to examine a number of myeloma proteins and monoclonal antibodies. We show that by purifying the same antiserum by adsorption with either the immunizing peptide or with intact immunoglobulin, one obtains preparations containing different specificities. Furthermore, rabbit antibodies against the synthetic peptide can affect the magnitude of the anti-bacterial levan response in vivo, as well as the expression of the A48Id in this response.

MATERIALS AND METHODS

Synthesis of Peptide, Immunization, and Purification of Rabbit Antibodies

The ID 32-44 peptide was synthesized and coupled to chicken egg ovalbumin according to published techniques (Koketsu and Atassi, 1974; Atassi et al., 1981; 4.94 moles peptide/mole ovalbumin). The ID 32-44 peptide with a C-terminal extension of Lys-Lys-Gly (ID 32-44-KKG) was also synthesized and coupled to BSA (13.5 moles peptide/mole BSA). The ID 32-44-KKG peptide is more efficiently coupled to a carrier and was used in purification and assays. Only the ID 32-44 conjugate was used for immunization. The HIV peptide was provided by Dr. R. Kennedy (Southwest Foundation for Biomedical Research, San Antonio, TX) and was originally designated HTLV III 735-752 (Kennedy et al., 1986). This peptide represents part of the viral envelope glycoprotein, its complete sequence is YDRPEGIEEEGGERDRDRSGC.

New Zealand White rabbits (Pocono Rabbit Farms, Canadensis, PA) were injected intradermally first with 500 µg of ID 32-44-ovalbumin emulsified in CFA followed at monthly intervals by 200 µg in CFA. Rabbits were bled 10 days following the second or subsequent injections. Rabbit antisera were purified by passage over a column containing either ID 32-44-KKG-BSA or the monoclonal antibody LPS11-6 coupled to cyanogen bromide-activated Sepharose 4B (Pharmacia).

Myeloma Proteins and Monoclonal Antibodies

The myeloma proteins ABPC48, UPC10, XRPC24, MOPC167, and TEPC15 are originally from the collection of Dr. M. Potter (National Institutes of Health, Bethesda, MD). The myeloma proteins J606 and J558 are from the Salk Institute (La Jolla, CA). The BALB/c anti-A48Id hybridomas IDA3, 10, 16, 17, and 23 (Legrain et al., 1981) are from Dr. P. Legrain (Institut Pasteur, Paris, France). All other monoclonal antibodies (mAbs) were produced in our laboratory and purified by affinity chromatography over a column containing a monoclonal rat anti-mouse kappa chain antibody (187.1, from Dr. M. Scharf, Albert Einstein College of Medicine, New York, NY).

The mAb 2-1-3 has been previously described (Victor-Kobrin et al., 1985). LPS11-6, NWSM9-3 and NWSM17-3 are monoclonal autoantibodies derived from B cell mitogen-stimulated BALB/c spleen cells. LPS11-6 binds cardiolipin, myelin basic protein (MPB), thyroglobulin (Tg) and intrinsic factor. NWSM9-3 binds Tg, and NWSM17-3 binds MBP. The mAb 10N109-1 was derived from B-cell mitogen-stimulated BALB/c spleen cells from mice treated at birth with IDA10 and selected for expression of the VhX24 gene family (Bonilla, et al., submitted, and manuscripts in preparation). V gene usage was determined by Northern blotting or nucleotide sequence analysis (see Figure 2 legend).

Radioimmunoassays

Figure 1. Microtiter plates (Falcon) were coated with 50 ul ID 32-44-KKG-BSA at 10 ug/ml for 3 hr. at 37 deg. C. After washing with PBS and blocking for 30 min. at R.T. with 1% BSA in PBS and washing again, 50 ul rabbit antibodies at 10 ug/ml were added with or without inhibitor and plates incubated at R.T. for 3 hr. After washing with PBS, 50,000 CPM in 50 ul of 125I-labelled donkey-anti-rabbit immunoglobulin antibodies (Amersham) diluted in 1% BSA in PBS were added and plates incubated for 3 hr. at R.T. After extensive washing, plates were cut apart and individual wells counted in a LKB 1272 CliniGamma radiation counter.

Figures 2-5. Plates were coated with rabbit antibodies at 10 ug/ml, followed by various myeloma proteins or monoclonal antibodies at 10 ug/ml in the presence or absence of varying concentrations of inhibitors. Binding of these antibodies to the plates was detected using a 125I-labelled monoclonal rat-anti-mouse kappa chain antibody in 1% BSA in PBS.

Plaque-Forming Cell Assays

BALB/c mice were injected i.p. at birth with various antibody preparations and immunized i.p. at age one month with 50 µg bacterial levan (provided by C. Moreno, Hammersmith Hospital,

London, England). Five days later, spleens were taken and levan-specific antibody-producing cells were enumerated in the absence or presence of syngeneic polyclonal anti-A48Id antibodies. Preparation of bacterial levan from A. levanicum, its coupling to sheep red blood cells, and plaque detection were performed according to previously published methods (Hammerling and Westphal, 1967; Lieberman et al., 1976).

RESULTS

Serum from a rabbit immunized with an ID 32-44-ovalbumin conjugate was affinity-purified over a column containing an ID 32-44-KKG-BSA conjugate. (See Materials and Methods; henceforth we will abbreviate rabbit anti-ID 32-44 antibodies as RaIDab.) Figure 1 shows that the ID 32-44-BSA conjugate in solution inhibits the binding of RaIDab to ID 32-44-BSA-coated plates. The specificity of RaIDab is demonstrated by the lack of inhibition by a BSA-conjugated peptide representing a portion of HIV enve-

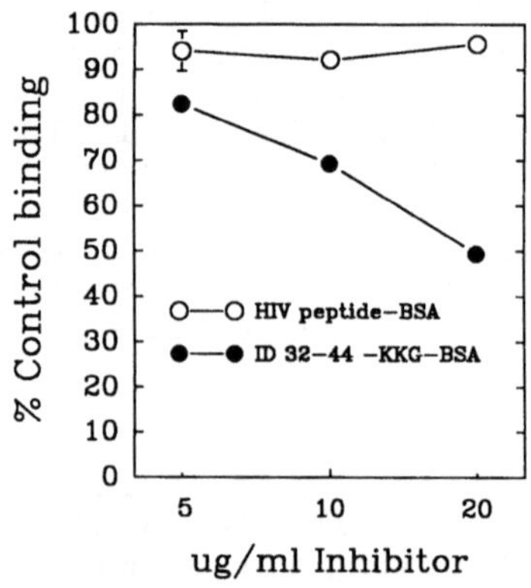

Fig. 1. RaIDab are specific for the immunizing peptide. ID 32-44-KKG inhibits the binding of RaIDab (detected with radiolabelled anti-rabbit immunoglobulin antibodies) to ID 32-44-KKG-BSA-coated plates, while a BSA conjugate of a peptide representing part of the HIV envelope glycoprotein and having no homology with ID 32-44 (Kennedy et al., 1986, see Materials and Methods) does not. Control CPM = 8,599 ± 284. Data points are mean ± S.E.M. for triplicate determinations.

lope glycoprotein and having no homology with ID 32-44.

Figure 2A shows the reaction of RaIDab with a variety of myeloma proteins and monoclonal antibodies. The data are ordered for clarity. Several of these antibodies were included as representatives of some of the Vh gene families (see Figure 2 legend). The antibodies listed in Figure 2A, from ABPC48 through 2-1-3 and 10N109-1, all express a gene of the VhX24 gene family. The antibodies LPS11-6 and NWSM9-3 use Vh7183, and NWSM17-3 uses VhQ52. The mAbs LPS11-6 and M167 react very well with RaIDab, while mAbs bearing VhX24 and other Vh gene families bind to variable degrees. Figure 2B compares some of the known sequences of Vh residues 32-44 for the antibodies we studied.

The two germline encoded protein sequences for residues 32-44 are represented by U10 and XRPC24 in Figure 2B. Neither of these antibodies binds strongly to RaIDab, although U10 possesses the precise sequence of ID 32-44. Partial sequencing of the LPS11-6 Vh showed that it belongs to the Vh7183 family, while Northern blotting indicates that the NWSM9-3 Vh also belongs to this family (data not shown). Homology between genes of the VhX24 and Vh7183 families can be as high as 75% (Brodeur and Riblet 1984). PY102 also expresses a Vh7183 gene, yet does not react with RaIDab (Figure 2).

Examination of the sequences of several of the antibodies studied suggests that a majority of RaIDab may be interacting with amino acids 40-44. TEPC15 has a Thr to Pro substitution relative to M167 at position 40, which abolishes reactivity with RaIDab. NWSM17-3 also has Pro at position 40, and has only 40% of the binding of M167. Differences in residues 32-39 and 44 between PY102 and NWSM17-3 may be responsible for their different binding to RaIDab. The relatively weak interaction of mAbs having high sequence homology with ID 32-44 with RaIDab must be due to differences in the accessibility of these residues to RaIDab in the context of the overall three-dimensional structure of the antibody variable region.

In Figure 3 we compare the binding of RaIDab to various antibodies in the presence and absence of 20 ug/ml of ID 32-44-BSA. In all cases where binding to RaIDab is observed, even where the interaction seems weak (e.g., with A48), the presence of ID 32-44 significantly reduces antibody binding. The HIV peptide-BSA conjugate at 20 ug/ml did not inhibit the binding of RaIDab to any of the antibodies tested (not shown). Thus, reaction of RaIDab with these antibodies is specific.

We also tested RaIDab for the ability to inhibit, in a solid-phase radioimmunoassay, the binding of various monoclonal antibodies to antigen-coated plates. RaIDab at a concentration of 20 ug/ml were able to inhibit the binding of M167 to PC-BSA, and 10N109-1 to rye levan by 28% and 35%, respectively. Rabbit anti-idiotypic antibodies against a monoclonal anti-influenza hemagglutinin antibody (PY206, Moran et al., 1984) did not inhibit. RaIDab were not able to significantly inhibit the antigen-binding of any other monoclonal antibodies tested (data not shown). The failure of RaIDab to inhibit the antigen binding of other antibodies with which it reacts well may be due either to

A

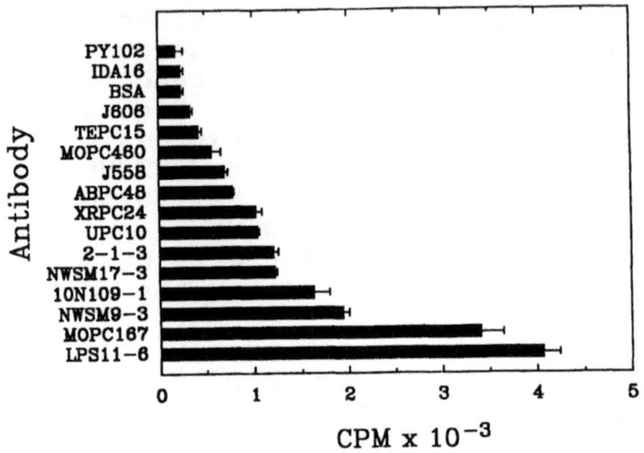

B

	CDR1							FR2					
	32	33	34	35	36	37	38	39	40	41	42	43	44
ID32-44	Tyr	Trp	Met	Ser	Trp	Val	Arg	Gln	Ala	Pro	Gly	Lys	Gly
PY102	---	Ala	---	---	---	---	---	---	Ser	---	Glu	---	Arg
J606	---	---	---	Asn	---	---	---	---	Ser	---	Gln	---	---
TEPC15	Phe	Tyr	---	Glu	---	---	---	---	Pro	---	---	---	Arg
MOPC460	Gly	Tyr	---	Asn	---	Ile							
J558	---	Tyr	---	Lys	---	---	Lys	---	Ser	His	---	---	Ser
ABPC48	---	---	---	---	---	---	---	---	---	---	---	Arg	---
XRPC24	---	---	---	---	---	Ala	---	---	---	---	---	---	---
UPC10	---	---	---	---	---	---	---	---	---	---	---	---	---
2-1-3	---	---	---	---	---	---	---	Arg	---	---	---	---	---
NWSM17-3	---	Gly	Val	His	---	---	---	---	Pro	---	---	---	---
MOPC167	Phe	Tyr	---	Glu	---	---	---	---	Thr	---	---	---	Arg

Fig. 2. A. Binding of RaIDab to different monoclonal antibodies. In this RIA The binding of monoclonal antibodies to RaIDab-coated plates was detected with a radiolabelled anti-mouse kappa chain antibody. Values given are mean ± S.E.M. for triplicate determinations. Vh gene families represented: IDA16, VhQ52 (Bona et al., 1985); 10N109-1, VhX24 (Bonilla et al., submitted); LPS11-6, Vh7183 (Bonilla et al., unpublished). For remainder see below.
B. ID 32-44 and related sequences are compared. References for sequences (Vh gene family in parentheses): PY102 (7183), Zaghouani et al., submitted; J606 (J606), MOPC460 (36-60), and J558 (J558), Kabat et al., 1987; MOPC167 (S107) and TEPC15 (S107), Rudikoff and Potter, 1976; ABPC48 (X24) and UPC10 (X24), Aufray et al., 1981; 2-1-3 (X24), Barak, 1988; XRPC24 (X24), Hartman and Rudikoff, 1984; NWSM17-3 (Q52), Bonilla, et al., unpublished data; 10N109-1 (X24), Bonilla, et al., submitted.

a small contribution of CDR1 to antigen binding in these cases, or to a much greater affinity of the monoclonal for antigen relative to the affinity of RaIDab for the monoclonal.

A series of BALB/c monoclonal anti-A48Id antibodies (the IDA series) have been obtained by Legrain et al. (1981). To explore the involvement of the Vh region represented by ID 32-44 in the formation of idiotopes recognized by these monoclonal anti-Ids, we attempted to inhibit the binding of IDA3, 10, 16, 17, or 23 to A48 with unconjugated peptide. Here we used a "sandwich" binding assay (Legrain and Buttin, 1983; Victor-Kobrin et al., 1985) where IDA antibody, in the presence or absence of peptide, was added to A48-coated plastic plates; binding of the anti-Id was detected with 125I-labelled A48. In these experiments, ID 32-44 was not able to inhibit the binding of any of the IDA antibodies to A48 (data not shown). These results indicate that either Vh residues 32-44 do not contribute to the idiotopes recognized by the five IDA antibodies we tested, or that the presence of a sequence homologous to ID 32-44 cannot confer binding to a particular IDA antibody in the absence of other residues which comprise the idiotope.

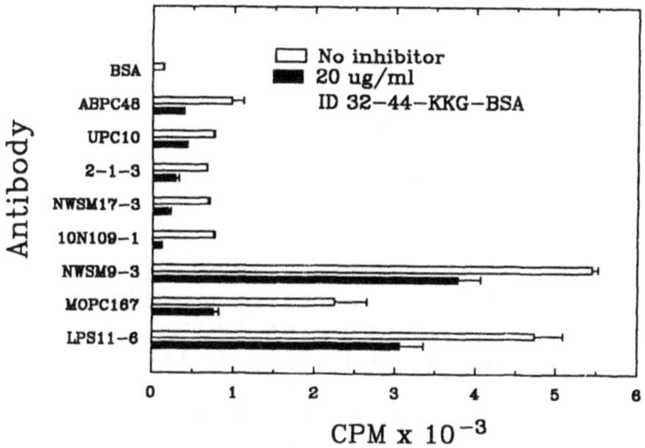

Fig. 3. Inhibition of the binding of RaIDab to monoclonal antibodies by ID 32-44-KKG-BSA. The experimental design is the same as in Figure 2 with ID 32-44-KKG-BSA mixed with monoclonal antibodies. Values are mean ± S.E.M. for triplicate determinations.

The polyclonal RaIDab, no doubt, consist of an array of more or less subtly different specificities. The question thus arises, "is reactivity of RaIDab with different mouse antibodies due to several different subpopulations having different specificities within the RaIDab, or is it due to a majority of the RaIDab which share a specificity and recognize the same structure on reactive mouse antibodies?" In order to address this question we purified rabbit antiserum by adsorption on a column containing the intact antibody LPS11-6 coupled to sepharose (this preparation is designated RaIDab-L).

In Figure 4 we see the binding of RaIDab-L to various antibodies. The data are arranged in the same order as in Figure 2 to facilitate visual comparison. RaIDab-L binds well to LPS11-6, M167, and NWSM9-3 as does RaIDab. On the other hand, 2-1-3, NWSM17-3, and 10N109-1 do not bind as well to RaIDab-L as they do to RaIDab. The most striking observation in Figure 4, however, is the markedly increased binding of RaIDab-L to A48. The failure of RaIDab-L to bind to U10 and 2-1-3, mAbs expressing the same Vh gene family, and sharing antigen binding and idiotypy (Victor-Kobrin et al., 1985) indicates again that even where strong structural similarity exists in a particular region of the protein, very subtle differences in structure strongly influence the ability of that region to interact with other molecules.

In Figure 5 we see that the reaction of both RaIDab and RaIDab-L with various antibodies can be inhibited by free ID 32-44. Although the absolute amounts of RaIDab or RaIDab-L bound to a particular mAb varies (see Figure 5 legend), the pattern of inhibition for both antibody populations is quite similar. These data show that the binding of RaIDab-L to the mAbs studied is due to selection of peptide-reactive rabbit antibodies, not to selection of rabbit antibodies which bind some other antigenic determinant which may be shared by LPS11-6 and other mAbs.

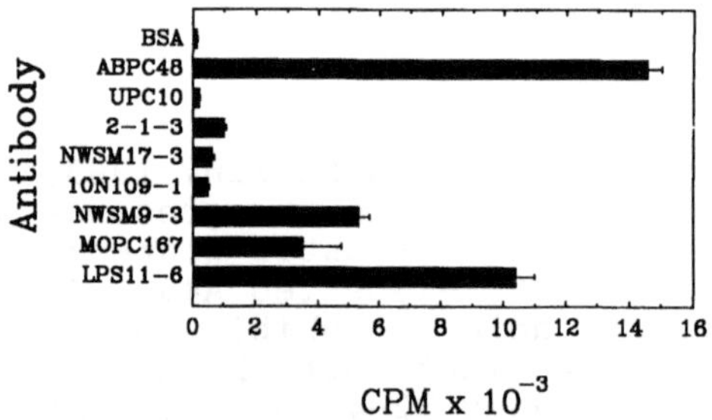

Fig. 4. Binding of rabbit antiserum purified with intact LPS11-6 (RaIDab-L) to various monoclonal antibodies. The experimental design is the same as in Figure 2. Values are mean ± S.E.M. for triplicate determinations.

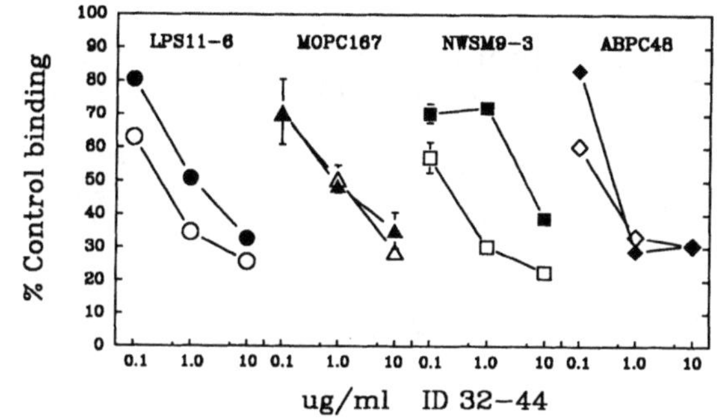

Fig. 5. Inhibition of binding of RaIDab and RaIDab-L to various monoclonal antibodies by free Id 32-44. The design is the same as in Figure 3 with unconjugated ID 32-44 mixed with the monoclonal antibodies. Open symbols show inhibition of RaIDab binding, closed symbols show inhibition of RaIDab-L. Control CPM for RaIDab are: LPS11-6, 13,381 ± 733; M167, 2,105 ± 346; NWSM9-3, 15,912 ± 1,299; A48, 5,375 ± 135. Control CPM for RaIDab-L are: LPS11-6, 6,472 ± 134; M167, 1,310 ± 131; NWSM9-3, 6,391 ± 384; A48, 4,882 ± 89. Values are mean ± S.E.M. for triplicate determinations.

We have previously shown that A48Id+ levan-specific clones are a minor, or "silent" component of the BALB/c anti-levan response (Hiernaux et al., 1981; Rubinstein et al., 1982, 1983). Treatment of BALB/c mice with syngeneic polyclonal anti-A48Id antibodies at birth increases the fraction of A48Id+ to the extent that they dominate the anti-levan response. Can treatment of mice with RaIDab produce a similar effect? Table I shows the number of anti-bacterial levan plaques/spleen of BALB/c mice injected at birth with various antibody preparations and immunized at age one month with bacterial levan. The first three groups in the table are the controls: untreated mice; mice injected with normal rabbit Ig; and mice injected with rabbit anti-Id antibodies against an arbitrary monoclonal antibody. The experimental groups are mice injected with RaIDab or RaIDab-L. The data in the last group, treatment with BALB/c polyclonal anti-A48Id, are from an earlier work (Hiernaux et al., 1981) and are included for comparison. An analysis of variance was performed using both the log(BL-specific PFC/spleen) and the percentage of A48Id+ BL-specific plaques as dependent variables. A significant group effect was detected only for the numbers of BL-specific plaques, $F = 38.52$, $p < 0.0001$. For the percentage of A48Id+ BL-specific plaques these values are $F = 1.64$, $p < 0.22$. Table I also shows the results of multiple comparisons according to the Tukey test for the log(BL-specific PFC/spleen). The means for both experimental groups are different from the means of the control groups with alpha 0.05. It is interesting to note that the effect of treatment with RaIDab or RaIDab-L is to increase the number of BL-specific plaques. It has been observed pre-

Table I
Effect of neonatal injection of various antibody preparations on A48Id expression by BALB/c anti-levan antibodies.

Treatment at birth	n	BL-specific plaques	BL-specific plaques, % A48Id+	Group
Normal	4	3.11 ± 0.21 (1288) (a)	8 ± 19 (b)	A (c)
100 ng normal rabbit Ig	3	3.43 ± 0.12 (2692)	8 ± 11	A
100 ng rabbit anti-PY206 (d)	4	3.12 ± 0.14 (1318)	11 ± 10	A
100 ng RaIDab	7	4.25 ± 0.21 (17783)	20 ± 13	B
100 ng RaIDab-L	2	3.84 ± 0.01 (6918)	34 ± 8	B
100 ng syngeneic polyclonal anti-A48 (e)	5	2.86 ± 0.22 (724)	65 ± 17	

(a) Values are mean ± S.D. of log(BL-specific PFC/spleen). The geometric mean of PFC/spleen is given in parentheses.
(b) Values are mean ± S.D. of the percentage of BL-specific plaques inhibited by syngeneic polyclonal anti-A48Id antibodies.
(c) Statistical grouping with respect to the numbers of BL-specific plaques. Treatment groups with the same letter are not significantly different (see text).
(d) PY206 is a monoclonal antibody specific for influenza virus hemagglutinin (Moran et al., 1984).
(e) Data for this group are from Hiernaux et al., 1981.

viously that treatment with syngeneic polyclonal anti-A48Id antibodies leads to a decrease in total anti-BL PFC with the simultaneous rise in the fraction which are A48Id+ (Hiernaux et al., 1981). Although the percentage of A48Id+ plaques in experimental versus control groups did not reach statistical significance, the trend is toward increasing amounts of A48Id following treatment with RaIDab or RaIDab-L.

DISCUSSION

We have shown that antibodies induced by immunization with a synthetic peptide representing a portion of an immunoglobulin Vh segment, and purified by affinity chromatography with the peptide, specifically recognize peptide-related structures in intact antibodies having regions homologous (although not identical) to the peptide. Other authors have reported similar observations in studies of the idiotypy of dextran-specific antibodies (McMillan et al., 1983; Seiden et al., 1986) and human rheumatoid factors (Chen et al., 1984a, 1984b, 1985a, 1985b; Goldfien et al., 1985; Go i et al., 1985). We have demonstrated that peptide-induced antibodies can be effectively purified using intact immunoglobulin, and that a comparison of the reactivity of peptide-purified versus antibody-purified antisera yields important information about the specificities generated by peptide immunization, and about the actual expression of peptide-associated structures in native immunoglobulins.

Comparison of Figures 2 and 4 shows that RaIDab are heterogeneous, and that different subpopulations of antibodies can be distinguished by purifying antiserum using peptide or intact antibody coupled to Sepharose 4B. The RaIDab used in Figures 2 and 3 were obtained following the third injection of ID 32-44-OVA. With this preparation, the binding to A48 is 20% relative to LPS11-6. The RaIDab and RaIDab-L used in Figures 4 and 5 were obtained following the fourth injection of immunogen. The binding obtained with RaIDab and A48 relative to LPS11-6 in Figure 5 is 42%, while with RaIDab-L and A48 relative to LPS11-6 is 130% and 70% in Figures 4 and 5, respectively. Thus, the different reactivities of RaIDab and RaIDab-L cannot be solely attributed to differences in the response of the rabbit to the third and fourth immunizations (although some difference is apparent).

McMillan et al. (1983) have shown that anti-peptide antibodies can distinguish two peptides which differ in two of thirteen residues and represent the private idiotopes of the dextran-binding myeloma proteins J558 and M104E). Chen et al. (1985a) prepared antibodies against a peptide representing CDR2 of a human rheumatoid factor light chain. They found that anti-peptide antibodies reacted only with rheumatoid factor light chains having complete sequence identity in the region represented by the peptide. Both of these studies used immunoblotting techniques and revealed binding of unfractionated rabbit antibodies with radiolabelled protein A. In the experiments we report here, anti-peptide antibodies react with a variety of intact immunoglobulins which bear homologous sequences, but in many cases not identical, to the immunizing peptide. The immunoblot technique lacks the quantitative advantage, and may detect only specificities dependent on the linear arrangement of amino acids in the area of interest, while missing antibodies which bind to the peptide only when it assumes a conformation that it cannot achieve when immobilized on nitrocellulose.

The selection of a binding activity to a particular target from a preparation of anti-peptide antiserum by affinity chromatography with the peptide has been used as a criterion of the specificity of the binding (Chen et al., 1984a, 1984b). While a positive result, i.e., retention of binding activity by the column, may be interpreted as identity between peptide and target, a negative result cannot be interpreted as lack of such an identity. For example, U10 has the precise sequence of ID 32-44, yet reacts rather poorly with RaIDab. Either the rabbit antiserum contains antibodies which can bind to U10 but cannot be affinity purified with peptide, or which can only bind to U10 when not in its native conformation. In the first case, U10 would react in an immunochemical assay using unfractionated rabbit antibodies, but the U10-binding activity would not be removed by affinity chromatography with the peptide. In the second case, U10 might react with antiserum when denatured. In this instance we learn that U10 bears the sequence of interest, but we do not know if this actually represents an idiotope expressed in the native state. We argue that where studies of idiotypy are concerned, a comparison of reagents obtained by different purification methods, and assays where target structures are intact and in solution (Figures 2-5) will inform us more clearly of the presence or absence of idiotopes as they are viewed by their anti-idiotypic opposites.

We have also demonstrated that antibodies against a synthetic peptide can influence in vivo an immune response mediated in part by antibodies which bear structures homologous to the peptide. The immune response to bacterial levan (BL) consists of two components, one directed against the beta(2-1) fructose linkages, and one directed against the beta(2-6) linkages (Bona et al., 1978). The former set of antibodies express genes of the VhJ606 gene family and are A48Id- (Lieberman et al., 1975; Brodeur and Riblet 1984). The latter set express VhX24 genes and some of these are A48Id+ (Hiernaux et al., 1981; Victor-Kobrin et al., 1985). Syngeneic polyclonal anti-A48Id antibodies injected at birth cause an increase in the expression of the A48Id in the BL response at age one month, while causing an overall decrease in the numbers of BL-specific antibody producing cells (Hiernaux et al., 1981; Table I). Neonatal injection of an identical dose of RaIDab or RaIDab-L causes a significant increase in the number of BL-specific B cells, while slightly raising the level of A48Id expression, but not sufficiently for a significant statistical difference for this parameter in experimental groups versus controls. The different effects of syngeneic versus these xenogeneic antibodies are related either to the different specificities of the two preparations (multiple Vh and Vl determinants in comparison to ID 32-44 related structures) or to effects of the different characters of mouse and rabbit C regions.

Synthetic peptides are powerful tools with which to identify, in the molecular landscape, the prominences upon which the attention of the immune system is focused. We have applied these techniques to a study of idiotypy in the A48Id system in an attempt both to understand idiotypic cross-reactivity among (mostly) structurally disparate antibodies, A48 and M167, and to examine closely the types of information that can be acquired by variations from approaches used by others. We have also used anti-peptide antibodies to influence the immune response to bacterial levan in vivo. Synthetic peptides will be used increasingly, no doubt, to gain important insights into immune responses and their regulation.

ACKNOWLEDGEMENTS

This work was supported by American Cancer Society grant IM-275D, National Science Foundation grant DCB-8408660, Welch Foundation grant Q994, and by the award to M.Z. Atassi of the Robert A. Welch Chair of Chemistry.

REFERENCES

Atassi, Z., Kazim, A. L. and Sakata, S. (1981) High yield coupling of peptides to protein carriers. Biochim. Biophys. Acta, 670, 300-302.

Auffray, C., Sikorav, J. L., Ollo, R. and Rougeon, F. (1981) Correlation between D region structure and antigen-binding specificity: evidences from the comparison of closely related immunoglobulin Vh sequences. Ann. Immunol. (Inst. Pasteur), 132D, 77-88.

Barak, Z. (1988) Immunochemical and molecular analysis of beta(2-6) fructosan binding monoclonal antibodies bearing the A48 regulatory idiotope. Doctoral dissertation, The City University of New York.

Bona, C., Lieberman, R., Chien, C. C., Mond, J., House, S., Green, I. and Paul, W. E. (1978) Immune response to levan. I. Kinetics and ontogeny of anti-levan and anti-inulin antibody response and expression of cross-reactive idiotype. J. Immunol. 120, 1436-1442.

Bona, C., Victor-Kobrin, C., Manheimer, A., Legrain, P., Buttin, G., Yancopoulos, G. and Alt, F. (1985) Genetic and molecular aspects of regulatory idiotopes. In Monoclonal Antibodies '84: Biological and Clinical Applications. Pinchera, A., Doria, G., Dammaco, F. and Bargellesi, A. eds., Editrice Kurtis s.r.l., Milan, pp. 55-75.

Bonilla, F. A., Zaghouani, H., Rubin, M. and Bona, C. (1988) Antigen binding, idiotypy, and Vk gene usage among hybridomas selected for expression of the VhX24 gene family: immunochemical and structural analysis. (Submitted).

Brodeur, P. M. and Riblet, R. (1984) The immunoglobulin heavy chain variable region (Igh-V) locus in the mouse. I. One hundred Igh-V genes comprise seven families of homologous genes. Eur. J. Immunol., 14, 922-930.

Chen, P. P., Fong, S., Normansell, D., Houghten, R. A., Karras, J. G., Vaughan, J. H. and Carson, D. A. (1984) Delineation of a cross-reactive idiotype on human autoantibodies with antibody against a synthetic peptide. J. Exp. Med., 159, 1502-1511.

Chen, P. P., Go i, F., Fong, S., Jirik, F., Vaughan J., H., Frangione, B. and Carson, D. A. (1985) The majority of human monoclonal IgM rheumatoid factors express a "primary structure-dependent" cross-reactive idiotype. J. Immunol., 134, 3281-3285.

Chen, P. P., Go i, F., Houghten, R. A., Fong, S., Goldfien, R., Vaughan, J. H., Frangione, B. and Carson, D. A. (1985) Characterization of human rheumatoid factors with seven antiidiotypes induced by synthetic hypervariable region peptides. J. Exp. Med., 162, 487-500.

Chen, P. P., Houghten, R. A., Fong, S., Rhodes, G. H., Gilbertson, T. A., Vaughan, J. H., Lerner, R. A. and Carson D. A. (1984) Anti-hypervariable region antibody induced by a defined peptide: an approach for studying the structural correlates of idiotypes. Proc. Natl. Acad. Sci. USA, 81, 1784-1788.

Goldberg, B., Paul, W. E. and Bona, C. (1983) Idiotype-antiidiotype regulation. IV. Expression of common regulatory idiotopes on fructosan-binding and non-fructosan-binding monoclonal immunoglobulin. J. Exp. Med., 158, 515-528.

Goldfien, R. D., Chen, P. P., Fong, S. and Carson, D. A. (1985) Synthetic peptides corresponding to third hypervariable region of human monoclonal IgM rheumatoid factor heavy chains define an immunodominant idiotype. J. Exp. Med., 162, 756-761.

Go i, F., Chen, P. P., Pons-Estel, B., Carson, D. A. and Frangione, B. (1985) Sequence similarities and cross-idiotypic specificity of L chains among human monoclonal IgM-kappa with anti-gamma-globulin activity. J. Immunol., 135, 4073-4078.

Hammerling, U. and Westphal, O. (1967) Synthesis and use of O-stearoyl polysaccharides in passive hemagglutination and hemolysis. Eur. J. Biochem., 1, 46-52.

Hartman, A. B. and Rudikoff, S. (1984) Vh genes encoding the immune response to beta(1,6)-galactan: somatic mutation in IgM molecules. EMBO J. 3, 3023-3030.

Hiernaux, J., Bona, C. and Baker, P. J. (1981) Neonatal treatment with low doses of anti-idiotypic antibody leads to the expression of a silent clone. J. Exp. Med., 153, 1004-1008.

Jerne, N. K. (1974) Towards a network theory of the immune system. Ann. Immunol. (Inst. Pasteur), 125C, 373-389.

Kabat, E. A., Wu, T. T., Reid-Miller, M., Perry, H. M. and Gottesman, K. S. (1987) Sequences of Proteins of Immunological Interest. Fourth edition. U.S. Department of Health and Human Services, Public Health Service, National Institutes of Health.

Kennedy, R. C., Henkel, R. D., Pauletti, D., Allan, J. S., Lee, T. H., Essex, M., and Dreesman, G. R. (1986) Antiserum to a synthetic peptide recognizes the HTLV III envelope glycoprotein. Science, 231:1556-1559.

Koketsu, J. and Atassi, M. Z. (1974) Immunochemistry of sperm-whale myoglobin-XVI: accurate delineation of the single region in sequence 1-55 by immunochemical studies of synthetic peptides. Some conclusions concerning antigenic structures of proteins. Immunochemistry, 11, 1-8.

Legrain, P. and Buttin, G. (1983) ABPC 48 cross-reactive idiotopes in BALB/c mice. Natural and levan-induced expression. J. Exp. Med., 158:872-884.

Legrain, P. and Buttin, G. (1985) The Vk gene expressed by BALB/c ABPC48 cross-reactive idiotypes induced by antiidiotypic immunization is identical to that of BALB/c anti-oxazolone and A/J anti-arsonate antibodies. J. Immunol., 134, 3468-3473.

Legrain, P., Voegle, D., Buttin, G. and Cazenave, P. A. (1981) Idiotype-anti-idiotype interactions and the control of the anti-beta(2-6) polyfructosan response in the mouse: specificity and idiotypy of anti-ABPC48 anti-idiotypic monoclonal antibodies. Eur. J. Immunol., 11, 678-685.

Lieberman, R., Potter, M., Humphrey, W., Jr. and Chien, C. C. (1976) Idiotypes of inulin-binding antibodies and myeloma proteins controlled by genes linked to the allotype locus of the mouse. J. Immunol., 117, 2105-2111.

Lieberman, R., Potter, M., Humphrey, W., Jr., Mushinski, E. B. and Vrana, M. (1975) Multiple individual and cross-reactive idiotypes on 13 levan-binding myeloma proteins of BALB/c mice. J. Exp. Med., 142, 106-118.

McMillan, S., Seiden, M. V., Houghten, R. A., Clevinger, B., Davie, J. M. and Lerner, R. A. (1983) Synthetic idiotypes: the third hypervariable region of murine anti-dextran antibodies. Cell, 35, 859-863.

Moran, T., Liu, Y. C., Schulman, J. L. and Bona, C. A. (1984) Shared idiotopes among monoclonal antibodies specific for A/PR/8/34 (H1N1) and X-31 (H3N2) influenza viruses. Proc. Natl. Acad. Sci. USA, 81, 1809-1812.

Ollo, R., Auffray, C., Sikorav, J. L. and Rougeon, F. (1981) Mouse heavy chain variable regions: nucleotide sequence of a germline Vh gene segment. Nucl. Acids Res., 9, 4099-4109.

Potter, M., Newell, J. B., Rudikoff, S. and Haber, E. (1982) Classification of mouse Vk groups based on the partial amino acid sequence to the first invariant tryptophan: impact of 14 new sequences from IgG myeloma proteins. Mol. Immunol., 19, 1619-1630.

Reale, M. A., Manheimer, A. J., Moran, T. M., Norton, G., Bona, C. A. and Schulman, J. L. (1986) Characterization of monoclonal antibodies specific for sequential influenza A/PR/8/34 virus variants. J. Immunol., 137, 1352-1358.

Rubinstein, L. J., Goldberg, B., Hiernaux, J., Stein, K. E. and Bona, C. (1983) Idiotype antiidiotype regulation. V. The requirement for immunization with antigen or monoclonal antiidiotypic antibodies for the activation of beta2-6 and beta2-1 polyfructosan-reactive clones in BALB/c mice treated at birth with minute amounts of anti-A48 idiotype antibodies. J. Exp. Med., 158, 1129-1144.

Rubinstein, L. J., Yeh, M. and Bona, C. (1982) Idiotype-anti-idiotype network. II. Activation of silent clones by treatment at birth with idiotypes is associated with the expansion of idiotype-specific helper T cells. J. Exp. Med., 156, 506-521.

Rudikoff, S. and Potter, M. (1976) Size differences among immunoglobulin heavy chains from phosphorylcholine-binding proteins. Proc. Natl. Acad. Sci. USA, 73, 2109-2112

Seiden, M. V., Heuckeroth, R., Clevinger, B., McMillan, S., Lerner, R. and Davie, J. M. (1986) Hypervariable region peptides variably induce specific anti-idiotypic antibodies: an approach to determining antigenic dominance. J. Immunol., 136, 582-587.

Victor-Kobrin, C., Bonilla, F. A., Bellon, B. and Bona, C. A. (1985) Immunochemical and molecular characterization of regulatory idiotopes expressed by monoclonal antibodies exhibiting or lacking beta2-6 fructosan binding activity. J. Exp. Med., 162, 647-662.

Zaghouani, H., Bonilla, F. A., Meek, K. and Bona, C. (1988) Molecular basis for expression of the A48 regulatory idiotope on antibodies encoded by V genes from various families. (Submitted).

ENHANCEMENT OF PRIMARY AND SECONDARY IMMUNE RESPONSES BY INTERFERON-GAMMA

Tilahun Yilma, Sally Owens, Elizabeth H. Fennie, and Kevin P. Anderson

Department of Veterinary Microbiology and Immunology, University of California, Davis, CA 95616

Department of Pharmacological Sciences, Genentech,Inc., South San Francisco, CA 94080

ABSTRACT

We have investigated the effects of interferon-gamma (IFN-γ) administered with "G" glycoprotein of vesicular stomatitis virus (VSV), on the neutralizing antibody response. Treatment of mice or cattle with recombinant DNA-derived IFN-γ at the time of primary immunization with "G" glycoprotein enhanced the secondary virus-neutralizing antibody response that followed a booster administration of the same antigen without IFN-γ treatment. Enhancement was statistically significant, and occurred at relatively low doses of IFN-γ in the absence of any additional adjuvants. Cattle treated with IFN-γ at the time of primary immunization were also more resistant to VSV challenge than those immunized without IFN-γ treatment. Such treatment in conjunction with primary immunization may therefore provide a practical means of enhancing protection from viral challenge without inflammatory adjuvants or boosters.

INTRODUCTION

Interferon-gamma (IFN-γ) is a secreted polypeptide product of stimulated T-lymphocytes which has potent immunomodulatory properties as well as antiviral activity (Sonnenfeld et al., 1978; Gresser et al., 1979). Treatment of macrophages and other specific cell types with recombinant DNA-derived IFN-γ can induce or enhance expression of class II major histocompatibility (MHC) products (King and Jones, 1983; Wong et al., 1983; Skoskiewicz et al., 1985; Chang and Lee, 1986), and treatment of purified resting B-lymphocytes can stimulate proliferation and differentiation into immunoglobulin-secreting cells (Leibson et al., 1984; Sidman et al., 1984; Nakagawa et al., 1985; Romagnani, et al., 1986.). In addition, *in vivo* administration of recombinant DNA-derived murine IFN-γ in conjunction with immunization has been reported to enhance immune responses to specific antigens (Nakamura et al., 1984; Playfair and de Souza, 1987). Vesicular stomatitis virus (VSV) is an enveloped, negative-strand RNA virus which causes a severe debilitating vesicular disease in cattle, horses and swine. We have investigated the effects of *in vivo* IFN-γ treatment on the antibody response of mice and cattle to a purified component of the VSV virion, the envelope "G" glycoprotein, because of its excellent immunogenic properties and ease of purification (Volk et al., 1982; Yilma et al.,1985). In addition, antibodies raised to "G" glycoprotein neutralize infectivity of VSV both in vitro (Kelley et al., 1972) and in vivo (Dietschold, et al., 1974). Therefore, an in vitro virus neutralization test can be used to monitor levels of protective antibodies, and the in vivo protective effects of immunization can be evaluated.

Our results demonstrate that administration of recombinant DNA-derived IFN-γ at the time of initial immunization with purified "G" glycopproteincan enhance the magnitude of a secondary virus neutralizing antibody response following a booster administration of the same antigen without IFN-γ treatment. This enhancement occurred even in animals with no significant enhancement of the primary neutralizing antibody response. Cattle treated with IFN-γ at the time of primary immunization were also more resistant to a subsequent viral challenge than those immunized without IFN-γ treatment.

MATERIALS AND METHODS

Animals. Female DBA/2 mice were obtained from Charles River Laboratories, quarantined for 2 weeks prior to initiation of the study, and shown to be negative in screening for common pathogens. Healthy yearling bovine heifers (200-300 kg) of mixed breed, seronegative to VSV, were quarantined for 4 weeks prior to use and then maintained, challenged and disposed of according to United States Department of Agriculture regulations.

Immunogen. VSV "G" glycoprotein was purified from VSV serotype New Jersey (VSV_{NJ}) as previously described (Volk et al., 1982; Yilma et al., 1985), and was >95% pure as judged by polyacrylamide gel electrophoresis. Control animals were immunized with purified hen egg ovalbumin obtained from Sigma Chemical Co.

Virus. Plaque-purified VSV used for challenge was propogated and titered in BHK-21 cells as previously described (Volk et al., 1982; Yilma et al., 1985).

IFN-γ. Murine and bovine IFN-γ were produced in E. coli using recombinant DNA technology. Cloning, expression, and purification of these proteins has been previously described (Burton et al., 1985; Gray and Goeddel, 1983). The purity of both preparations was greater than 95% as determined by polyacrylamide gel electrophoresis. Specific activities of murine and bovine IFN-γ were $1 \times 10_7$ units/mg and $2 \times 10_6$ units/mg respectively. Activity was determined by cytopathic effect inhibition assays using encephalomyocarditis virus on L929 cells for murine IFN-γ, and VSV on Madin-Darby bovine kidney cells for bovine IFN-γ (Gray and Goeddel, 1983, Czarniecki et al., 1985).

Immunizations. Mice were immunized by intramuscular injection of 0.5μg of purified VSV "G" glycoprotein or ovalbumin in a total volume of 0.1 ml sterile saline into the left hindquarter. IFN-γ was administered at the same site on day -2, 0, and +2 relative to immunization. On day 0, IFN-γ and VSV "G" glycoprotein were administered as a single injection. Mice receiving no IFN-γ treatments were injected with saline on days -2 and +2 relative to immunization. Booster immunizations on day 28 were identical except that no IFN-γ treatment was included. IFN-γ treatment and immunization of cattle followed a similar course to that of the mice, except 112 μg of VSV "G" protein in 1.0 ml of phosphate buffered saline (PBS) was used for primary immunization. Cattle were given 50 μg of VSV "G" glycoprotein for booster immunization. Mice were bled at indicated intervals via the tail vein, and cattle were bled via the jugular vein.

Serum neutralization. Serum neutralizing antibody titers were determined and expressed as the $\log_{10}$ reciprocal of the greatest dilution of serum which resulted in complete protection against the cytopathic effect of 100 tissue culture 50% infectious doses (TCID-50) of VSV_{NJ} in Madin-Darby bovine kidney cells.

VSV challenge. Two cows from each group were challenged by intradermal injection of VSV at specified sites on the tongue (Yilma et al., 1985; Mackett et al., 1985). This technique allows estimation of the degree of protection resulting from immunization since lesions at each injection site develop independently. Two inocula of 100 plaque forming units (pfu) of VSV_{NJ} were injected at 2 distinct sites on the upper portion of the tongue of each animal (upper quadrants), and two inocula of 10_3 pfu of VSV_{NJ} were injected at two additional sites on the lower portion of the tongue (lower quadrants). Log_{10} VSV neutralizing titers of sera obtained from animals treated with IFN-γ were compared to log_{10} titers of sera from animals immunized but not treated with IFN-γ on the same bleed date.

RESULTS

Murine IFN-γ treatment did not significantly alter the primary antibody response of mice to VSV "G" glycoprotein. VSV neutralizing titers of pooled sera from each treated group as well as the control group were only slightly elevated above background levels for 4 weeks following immunization (Table I). All groups of mice responded to the boost at 28 days with elevated levels of neutralizing serum titers, but the secondary antibody response was markedly enhanced in those mice which had been treated with IFN-γ (Table I). Fourteen days after booster administration, the VSV neutralizing titers of pooled sera from mice given three doses of either 50, 500 or 5,000 units of IFN-γ at the time of initial immunization were enhanced threefold, eight-fold and six-fold, respectively, when compared to the titer of pooled sera of control animals receiving no IFN-γ treatment. A similar trend was observed at the same doses of IFN-γ when the amount of "G" glycoprotein used as immunogen was reduced by a factor of 10 (data not shown). VSV neutralizing titers were determined for sera of individual animals that were treated with doses of 500 units of IFN-γ and those not treated with IFN-γ. On all bleed dates after the day 28 boost, statistically significant increases in antibody titers were observed in those animals treated with IFN-γ in conjunction with primary immunization compared to control animals. These results were confirmed in an independent experiment performed identically except that VSV "G" glycoprotein and IFN-γ were administered subcutaneously rather than by the intramuscular route.

Table I. Log_{10} VSV-Neutralizing Titers in Pooled Serum of Immunized Mice*

		Log_{10} VSV-Neutralizing Titers on Indicated Days							
Immunogen	Adjuvant	0	7	14	21	28	35	42	126
G	Saline	<1.00	1.00	1.60	1.90	1.90	2.81	2.51	2.08
G	50 U IFN	1.00	1.18	1.60	1.78	1.78	2.98	2.98	1.90
G	500 U IFN	<1.00	1.00	1.60	1.08	2.08	3.11	3.41	2.81
G	5000 U IFN	1.00	1.78	1.90	1.90	2.20	3.28	3.28	2.81
G	CFA	<1.00	1.18	2.08	2.81	3.11	4.31	4.31	4.19
OVA	Saline	<1.00	<1.00	1.18	<1.00	1.48	1.18	1.18	<1.00
OVA	500 U IFN	<1.00	<1.00	<1.00	<1.00	1.30	1.18	<1.00	<1.00
OVA	CFA	1.18	1.30	1.48	<1.00	1.30	<1.00	<1.00	<1.00

*Groups of six mice were immunized with G glycoprotein or ovalbumin on days 0 and 28, and bled via the tail vein on the indicated days relative to primary immunization. Murine IFN-γ or saline was given at the same site on days -2, 0, and +2 relative to immunization. Pooled sera from each bleed were assayed.

In order to evaluate the adjuvant effects of IFN-γ treatment on immunization in a natural host for VSV, serum VSV neutralizing titers were monitored in yearling heifers immunized by intramuscular injection of 112 μg of "G" glycoprotein and treated with bovine IFN-γ. In all cattle immunized with "G" glycoprotein a primary neutralizing antibody response was observed. Animals receiving the highest doses of bovine IFN-γ (5 x 10_6 units/animal; comparable to 500 units/ animal in mice when considered on a per kg basis) exhibited enhanced VSV neutralizing titers compared to animals not treated with bovine IFN-γ . A 5-fold difference observed on day 18 after vaccination was statistically significant. Animals receiving lower doses of bovine IFN-γ (1 x 10_6 or 2 x 10_5 units/animal) showed no significant enhancement of serum VSV neutralizing titers following primary immunization.

After serum VSV neutralizing titers in cattle had reached a maximum and declined to relatively low levels (53 days after immunization), two animals from each group were challenged by intradermal injection of live VSV into the tongue. The resultant lesions give an estimate of the degree of protection afforded by immunization (Yilma et al., 1985; Mackett et al., 1985). Animals not treated with IFN-γ at the time of "G" glycoprotein immunization (G13 and G14) developed characteristic lesions within 48 hours at all sites injected with 10^2 pfu of VSV_{NJ} (Table II).. In contrast, animals treated with the highest dose of bovine IFN-γ at the time of immunization with "G" glycoprotein (G01 and G02; 5 x 10_6 units/animal) did not develop any lesions at sites injected with 10_2 pfu of virus. In animals treated with lower doses of IFN-γ (G05, G06, G09, G10), vesicular lesions developed at 50 percent of sites injected with 10_2 pfu of VSV_{NJ}. One of the 2 control animals in this experiment was immunized with ovalbumin in complete Freund's adjuvant (G22) and failed to develop a lesion at one site of injection, possibly as a result of non-specific immunity from the use of complete Freund's adjuvant. Animals injected with saline in conjunction with immunization, as in previous challenge experiments using this system, invariably developed lesions at all sites injected with 10^2 pfu of VSV_{NJ} (Yilma et al., 1985; Mackett et al., 1985).

Table II. Protection from VSV Challenge in Cattle Treated with Bovine IFN-γ with immunization by VSV G Protein*

Animal	Immunogen	Adjuvant	Log_{10} Titer at challenge	Lesions Present after challenge	
				with 10^2 pfu	with 10^3 pfu
13	G	Saline	1.90	2	2
14	G	Saline	2.20	2	2
09	G	$2x10^5$ U IFN	2.51	1	2
10	G	$2x10^5$ U IFN	1.90	0	1
05	G	$1x10^6$ U IFN	2.20	2	2
06	G	$1x10^6$ U IFN	1.60	1	2
01	G	$5x10^6$ U IFN	2.81	0	2
02	G	$5x10^6$ U IFN	2.20	0	2
17	G	CFA	3.11	0	0
18	G	CFA	3.28	0	0
21	OVA	CFA	<1.00	2	2
22	OVA	CFA	<1.00	1	2

*Animals immunized as shown were challenged with VSV by intradermal inoculation at four sites on the tongue (two sites received 10^2 pfu and two received 10^3 pfu).

Since lesions developed at all sites in animals immunized using soluble "G" glycoprotein without IFN-γ treatment, the lack of lesions in some animals immunized with "G" glycoprotein and treated with lower doses of IFN-γ may indicate partial protection resulting from this treatment. This partial protection occurred even though no enhancement of the primary antibody response had been observed. Only 3 animals failed to develop lesions at sites inoculated with 10^3 pfu of virus. Neither of the two animals immunized with "G" glycoprotein in complete Freund's adjuvant (G17 and G18) developed any lesions in response to virus challenge and one animal receiving doses of 2 x 10^5 units of bovine IFN-γ in conjunction with "G" glycoprotein immunization (G10) developed a lesion at only one site of inoculation.

To determine the effect of bovine IFN-γ treatment at the time of primary immunization on the secondary antibody response, serum VSV neutralizing titers were monitored in remaining unchallenged animals of each group following a booster administration of "G" glycoprotein. A booster administration of 50 μg of "G" glycoprotein given 124 days after primary immunization enhanced VSV neutralizing titers of all animals except those which were initially immunized with "G" glycoprotein in complete Freund's adjuvant. Titers in these animals remained elevated after the primary immunization and were essentially unchanged following the booster immunization. The magnitudes of the peak titers observed after the booster administration were apparently dependent upon the dose of bovine IFN-γ used at the time of primary immunization. Control animals which did not receive bovine IFN-γ treatment at the time of primary immunization exhibited the lowest titers following the boost, while peak titers in animals which were treated with bovine IFN-γ increased as the dose of IFN-γ increased. The highest peak titer occurred in an animal which received the highest dose of IFN-γ.

DISCUSSION

The experiments presented demonstrate that IFN-γ, when administered in conjunction with a specific immunogen, can potentiate an enhanced secondary antibody response to that immunogen many weeks later. The optimum dose of IFN-γ for enhancement of the secondary antibody response was 500-5000 units per mouse administered on days -2, 0, and +2 relative to primary immunization. Similar enhancement of secondary antibody titers was observed when the dose of immunogen used and the route of adminstration of the immunogen and IFN-γ were altered. A dose-dependent enhancement of the secondary VSV neutralizing antibody response was also observed in cattle treated with IFN-γ in conjunction with primary immunization. In addition to the enhanced secondary response, a statistically significant enhancement of the primary VSV neutralizing antibody response was noted in cattle treated with the highest dose of IFN-γ (5 x 10_6 units per animal).

Two previously reported activities of IFN-γ may influence generation of memory cells responsible for the enhanced secondary antibody response. Several investigators (Leibson et al., 1984; Sidman et al., 1984; Nakagawa et al., 1985; Romagnani, et al., 1986.) have reported that in vitro proliferation of resting B cells and their differentiation into antibody-secreting cells can be stimulated in the presence of low concentrations of IFN-γ. An enhancement of the number of antibody- secreting cells could lead to an enhancement in the number of residual antigen reactive memory cells which could function in a secondary response. Alternatively, the number of antibody-secreting cells (and hence specific memory cells) generated might be increased through an indirect mechanism involving the reported ability of IFN-γ to enhance expression of class II MHC products (King and Jones, 1983; Wong et al., 1983; Skoskiewicz et al., 1985; Chang and Lee, 1986). An effective antibody response requires recognition of antigen by specific helper T cells in association with MHC class II products on the surface of antigen-presenting cells such as macrophages (Dialynas et al., 1983;

Unanue and Allen, 1987; Schwartz, 1985). Stimulation of MHC class II products on antigen presenting cells by IFN-γ treatment could therefore enhance the efficiency of antigen presentation and lead to increased numbers of antibody- secreting cells.

The ability of immunized animals to mount an enhanced secondary antibody response as a result of IFN-γ treatment in conjunction with primary immunization correlated with reduced susceptibility of IFN-γ treated animals to a challenge with VSV. It is unlikely, however, that this protection was the result of elevated antibody titers at the time of challenge. VSV neutralizing antibody titers were not significantly higher in IFN-γ treated animals than in untreated animals, and in all cases the titers were below those considered to be protective. Previous experiments have indicated that cattle immunized with soluble VSV "G" glycoprotein without adjuvant must have $\log_{10}$ serum neutralizing titers of at least 3.1 to provide protection from a challenge of 10_2 pfu of VSV (Yilma et al., 1985; Mackett et al., 1985). In our studies, IFN-γ treated animals with $\log_{10}$ serum titers of 2.8, 2.2, or 1.9 at the time of challenge were completely protected from challenge with 10_2 pfu of VSV. Animals treated with saline rather than IFN-γ had similar $\log_{10}$ titers of 1.9 and 2.2 at the time of challenge, but developed lesions in response to challenge with the same amount of virus.

These data provide supportive evidence for the conclusion that protection from VSV challenge is a result of the enhanced secondary antibody response potentiated by IFN-γ treatment and triggered by the virus challenge. However, we cannot rule out the possibility that an immune function other than the humoral response (e.g. cell mediated immunity) may be enhanced by IFN-γ treatment. Previous publications have demonstrated that IFN-γ can suppress, as well as enhance, in vitro immune responses (Lucero et al.,1980; Johnson and Torres, 1983; Mond, et al., 1985; Rabin et al., 1986). In addition, some in vivo experiments using crude preparations of IFN-γ suggest that the timing of treatment relative to immunization and the concentration used can determine whether in vivo antibody production is suppressed or enhanced with IFN-γ (Sonnenfeld et al., 1977; Sonnenfeld et al., 1978). No suppression of the in vivo antibody response to VSV "G" protein was observed at any of the doses examined in this report, demonstrating that highly purified, recombinant DNA-derived IFN-γ is not immunosuppressive when administered according to the protocol described here. However, further work will be necessary to determine if highly purified IFN-γ can suppress an in vivo antibody response when administered using a different route or schedule of administration.

The results of our experiments suggest that IFN-γ treatment might prove useful as an adjuvant for immunization in man as well as animals. Cattle treated with IFN-γ in conjunction with immunization, showed an enhanced primary humoral response, and more importantly, IFN-γ treatment at the time of initial immunization potentiated the ability of both mice and cattle to mount a secondary antibody response against a subsequent exposure to the same antigen. Protection from VSV challenge correlated with the ability of IFN-γ treated animals to mount an enhanced secondary antibody response suggesting that the viral challenge itself may serve as the stimulus for an enhanced secondary response.

No signs of inflammation were observed in either mice or cattle treated with IFN-γ, and relatively low doses were required to elicit enhanced antibody titers. Optimum enhancement of the secondary antibody response in mice occured at a dose of approximately 15 (OAmg per square meter. This dose is far below that where minimal side effects such as chills and fatigue have been observed in human volunteers following injection (50-100 μg/square meter), and is less than the maximum tolerated dose in humans by a factor of 30 (Foon et al., 1985; Kurzrock et al., 1986). IFN-γ treatment in conjunction with primary immunization therefore has the potential for use as an effective, non-inflammatory, immune adjuvant. Whether IFN-γ can enhance the activity of currently available adjuvants or work synergistically with other cytokines and differentiation factors has yet to be explored.

ACKNOWLEDGMENTS

We thank P. Gribling, R. Harny and S. Lieb for excellent technical assistance, and Dr. A. Ammann, R. Shalaby, and Dr. M. Shepard for helpful discussion. Bovine IFN-γ was generously provided by Dr. L.E. Burton. We also thank the Journal of Immunology for allowing us to reprint material previously published in the Journal.

REFERENCES

Burton, L.E., P.W. Gray, D.V. Goeddel, and E. Rinderknecht. 1985. Modifications of recombinant human and murine interferon-gamma and their biochemical characteristics. In The Biology of the Interferon System, 1984. H. Kirschner and H. Schellekens (Eds.) Elsevier, Amsterdam. p. 403.

Chang, R. and S.H. Lee, J. Immun. 1986. Effects of interferon-gamma and tumor necrosis factor-alpha on the expression of an Ia antigen on a murine macrophage cell line. J. Immunol. 137:2853.

Czarniecki, C.W., E.B. Hamilton, C.W. Fennie, and R.I. Wolf. 1985. In vitro biological activities of Escherichia coli-derived bovine interferons -alpha, -beta, and -gamma. J. Interferon Res. 6:29.

Dialynas, D.P., D.B. Wilde, P. Marrack, A. Pierres, K.A. Wall, W. Havran, G. Otten, M.R. Loken, M. Pierres, J. Kappler, and F.W. Fitch. 1983. Characterization of the murine antigenic determinant, designated L3T4a, recognized by monoclonal antibody GK1.5: Expression of L3T4a by functional T cell clones appears to correlate primarily with class II MHC antigen-reactivity. Immunological Rev. 74:29.

Dietzschold, B., L.G. Schneider, and J.H. Cox. 1974. Serological characterizations of the three major proteins of vesicular stomatitis virus. J. Virol. 14:1.

Gray, P.W. and D.V. Goeddel. 1983. Cloning and expression of murine interferon-gamma. Proc. Natl. Acad. Sci. U.S.A. 80:5842.

Gresser, I., J. DeMaeyer-Guignard, M.G. Tovey, and E. DeMaeyer. 1979. Electrophoretically pure mouse interferon exerts multiple biologic effects. Proc. Natl. Acad. Sci. U.S.A. 76:5308.

Johnson, H.M. and B.A. Torres. 1983. Recombinant mouse interferon-gamma regulation of antibody production. Infect. Immun. 41:546.

Kelley, J.M., S.U. Emerson, and R.R. Wagner. 1972. The glycoprotein of vesicular stomatitis virus is the antigen that gives rise to and reacts with neutralizing antibody. J. Virol. 10:1231.

King, D.P. and P.P. Jones. 1983. Induction of Ia and H-2 antigens on a macrophage cell line by immune interferon. J. Immun. 131:315.

Kurzrock, R., J.R. Quesada, M. Talpaz, E.M. Hersh, J.M. Reuben, S.A. Sherwin,and J.U. Gutterman. 1986. Phase I study of multiple dose intramuscularly administered recombinant gamma interferon. J. Clin. Onc. 4:1101.

Leibson, H.J., M. Gefter, A. Zlotnik, P. Marrack, and J.W. Kappler. 1984. Role of gamma-interferon in antibody-producing responses. Nature 309:799.

Lucero, M.A., J. Weitzerbin, S. Stefanos, C. Billardon, E. Falcoff, and W.H. Fridman. 1980. Immunosuppressive properties of purified immune T-interferon. Cell. Immunol. 54:58.

Mackett, M., T. Yilma, J.K. Rose, and B. Moss. 1985. Vaccinia virus recombinants: expression of VSV genes and protective immunization of mice and cattle. Science 227:433.

Mond, J.J., F.D. Finkelman, C. Sarma, J. Ohara, and S. Serrate. 1985. Recombinant interferon-gamma inhibits the B cell proliferative response stimulated by soluble but not by sepharose-bound anti-immunoglobulin antibody. J. Immunol. 135:2513.

Nakagawa, T., T. Hirano, N. Nakagawa, K. Yoshizaki, and T. Kishimoto. 1985. Effect of recombinant IL-2 and IFN-gamma on proliferation and differentiation of human B cells. J. Immunol. 134:959.

Nakamura, M., T. Manser, G.D.N. Pearson, M.J. Daley, and M.L. Gefter. 1984. Effect of IFN-gamma on the immune response in vivo and on gene expression in vitro. Nature 307:381.

Playfair, J.H.L. and J.B. de Souza. 1987. Recombinant gamma interferon is a potent adjuvant for a malaria vaccine in mice. Clin. Exp. Immunol. 67:5.

Rabin, E.M., J.J. Mond, J. Ohara, and W.E. Paul. 1986. Interferon-gamma inhibits the action of B cell stimulatory factor (BSF)-1 on resting B cells. J. Immunol. 137:1573.

Romagnani, S., M.G. Giudizi, R. Biagiotti, F Almerigogna, C. Mingari, E. Maggi, C. Liang, and L. Moretta. 1986. B cell growth factor activity of interferon-gamma. Recombinant human interferon-gamma promotes proliferation of anti-IgM-activated human B lymphocytes. J. Immunol. 136:3513.

Schwartz, R.H. 1985. T-lymphocyte recognition of antigen in association with gene products of the major histocompatibility complex. Ann. Rev. Immunol. 3:237.

Sidman, C.L., J.D. Marshall, L.D. Schultz, P.W. Gray, and H.M. Johnson. 1984. gamma-Interferon is one of several direct B cell-maturing lymphokines. Nature 309:801.

Skoskiewicz, M.J., R.B. Colvin, E.E. Schneeberger, and P.S. Russell. 1985. Widespread and selective induction of major histocompatibility complex- determined antigens in vivo by gamma interferon. J. Exp. Med. 162:1645.

Sonnenfeld, G., A.D. Mandel, and T.C. Merigan. 1977. Cell. Immunol. 34:193.

Foon, K.A., S.A. Sherwin, P.G. Abrams, H.C. Stevenson, P. Holmes, A.E. Maluish, R.K. Oldham, and R.B. Herberman. 1985. A phase I trial of recombinant gamma interferon in patients with cancer. Cancer Immunol. Immunother. 20:193.

Sonnenfeld, G., A.D. Mandel, and T.C. Merigan. 1978. Time and dosage dependence of immunoenhancement by murine type II interferon preparations. Cell. Immun. 140:285.

Unanue, E.R. and P.M. Allen. 1987. The basis for the immunoregulatory role of macrophages and other accessory cells. Science 236:551.

Volk, W.A., R.M. Snyder, D.C. Benjamin, and R.R. Wagner. 1982. Monoclonal antibodies to the glycoprotein of vesicular stomatitis virus: comparative neutralizing activity. J. Virol. 42:220.

Wong, G.H.W., I. Clark-Lewis, J.L. McKinn-Breschkin, A.W. Harris, and J. Schrader. 1983. Interferon-gamma induces enhanced expression of Ia and H-2 antigens on B lymphoid, macrophage and myeloid cell lines. J. Immun. 131:788.

Yilma, T., R.G. Breeze, S. Ristow, J.R. Gorham and S.R. Lieb. 1985. Immune responses of cattle and mice to the G glycoprotein of vesicular stomatitis virus. Adv. Exp. Med. Biol. 150:101.

INTERLEUKIN 1 AND ITS SYNTHETIC PEPTIDES AS ADJUVANTS FOR POORLY IMMUNOGENIC VACCINES

G. Antoni, R. Presentini, F. Perin, L. Nencioni, L. Villa, S. Censini, P. Ghiara, G. Volpini, P. Bossu', A. Tagliabue, and D. Boraschi

Sclavo Research Center, 53100 Siena, Italy

INTRODUCTION

A large variety of substances have the property of regulating the immune system, by increasing or depressing the response to antigens after interaction with suitable cellular components. Recent reviews describe numerous natural and synthetic peptide endowed with immunomodulatory activity (Werner et al., 1986; Werner, 1987).

Many immunostimulatory compounds present a particular interest, because of their possible use as adjuvants for vaccines. In most cases the adjuvant effect appears mediated by interleukin 1 (IL 1), which may represent the natural adjuvant produced by the organism to increase response to infection and antigenic stimulus (Dinarello, 1988). In fact crude IL 1 preparations have been described to possess immunoadjuvant activity (Staruch and Wood, 1983).

IL 1 is involved in a complex panel of activities which include, beside the enhancement of immune response to antigens, also a series of inflammatory events, like pyrogenicity, hyperalgesia and production of acute phase proteins. This inflammatory capacity and the possible involvement in the pathogenesis of rheumatic diseases make difficult the possibility of the clinical use of IL 1 as immunopotentiator.

On the basis of the hypothesis that different moieties of the molecule could be responsible of the different effects, we have undertaken a study intended to evaluate the biological activities of selected peptide fragments. We thus identified and synthesized a short fragment, corresponding to the sequence 163-171 of human IL 1 β (VQGEESNDK), which in fact showed IL 1-like immunomodulatory properties both *in vitro* and *in vivo*, but proved devoid of the inflammatory effects of the entire molecule (Antoni et al., 1986; Nencioni et al., 1987; Boraschi et al., 1988). In particular, the peptide, like the entire IL 1 molecule, showed strong adjuvant properties for both T-dependent and T-independent antigens.

We also examined in the same system the activity of shorter and longer fragments around the peptide 163-171, in order to identify the minimal sequence corresponding to the maximal biological activity.

STRUCTURAL ANALYSIS OF IL 1

Two forms of IL 1 have been described so far, named IL 1 α and IL 1 β respectively, whose sequences have been deduced after cloning and expression of the corresponding complementary DNAs (Auron et al., 1984; March et al., 1985). The two proteins present a low but significant degree of homology (March et al., 1985) and practically overlapping spectra of biological properties (Rupp et al., 1986). Our studies have taken in particular consideration IL 1 β, which represents the predominant circulating form of IL 1 (Dinarello, 1988).

The research of regions of a protein molecule potentially responsible of the biological activity can be performed by examining which portions of the molecule are better exposed on the external surface, thus being better accessible for the interaction with other molecules of the environment. At the beginning of our studies the tridimensional structure of IL 1 β was not yet known. Therefore we have used predictive methods, based on the examination of the amino acid sequence of the protein.

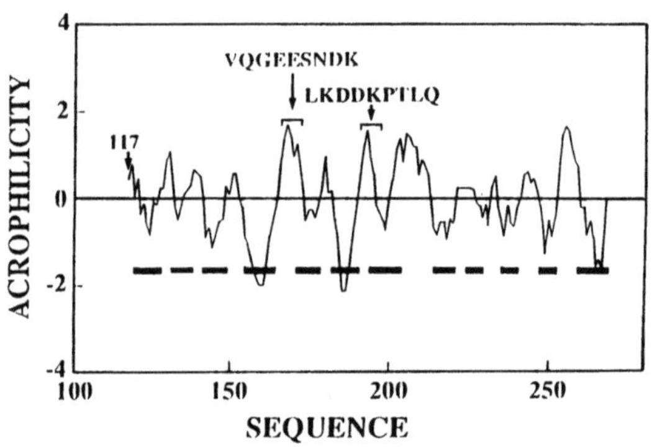

Figure 1. Acrophilicity profile of human interleukin 1 β. The profile is calculated on segments of six residues (Hopp, 1984; Hopp, 1986). The plot is limited to the sequence 117-269, representing the biologically active mature protein (March et al., 1985). The selected peptides, indicated on the figure, have been synthesized by solid phase as described (Antoni et al., 1986) and purified by HPLC (Perin et al., 1987). The black segments represent fragments corresponding to beta sheets (Priestle et al., 1988).

IL 1 β has thus been examined for determination of the acrophilicity profile, as proposed by Hopp (1984, 1986). This method appeared particularly promising for the research of well exposed fragments of protein molecules. The acrophilicity profile of human IL 1 β is shown in figure 1. Two peaks of high acrophilicity were selected, and the corresponding peptides (163-171, VQGEESNDK and 189-197, LKDDKPTLQ) were synthesized and examined for biological activity. We also performed similar studies on murine IL 1 α, selecting also for this molecule two peptides (174-182, YSSGGDDSK and 199-208, QGEDQPVLLK), corresponding to peaks endowed with high acrophilicity and superimposable to those of human IL 1 β when the two sequences are suitably aligned (Antoni et al., 1986).

The tridimensional structure of human IL 1 β has now been resolved (Priestle et al., 1988). The protein is formed of twelve beta sheets, separated by loops forming well exposed corners. As shown in figure 1, each peak of positive acrophilicity corresponds to a loop between two beta sheets, to confirm the validity of the adopted method of structural prediction.

BIOLOGICAL ACTIVITY OF THE PEPTIDE FRAGMENT 163-171 OF IL1 β

The biological properties of the selected peptides were evaluated in different tests (Antoni et al., 1986). Fragment 163-171 showed a significant T cell activation capacity, as it was able to induce IL 2 production in spleen cells. Experiments still in progress are also demonstrating a stimulatory capacity of the peptide on the production of IL 4 in both spleen lymphocytes and the T helper clone D10.G4.1 (manuscript in preparation). However, at variance with the entire IL 1 molecule, the peptide did not show inflammatory effects as it was unable to induce the production of prostaglandin E_2 in human fibroblasts and it was devoid of pyrogenic properties *in vivo*. None of the other peptides synthesized appeared to be active in any of the assays performed.

In consideration of these favourable properties of the peptide 163-171, we also investigated its immunostimulatory activity *in vivo*. The peptide showed a significant adjuvanticity for both T-dependent (sheep red blood cells, SRBC) and T-independent (type III pneumococcal polysaccharide, SIII) antigens. This effect was evaluated in the hemolytic plaque assay, as increase in the frequency of antibody-producing spleen cells specific for the antigen in mice immunized with SRBC (both on primary and secondary response) and it was also highly significant in mice receiving the T-independent antigen SIII, a typical poorly immunogenic antigen (Nencioni et al., 1987).

However the *in vivo* adjuvant activity of the peptide, as evaluated in the above cited experiments, appeared much lower than that of the entire IL 1, as the dose of peptide needed to obtain the same effect as IL 1 resulted several orders of magnitude higher. We first ascribed the low activity of the peptide to the need for other IL 1 moieties to reach the correct structural conformation or to achieve the interaction with the IL 1 receptor. However we later focused on the presence of trifluoroacetate anion (TFA, 1/2 mole per mole of peptide) in the HPLC purified preparations of peptide used for the experiments, and hypothesized an inhibitory effect of TFA on the in vivo activity of the peptide (Boraschi et al., 1988).

Therefore we have transformed the peptide to the hydrochloride salt, by chromatography on a column of Amberlite IRA 401, Cl^- form. The peptide hydrochloride obtained (2 equivalents of Cl^- per mole of peptide) showed an extraordinary increase of specific adjuvant activity, reaching an effect fully comparable, in molar terms, to that of recombinant IL 1 (figure 2).

The different efficacy of the hydrochloride salt compared to the trifluoroacetate appeared dependent on an inhibitory effect of the TFA ion in the in vivo assay, as the addition of TFA to the peptide hydrochloride or to recombinant IL 1 β strongly decreased their adjuvant effects.

INHIBITORY EFFECT OF A MONOCLONAL ANTIBODY PREPARED WITH THE PEPTIDE 163-171.

The particular role of the peptide 163-171 as the fragment mainly responsible of the adjuvanticity of the IL 1 β molecule was also confirmed indirectly, by evaluating the interaction of IL 1 and peptide with a monoclonal antibody specifically directed against this part of the molecule,

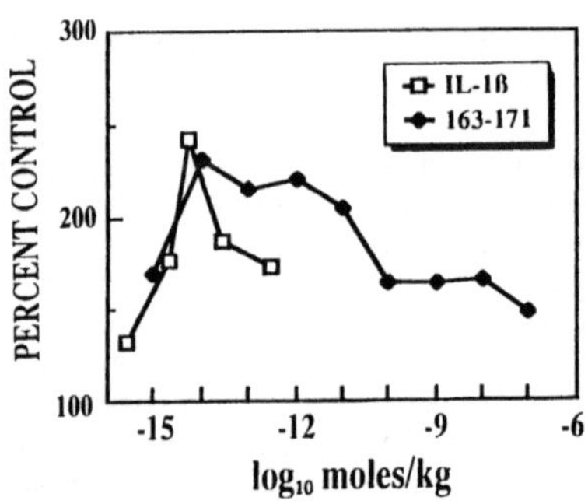

Figure 2. Adjuvant effect of peptide 163-171 hydrochloride and of recombinant human IL 1.
The PFC test was performed according to Cunningham and Szenberg, (1968), as previously described (Nencioni et al., 1987). The IL 1 used in these experiments was higly purified human recombinant IL 1 β (specific activity about $1\text{-}2\times10^7$ U/mg), corresponding to the mature fragment 117-269, produced in our laboratories by expression in E. Coli of cDNA.

Table 1. Inhibition of the immunostimulatory activity of IL 1 β and peptide fragment 163-171 by mAb Vhp20.

Added to adjuvant		Immunostimulation[a] with rIL 1 β	Immunostimulation[a] with fragment 163-171
saline		100.0	100.0
Vhp20[b]	10:1	68.4	39.5
	100:1	61.0	13.1
	1000:1	26.5	3.2

[a] Evaluation of adjuvant activity was performed as previously described (Nencioni et al., 1987), but using the peptide in the form of hydrochloride salt instead of trifluoroacetate (Boraschi et al., 1988). Immunostimulation is expressed as percent of the adjuvant activity achieved by optimal doses of IL 1 β (0.5 ng/Kg) and of the 163-171 peptide (10 ng/Kg).

[b] mAb Vhp20 was prepared by immunization of BALB/c mice with peptide 163-171 conjugated to keyhole limpet hemocyanin. Hybridoma supernatants were screened against the immunizing peptide, conjugated to bovine serum albumin, and against human recombinant IL 1 β. Positive clones were expanded in ascites and the mAb (IgG2a) was purified by affinity chromatography on Protein A-Sepharose (Pharmacia). The mAb Vhp20 was admixed to IL 1 β or to the 163-171 peptide at different molar ratios (mAb:adjuvant) and incubated overnight at +4°C before i.v. inoculation with SRBC.

obtained by immunization of BALB/c mice with the peptide covalently linked to keyhole limpet hemocyanin. The antibody, named as Vhp20, was able to interact with recombinant IL 1 β in radioimmunoassay and it also strongly inhibited the adjuvant activity of both the entire protein and the peptide fragment (table 1).

EVALUATION OF THE ADJUVANTICITY OF SHORTER AND LONGER FRAGMENTS AROUND THE SEQUENCE 163-171.

We have then synthesized several fragments of different length around the sequence 163-171, in order to find the minimal segment responsible of the adjuvant activity. The results obtained are summarized on table 2. The activity is clearly not increased by increasing the number of residues of the peptide, while shorter peptides (down to the sequence 165-171) appear at least as active as the sequence 163-171.

Table 2. Adjuvant effect of peptides of different length around the sequence 163-171 of human IL 1 β[a].

Peptide position	Peptide sequence	Adjuvant activity[b]
165-171	GEESNDK	++
164-171	QGEESNDK	++
163-171	VQGEESNDK	++
162-171	FVQGEESNDK	++
161-171	SFVQGEESNDK	+
160-171	MSFVQGEESNDK	n.t.
159-171	SMSFVQGEESNDK	++
163-176	VQGEESNDKIPVAL	-

[a] Peptides were synthesized by solid phase and purified by HPLC on a gradient of 0.1% TFA - methanol. Peptides were then transformed to the hydrochloride salt by ion exchange chromatograpy on a column of Amberlite IRA 401, Cl^- form.
[b] The adjuvant activity of the peptides was determined in vivo using SRBC as immunizing antigen.

The evaluation of other shorter fragments is currently in progress. Fragment 165-169 (GEESN) is of particular interest, as it represents the loop between the two beta sheets 4 and 5 (according to the numbering of Priestle, 1988).

CONCLUSIONS

On the basis of the above described results, the complexity of the biological activities of IL 1 could be possibly explained with the hypothesis that different fragments of the molecule are responsible of the different effects. In particular our experiments show that the peptide corresponding to the fragment 163-171 possesses immunostimulatory activity, but is devoid of pyrogenic and other inflammatory effects.

Besides the theoretical interest for a better understanding of the complex mechanism of action of the IL 1 molecule, our results also present practical implications; in fact, because of its high in vivo immunostimulatory activity in the absence of inflammatory effects, the peptide may find important clinical applications as adjuvant for poorly immunogenic vaccines and in the restoration of the immune response in different pathological conditions.

SUMMARY

The adjuvant activity of the peptide corresponding to the fragment 163-171 of human interleukin 1 β (VQGEESNDK) has been evaluated on the immune response to both T-dependent and T-independent antigens. The hydrochloride derivative of the peptide showed an effect quantitatively comparable in molar terms to that of the entire protein.

At variance with the entire IL 1 protein the peptide appeared devoid of inflammatory effects and therefore it may find clinical applications as adjuvant for poorly immunogenic vaccines or as immunorestorative agent.

The activity of other fragments in proximity of the sequence 163-171 was also evaluated. The shorter fragment 165-171 appeared as active as the 163-171 peptide.

REFERENCES

Antoni, G., Presentini, R., Perin, F., Tagliabue, A., Ghiara, P., Censini, S., Volpini, G., Villa, L., and Boraschi, D., 1986, A short synthetic peptide fragment of human interleukin 1 with immunostimulatory but not inflammatory activity ,J. Immunol., 137:3201.

Auron, P. E., Webb, A. C., Rosenwasser, L.J., Mucci, S. F., Rich, A., Wolff, S. M., and Dinarello, C. A.,1984, Nucleotide sequence of human monocyte interleukin 1 precursor cDNA, Proc. Natl. Acad. Sci. USA, 81:7907.

Boraschi, D., Nencioni, L., Villa, L., Censini, S., Bossu', P., Ghiara, P., Presentini, R., Perin, F.,Frasca, D., Doria, G., Forni, G., Musso, T., Giovarelli,M., Ghezzi, P., Bertini, R., Besedovsky, H. O., del Rey, A., Sipe, J. D., Antoni, G., Silvestri, S. and Tagliabue, A., 1988, In vivo stimulation and restoration of the immune response by the noninflammatory fragment 163-171 of human IL-1 β, J. Exp. Med., 168:675.

Cunningham, A. J., and Szenberg, A., 1968, Further improvements in the plaque technique for detecting single antibody-forming cells, Immunology, 14:599.

Dinarello, C. A., 1988, Biology of interleukin 1, FASEB J., 2:108.

Hopp, T. P., 1984, Protein antigen conformation: folding patterns and predictive algorithms; selection of antigenic and immunogenic peptides, in: "Synthetic Antigens," P. Neri, and S. Piazzi, eds., Annali Sclavo, Siena.

Hopp. T. P., 1986, Protein surface analysis. Methods for identifying antigenic determinants and other interaction sites, J. Immunol. Meth., 88:1.

March, C. J., Mosley, B., Larsen, A., Cerretti, D. P., Braedt, G., Price, V., Gillis, S., Henney, C. S., Kronheim, S. R., Grabstein, K., Conlon, P. J., Hopp, T. P., and Cosman, D., 1985, Cloning, sequence and expression of two distinct human interleukin-1 complementary DNAs, Nature, 315:641.

Nencioni, L., Villa, L., Tagliabue, A., Antoni, G., Presentini, R., Perin, F., Silvestri, S, and Boraschi, D., 1987, In vivo immunostimulatory activity of the 163-171 peptide of human IL-1 β, J. Immunol., 139:800.

Perin, F., Presentini, P., and Antoni, G., 1987, Single-step high-yield purification of a synthetic peptide fragment of

human interleukin 1 by preparative high-performance liquid chromatography, J. Chromatography, 397:365
Priestle, J. P., Schar, H. -P., and Grutter, M. G., 1988, Crystal structure of the cytokine interleukin 1 β, EMBO J. 7:339.
Rupp, E. A., Cameron, P. M., Ranawat, C. S., Schmidt, J. A., and Bayne, E. K., 1986, Specific bioactivities of monocyte-derived interleukin 1 α and interleukin 1 β are similar to each other on cultured murine thymocytes and on cultured human connective tissue cells, J. Clin. Invest., 78:836
Staruch, M. J., and Wood, D. D., 1983, The adjuvanticity of interleukin 1 in vivo, J. Immunol., 130:2191.
Werner, G. H., Floc'h, F, Migliore-Samour, D., and Jolles, P., 1986, Immunomodulating peptides, Experientia, 42:521.
Werner, G. H., 1987, Natural and synthetic peptides (other than neuropeptides) endowed with immunomodulating activities, Immunology Letters, 16:363.

PROTECTION INDUCED BY SYNTHETIC PEPTIDES CORRESPONDING TO THREE SEROTYPES IN FOOT AND MOUTH DISEASE VIRUS

Charles Gale*, Tim Doel**, Gerry Brooke*, Don White*, Grace Mulcahy**, and Richard DiMarchi*

*Lilly Research Laboratories
A Division of Eli Lilly and Company
Greenfield, IN 46140

**The Institute for Animal Health Research
Pirbright Laboratory
Woking, Surrey, GU24 ONF, England

Foot and Mouth Disease (FMD) is a highly contagious disease which primarily affects cloven-hoofed animals. It is a disease of major economic importance. Mortality from FMD is low, except in young animals where it can be quite high. The disease is debilitating, spreads rapidly and requires only a short incubation period for establishing visible disease symptoms in infected animals. An outbreak of the disease can be economically devastating. In countries where the disease is endemic, the commonly used method of control is vaccination with an inactivated virus preparation. FMD vaccines, while proven to be effective, are not without risk. Outbreaks of the disease have been linked to vaccines which were improperly inactivated. Infections have also been traced to the escape of virus from production facilities. For these reasons, the developments in synthetic approaches to vaccination have attracted the attention of scientists working towards the control of this disease (Bittle et al.,1982; Pfaff et al., 1985; DiMarchi, et al., 1986).

The foot and mouth virus belongs to the aphthovirus genus of the family Picornaviridae. Picornaviruses are small single-stranded RNA viruses with icosahedral symmetry with a diameter of approximately 25 nm. The virus is 8500 nucleotides in length and is encapsulated by four structural proteins VP1, VP2, VP3 and VP4. The surface protein, VP1, has been shown to contain the immunodominant sites for viral neutralization and binding of protective antibodies (LaPorte et al., 1973; Bachrach et al., 1975). On the basis of a series of chemical and enzymatic degradations of VP1, Strohamaier et al., 1982, suggested that two regions represented minimally by residues 146 to 154 and 201 to 213 were responsible for these activities. Studies by Bittle et al., 1982, revealed that residues 141 to 160 provided guinea pigs a neutralizing antibody response superior to that of VP1 protein.

We have previously reported the complete protection of guinea pigs and cattle against direct FMD virus challenge following immunization with a particular forty-residue peptide (DiMarchi et al., 1986). The protective peptide essentially consisted of the amino acid residues 141 to 158 and 200 to 213 of the VP1 surface protein of the FMD virus. The two sequences were connected by a diproline spacer and the peptide was used free of any carrier protein. The amino acid sequence was derived from that of the O_1K serotype and challenge was with the highly homologous O_1BFS serotype virus.

Challenge experiments have been conducted in guinea pigs with a range of peptides equivalent to either or both sequences 141 to 158 and 200 to 213 of VP1. The experimental observations are reported in Table 1. These results indicate that a carrier-free peptide is more effective. They also indicate a marked increase in potency when residues 200 to 213 are present. The induction of antibody appears to be dependent on both the size of the peptide and the relative position of each antigenic site within it (DiMarchi et al., 1986, 1988; Doel et al., 1988).

While the O-serotype represents a major problem in Western Europe and South America, it is not alone. Commonly, cattle are immunized with a trivalent vaccine which confers protection against A- and C-serotypes in addition to the O-serotype. Consequently, it remained to be shown that the analogous peptides of the A- and C-serotype would induce protection comparable to that observed with the O-serotype. For this purpose, the corresponding peptides were prepared by solid phase peptide synthesis as described previously by DiMarchi et al., 1986. Dunkin Hartley guinea pigs, that were randomly bred in a closed colony and weighed approximately 500 grams, were immunized by subcutaneous injection. Each peptide was delivered in 2 ml of a 1:1 emulsified mixture of 0.04 M phosphate buffer, pH 7 and an incomplete oil adjuvant.

Table 1. Protection in Guinea Pigs Vaccinated With Foot and Mouth Disease Synthetic Peptides

Peptide dose (n moles)	P19[a]	P19[a] KLH	P14[a]	P15[a]
150	5/5[b]	0/5	5/5	5/5
50	1/5	0/5	5/5	5/5
17	2/5	0/5	5/5	5/5
6	0/5	0/5	4/4	5/5
2	0/5	0/5	3/4	4/5
0.6	0/5	0/5	5/5	3/5
0.2	0/5	0/5	1/5	1/5

[a]Amino acid numbers refer to 0_1K sequence, other amino acids are listed in one letter code.

Sequences are: P19 = (141-158)PCG

P14 = (200-213)PPS(141-158)PCG

P15 = CC(200-213)PPS(141-158)PCG

[b]numerator = number protected, denominator = number tested

All test animals were challenged by intradermal inoculation of the left hind foot pad with a minimum of 3,000 guinea pig infective dose units of virulent virus 28 days after immunization. The criterion of protection was the absence of any secondary lesions over the next seven days. The results in each table are reported as the number of protected animals over the total number challenged. The results of the challenge with virus corresponding to the sequence of the immunizing peptide (homotypic) are shown in Table 2. The A_{24} and O_1K peptides were capable of providing complete protection at all three dose levels against their respective virus challenge. The O_1K results are in complete agreement with observations we have reported previously (DiMarchi et al., 1986; 1988). The C_3 peptide provided complete protection at the intermediate dose of 0.2 mg. At the two other doses, the level of protection was significant, but less than complete. These observations have been confirmed in subsequent testing.

Table 2. Homotypic Protection of Guinea Pigs With 40-Residue Peptide

Peptide	Peptide Dose			Total
Serotype	1 mg	0.2 mg	0.04 mg	Protected
A_{24}	5/5	5/5	5/5	15/15
O_1K	4/4	5/5	5/5	14/14
C_3	2/4	5/5	3/5	10/14

This particular C-virus proves to be a more stringent test of the peptide-vaccine efficacy in guinea pigs. It is not immediately clear why this C-subtype of the FMD virus should be more difficult to protect against. Unfortunately, virus vaccinates were not evaluated in our test. This could have provided a control for the virulence of the challenge virus. While equivalent infectious doses of each virus subtype were used, the speed of viral replication in the guinea pig may vary greatly. The antipeptide and antiviral antibody titers seemed to rise at comparable rates for each peptide serotype. In cattle challenge experiments, Black et al., 1986, have reported the A_{24} virus to be approximately 30 times more immunogenic than the O_1 virus, and the C_3 virus to be of intermediate immunogenicity. Consequently, our results may be specific to the guinea pig adapted virus, or more likely an inherent feature of the peptide immunogens relative to inactivated virus.

Table 3. Heterotypic Protection of Guinea Pigs With 40-Residue Peptide

			Viral Serotype		
		Dose	A_{24}	C_3	O_1K
Immunizing Peptide Serotype	A_{24}	0.2 mg	5/5	0/5	4/5
	C_3	0.2 mg	1/5	3/5	3/5
	O_1K	0.2 mg	1/5	0/5	5/5
	A_{24}	5.0 mg	5/5	1/5	3/5
	C_3	5.0 mg	1/5	5/5	3/5
	O_1K	5.0 mg	4/5	1/5	5/5

As an extension of this initial study, we chose to cross-challenge the animals with heterologous virus. It is generally recognized that there is little protective enhancement between virus strains included in vaccines Black et al., 1986. The results of the heterologous challenge are shown in Table 3.

It is immediately obvious that the results of the homotypic challenge experiment reported in Table 2 are confirmed in this experiment. Complete, homotypic protection was achieved with the A- and O-Serotypes. The C-serotype yielded complete protection at the 5 mg dose. Looking down any of the three viral serotype columns of Table 3 indicates the relative ease of protection against each challenge virus. Surprisingly, the O_1 challenge virus proved the easiest to protect against. A significant degree of cross protection was observed with both the A- and C-peptides. The C_3 challenge virus was the most difficult to protect against. Neither the A- nor O-peptide were capable of inducing a protective response against the C_3 challenge. The A_{24} challenge was effectively blocked by the high-dose O_1-peptide immunization. The C_3-peptide was virtually ineffective against the A_{24} challenge.

These observations support the contention that the O-virus lies intermediate between the A- and C-serotypes on evolutionary scales (Dopazo et al., 1988). In a practical sense, these data demonstrate that these peptide antigens are capable of inducing a broad degree of viral protection. It remains to be determined whether any degree of synergy can be derived from the simultaneous immunization of animals with peptides representing multiple serotypes.

REFERENCES

Bachrach, H. L., Moore, D. M., McKercher, P. D. and Polatnik, J. (1975), Immune and antibody responses to an isolated capsid protein of foot-and-mouth disease virus. J. Immunol. 115, 1635-1641.

Bittle, J. L., Houghten, R. A., Alexander, H., Shinnick, T. M., Sutcliffe, J. G., Lerner, R. A., Rowlands, D. J. and Brown, F. (1982), Protection against foot and mouth disease by immunisation with a chemically synthesised peptide predicted from the viral nucleotide sequence. Nature (London) 298, 30-33.

Black, L., Nicholls, M. J., Rweyemamu, M. M., Ferrari, R. and Zunino, M. A. (1986), Foot-and-mouth disease vaccination: a multifactorial study of the influence of antigen dose and potentially competitive immunogens on the response of cattle of different ages. Research in Vet. Sci. 40, 303-307.

DiMarchi, R., Brooke, G., Gale, C., Cracknell, V., Doel, T., and Mowat, N. (1986), Protection of cattle against foot-and-mouth disease by a synthetic peptide. Science 232, 639-641.

DiMarchi, R., Brooke, G., Gale, C., and Doel, T. (1988), Structure-function relationship in protection against foot and mouth disease (FMD) by a synthetic peptide. Proc. of the Tenth Am. Pep. Symposium 10, 531-533.

Doel, T. R., Gale, C., Brooke, G., and DiMarchi, R. (1988), Immunization against Foot-and-Mouth Disease with Synthetic Peptides Representing the C-terminal Region of VP1. J. Gen. Virol. 69, 2403-2406.

Dopazo, J., Sobrino, F., Palma, E. L., Domingo, E., and Moya, A. (1988), Proc. Natl. Acad. Sci. USA 85, 6811-6815.

Laporte, J., Grosclaude, J., Wantyghem, J., Bernard, S., and Rouze, P., (1973), Neutralisation en culture cellulaire du pouvoir infectieux de virus de la fievre aphteuse par des serums proovenant de pores immunises a l'aide d'une proteine virale purifiee. C. R. Acad. Sci. Paris 276, 3399-3402.

Pfaff, E., Leban, J., Kuhn, C., Bohm, H. O., Thiel, J. -J. and Schaller, H. (1985), Structural analysis of the FMDV antigenic determinant, 199-202. In R. A. Lerner, R. M. Chanock, and F. Brown (ed.), Vaccines 85. Cold Spring Harbor Laboratory, Cold Spring Harbor, N. Y.

Strohmaier, K., Franze, R., and Adam, K.-H. (1982), Location and characterisation of the antigenic portion of the FMDV immunising protein. J. Gen. Virol. 59, 295-306.

CHARACTERIZATION OF A SYNTHETIC ENVELOPE PEPTIDE OF SRV-1 AND SRV-2 VIRUS THAT SPECIFICALLY BINDS RHESUS ANTI-SRV-1 AND ANTI-SRV-2 NEUTRALIZING ANTIBODIES

A. Malley,[1] S. Shiigi,[1] E. Benjamini,[2] L. Werner,[2] C.Y. Leung[2], H-S. Kwang[2], M. Axthelm[1], and L.M. Hallick[3]

[1]Oregon Regional Primate Research Center, Beaverton, OR
[2]University of California Med. School at Davis, Davis, CA
and [3]Oregon Health Sciences University, Portland, OR

INTRODUCTION

Currently, 5 distinct Simian type D retrovirus serotypes have been defined: SRV-1 (Calif., New Eng.), SRV-2 (2R, 2C), SRV-3 (MPMV), SRV-4 (Calif. Cyno), and SRV-5 (Chinese). Neutralizing antibodies made against the indicated Simian viruses do not cross react. Thus, we may be able to produce synthetic envelope peptides specific for each Simian virus serotype and make antibodies against the neutralizing epitopes. The experiments presented here with SRV-1 and SRV-2 viruses represent our attempts to develop a model approach for the preparation of synthetic envelope peptide vaccines.

Nucleic acid sequences and predicted amino acid sequences of both SRV-1 and SRV-2 viruses have been defined (Power et al., 1986), and a great deal of homology exists between these two viruses. The major difference between these viruses is found in the envelope proteins of the two viruses. Upon comparison of the amino acid sequences of the envelopes, regions were identified that are unique to each virus envelope. These peptides were analyzed for their hydrophobicity and secondary structural features (Hopp and Woods, 1981; Chou and Fasman, 1974). The selected unique peptide regions were synthesized by Dr. Eli Benjamini's laboratory (Univ. of Calif. Med. School at Davis, Davis, Calif.). We tested the ability of these peptides to block rhesus monkey anti-SRV-1 or anti-SRV-2 antibody binding with virus-infected Raji cells or neutralization of the syncytium forming activity of virus.

METHODS

Immunofluorescence Assay (IFA)

Raji cells and virus-infected Raji cells (25 μl/microtiter plate well) are dispensed at a final concentration of 2.5 X 10^5 cells/well. An equal volume of serum is added, the plate covered, and mixed by gentle agitation. The suspension is incubated at 4°C on ice for 30 minutes, and the plates centrifuged at 1000 rpm for 5 minutes. Plates are washed twice, the cells resuspended, and 25 μl of fluorescent-labeled anti-IgG (1:200 dilution) is added to each well and incubated at 4°C on ice for 30 minutes. Plates are centrifuged at 1000 rpm for 5 minutes, washed twice, and 200 µl of 1% formaldehyde is added to each well. Samples are analyzed immediately or stored at 4°C after sealing with plate sealer. Cells are examined for staining with the fluorescent-labeled anti-IgG antibody in an EPICS C fluorescence activated cell sorter. Mean channel peak and the percentage of positive staining cells is noted. The mean channel peak times the percentage of positive staining cells is plotted against the units of antibody activity.

Neutralization Assay

Neutralizing activity of serum samples or affinity purified fractions were assayed for their ability to inhibit syncytium formation of Raji cells cultured in the presence of either SRV-1 or SRV-2 virus. Twenty-five microliters of a 1:5 dilution of monkey serum or isolated antibody fraction are combined with 25 µl of culture supernatant containing 5 to 8 syncytium-forming units of virus in duplicate wells of a microtiter plate. The diluent and cell culture supernatant medium consisted of RPMI 1640, 10% FCS, 1% L-glutamine, and 25 units of gentamicin. After a 30 minute incubation at room temperature, 50 µl of Raji cells (2 X 10^5/ml) are added to the wells. Cultures are fed with 100 µl of culture medium on days 3 and 6, and examined on days 9 and 12 for syncytium formation.

Affinity Methods

SRV-1 envelope peptides 147-162 and 150-157 were coupled to keyhole limpet hemocyanin (KLH) by the carbodiimide method (Kipp and Miller, 1980). The resulting peptide-KLH conjugates were purified by passage over Sephadex G-25 and Dowex 1 columns. The peptide-KLH conjugates were coupled to cyanogen bromide activated Sepharose 4B (Pfeiffer, Wylie, and Schuster, 1987), and unreactive sites were blocked with an excess of 0.2 mM glycine, pH 8.

Samples applied to the Sepharose-SRV-1 peptide adsorbents were run at a flow rate of 10 ml/hr, washed with 20 volumes of buffer, and bound material eluted with 0.2 M glycine-HCl, pH 2.3 (3 volumes). The neutralized eluted protein and non-adherent fraction was concentrated by negative pressure, and assayed for antibody activity.

The insoluble SRV-2 envelope peptide 88-103 was used to absorb anti-SRV-2 antibody activity. The SRV-2 peptide 88-103 was neutralized by treatment with 1M Tris buffer pH 9, followed by washing with PBS, and stored at 4°C until used. Samples applied to the SRV-2 peptide 88-103 in a microcentrifuge tube (1.5 ml volume) was mixed by vortex for 30 seconds, and placed on a rocking platform at 4°C for 1 hr. The suspension was centrifuged at 10,000 rpm for 10 minutes, and the supernatant carefully removed. The pellet was resuspended by vortex, and washed in PBS. The original supernatant and all washes were combined and represent the non-adherent fraction. Material bound with the SRV-2 peptide 88-103 was eluted with 0.2 M glycine-HCl pH 2.3, and the material eluted with glycine and the first 2 washes of the adsorbent with PBS were combined and represent the adherent fraction. Both adherent and non-adherent fractions were concentrated by negative pressure.

RESULTS AND DISCUSSION

Each of the seven unique SRV-1 envelope peptides were tested for their ability to specifically block rhesus anti-SRV-1 antibody binding with SRV-1 infected Raji cells (Table 1). None of the SRV-1 envelope peptides blocked rhesus anti-SRV-2 antibodies binding with SRV-2 infected Raji cells. On the other hand, the synthetic peptide 147-162 significantly reduced the level of binding of anti-SRV-1 antibodies with SRV-1 infected Raji cells. In addition, the synthetic peptide 186-207 appeared to have some ability to block anti-SRV-1 antibody interaction with SRV-1 infected Raji cells.

Although 9 unique SRV-2 envelope regions were defined upon comparison of the amino acid sequences of SRV-1 and SRV-2 envelopes, only three of these sequences were hydrophilic. Table 2 shows the reactivity of these peptides and their ability to block anti-SRV-2 antibody binding with SRV-2 infected Raji cells. The insoluble synthetic peptide 88-103 clearly is very effective in reducing the level of anti-SRV-2 antibody that binds in the IFA test with SRV-2 infected Raji cells. The loss of antibody activity is not

Table 1. Blocking of Anti-SRV-1 or Anti-SRV-2 Antibody with SRV-1 Specific Envelope Peptides

Peptide Sequence	Units of anti-SRV Antibody[a] Anti-SRV-1	Anti-SRV-2
None	1400	700
63-79	1200	700
112-126	N.T.[b]	700
147-162	260	700
186-207	580	700
194-207	900	900
320-334	1000	500
366-380	1000	700

[a]2 mg of synthetic peptide was incubated with 100 ul of antiserum for 1 hr at 37°C, and overnight at 4°C. These mixtures were centrifuged at 2000 rpm for 10 min at 4°C, and the supernatants tested for their ability to bind to SRV-1-infected or SRV-2-infected Raji cells. The units of antibody activity measured by immunofluorescence assay (IFA).
[b]N.T. - Not tested.

Table 2. Adsorption of Anti-SRV-2 antibody with Synthetic SRV-2C Envelope Peptides

	SRV-2C Envelope Peptide Sequence					
	83 - 103		88 - 103		233 - 249	
Source of Serum	Unads	Ads[a]	Unads	Ads	Unads	Ads
Rhesus SRV-2R	190[b]	210	120	4	110	300
Rhesus SRV-2R	480	140	300	21	400	400
Celebes SRV-2C	270	320	640	43	870	940
Rhesus Control	<1	<1	<1	<1	<1	<1
Celebes Control	<1	<1	<1	<1	<1	<1
Rhesus SRV-1	<1	<1	<1	<1	<1	<1

[a]2 mg of synthetic peptide was incubated with 100 ul of antiserum for 1 hr at 37°C, and overnight at 4°C. The mixtures were centrifuged at 2000 rpm for 10 min at 4°C, and the supernatants tested for their ability to bind to SRV-2 infected Raji cells.
[b]Units of anti-SRV-2 antibody measured by immunofluorescence assay (IFA).

simply due to absorption by the insoluble 88-103 peptide, as peptide 83-103 is also insoluble and is not very effective in reducing the level of antibody that binds with SRV-2 infected Raji cells. The loss of binding activity of the 83-103 synthetic peptide may be due to an alteration of the structure by the addition of 5 hydrophobic amino acids on the N-terminal side of peptide 88-103. In addition, SRV-2 peptide 88-103 does not bind either anti-TMVP or anti-Timothy grass pollen antibodies.

Table 3 shows that rhesus anti-SRV-1 neutralizing antibodies bind to the Sepharose 147-162 adsorbent and are specifically eluted. In contrast, rhesus anti-SRV-2 antibodies do not bind with this affinity adsorbent. Also, Table 3 shows that rhesus anti-SRV-2 neutralizing antibodies are

Table 3. Adsorption of Rhesus Anti-SRV-1 or Anti-SRV-2 Neutralizing Antibodies by Affinity Methods

Adsorbent	Serum Sample	Fraction	Activity[a]
Seph-SRV-1 147-162	anti-SRV-1	Original	1:16
		Non-adh	0
		Adh	1:16
	anti-SRV-2	Original	1:32
		Non-adh	1:32
		Adh	0
SRV-2 peptide 88-103	anti-SRV-1	Original	1:16
		Non-adh	1:16
		Adh	0
	anti-SRV-2	Original	1:32
		Non-adh	0
		Adh	1:32

[a]Neutralization tested the ability of various fractions to prevent 8 syncytium-forming units of virus from developing syncytium formation of the virus in Raji cells.

Table 4. Adsorption of Rhesus Anti-SRV-2 Antibodies by Affinity Methods

		Immunofluorescence[a]			Neutralization[b]		
Adsorbent	Antibody Specificity	Orig	Non-Adh	Adh	Orig	Non-Adh	Adh
SRV-2 peptide 88-103	Rh anti-SRV-2	340	<1	326	1:32	0	1:32
	Rh anti-SRV-1	1000	1000	<1	1:16	1:16	0

[a]Anti-SRV-1 and Anti-SRV-2 antibody were tested for their ability to bind SRV-2 infected Raji cells by IFA. Controls included SRV-1 infected and uninfected Raji cells and were negative.
[b]Neutralization assay tested the ability of various fractions to prevent 8 syncytium-forming units of virus from developing syncytium formation of the virus in Raji cells.

specifically bound to the insoluble synthetic SRV-2 envelope peptide 88-103, and eluted with 0.2 M glycine-HCl buffer. Table 4 compares the adsorption of rhesus anti-SRV-2 antibody bound to the SRV-2 envelope peptide 88-103 in both the IFA and neutralization tests. Again, it is apparent that the SRV-2 peptide 88-103 binds antibodies specific for the SRV-2 virus, and does not bind antibody directed against the SRV-1 virus. Although the IFA and neutralization tests might not measure exactly the same antibodies, it is apparent that these two tests show parallel reactivity with regard to the ability of the synthetic peptides to bind specific antibodies measured by these assays.

The data clearly indicates that SRV-1 envelope peptide 147-162 and SRV-2 peptide 88-103 represent major determinants that are involved in the formation of SRV-1 and SRV-2 neutralizing antibodies. Experiments are now in progress to induce anti-SRV-1 and SRV-2 neutralizing antibodies by immunization of rhesus monkeys with the defined synthetic envelope peptides. Once these animals produce satisfactory levels of neutralizing antibodies, these monkeys will be challenged with the appropriate virus to determine if these antibodies can protect the animals from retroviral infection. Our data strongly suggests that a comparison of predicted amino acid sequences of unique envelope regions for hydrophobicity and secondary structural features provides a useful means of identifying major antigenic determinants of type D retrovirus envelopes, and these methods may prove useful in attempts to identify similar determinants on other type D serotypes.

REFERENCES

Chou, P.Y., and Fasman, G.D., 1974, Prediction of protein conformation. Biochem. 13:222.

Hopp, T.P., and Woods, K.R., 1981. Prediction of protein antigenic determinants from amino acid sequences. Proc. Natl. Acad. Sci. 78:3824.

Kipp, D., and Miller, A., 1980, Preparation of protein-conjugated red blood cells with ECDI in: "Methods of Cellular Immunology," B.B. Mishell and S. Shiigi, eds. WH Freeman & Co., San Francisco.

Pfeiffer, N.E., Wylie, D.E., and Schuster, S.M., 1987, Immunaffinity chromatography utilizing monoclonal antibodies. Factors which influence antigen-binding capacity. J. Immunol. Methods 97:1.

Power, M.D., Marx, P.A., Bryant, M.L., Gardner, M.B., Barr, P.J., and Luciw, P.A. 1986, Nucleotide sequence of SRV-1, a type D simian acquired immune deficiency syndrome retrovirus. Science 231:1567.

specifically bound to the insoluble synthetic SRV-2 envelope peptide 88-107, and eluted with 0.2 M glycine-HCl buffer. Table 4 compares the adsorption of rhesus anti-SRV-2 antibody bound to the SRV-2 envelope peptide 88-107 in both the IFA and neutralization tests. Again, it is apparent that the SRV-2 peptide 88-107 binds antibodies specific for the SRV-2 virus, and does not bind antibody directed against the SRV-1 virus. Although the IFA and neutralization tests might not measure exactly the same antibodies, [illegible] specificity with regard to the ability of the synthetic peptide to bind specific antibodies recognized by these assays.

The work [illegible] indicates [illegible]

SYNTHETIC PEPTIDE REPRESENTING A T-CELL EPITOPE OF CRM197 SUBSTITUTES AS CARRIER MOLECULE IN A HAEMOPHILUS INFLUENZAE TYPE B (HIB) CONJUGATE VACCINE

Garvin S. Bixler, Jr., Ronald Eby, Kathy M. Dermody,
Robert M. Woods, Robert C. Seid, and Subramonia Pillai

Praxis Biologics, Inc.
300 East River Road, Rochester, NY 14623

ABSTRACT

The cross-reactive material (CRM197) of diphtheria toxin is considered to be advantageous as a carrier molecule in the formulation of a *Haemophilus influenzae* type b conjugate vaccine. In order to more precisely understand the function of the CRM197 in the vaccine, we have begun mapping the T-cell epitopes of the protein. A peptide which represents a segment of the primary sequence of CRM197 has been identified and found to stimulate diphtheria toxoid or CRM197-primed murine T-lymphocytes. In addition, the peptide is capable of priming T-cells *in vivo* for a subsequent *in vitro* T-cell response to itself or to the intact CRM197 molecule. The ability of the peptide to replace CRM197 as a carrier molecule was examined by immunizing mice with PRP, PRP-CRM197 conjugate, or PRP covalently coupled to the peptide. Antibodies to PRP were only detected in the PRP-CRM197 or PRP-peptide immunized groups. Both conjugates elicited primary and secondary antibody responses. Thus, a synthetic peptide representing a defined T-cell epitope of CRM197 has been functionally demonstrated based on its ability to act as a carrier molecule in a PRP conjugate vaccine.

INTRODUCTION

The capsular polysaccharide of *Haemophilus influenzae* type b (Hib) has been identified as the focus of bactericidal (Cates *et al*., 1985) or protective (Smith *et al*., 1973; Anderson *et al*., 1972) antibody and subsequently has been the basis for several commercial vaccine formulations. The purified capsular polysaccharide composed of polyribosyl ribitol phosphate (PRP) has been found to be efficacious in children greater than 18 months of age (Greenberg *et al*., 1987). Unfortunately, the age group (6-18 months) with peak incidence of Hib infection respond poorly to the purified PRP alone (Anderson *et al*., 1972; Peltola *et al*., 1977). Presumably, this reflects the inability of an immature system to deal effectively with thymus-independent polysaccharide antigens (Bixler and Pillai, 1989). Some of the difficulties associated with recognition of polysaccharides by an immature immune system can be overcome by covalently coupling the polysaccharide to an appropriate protein carrier molecule. This converts the polysaccharide from a T-independent to a T-dependent antigen. Several studies have shown that the resulting conjugates are efficacious at early ages (Chu *et* al., 1983; Anderson et al., 1985, 1987; Insel et al., 1986) which is a significant improvement over vaccines composed of polysaccharide alone.

The cross-reactive material (CRM197) of diphtheria toxin has been found to be advantageous as a carrier molecule for PRP. In order to more precisely understand the function of CRM197 as a carrier molecule, we have undertaken the task of delineating T-cell epitopes within the protein which may participate in the recognition of a PRP conjugate vaccine. In this report, using synthetic peptides we have described our initial studies which localized and partially defined a T-cell epitope of CRM197.

RESULTS AND DISCUSSION

From the amino acid sequence of CRM197, several regions of the protein were selected for evaluation of their capacity to stimulate T-cells obtained from protein-primed mice. The regions suspected of harboring putative T-cell epitopes were selected on the basis of computer analyses of the primary sequence of the protein (Chou and Fasman, 1974; Kyte and Doolittle, 1982; Hopp and Woods, 1983; DeLisi and Berzofsky, 1985).

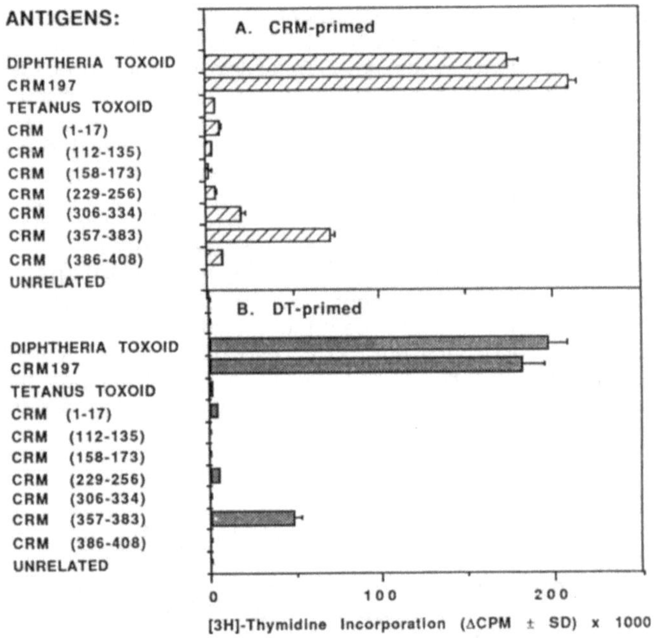

Figure 1. Responses of protein-primed murine T-lymphocytes to *in vitro* challenge with synthetic peptides of CRM197.

Six regions of the protein were focused upon and another region of the protein, 158-173, of approximately the same size was selected at random to serve as a control peptide. These peptides were synthesized (OCS Laboratories, Denton, TX) and their capacity as well as that of a peptide from an unrelated protein to stimulate T-cells assessed. The results of lymphocyte proliferation assays using T-cells obtained from CRM197 or diphtheria toxoid (DT)-primed SJL mice are summarized in Figure 1. Although a broad dose range (0.1 - 100 µg/ml) of peptides was used for in vitro challenge, only one CRM197 peptide, 357-383, consistently elicited a significant response in both CRM197 or DT-primed mice. This suggests that of the regions of CRM197 examined in SJL (H-2^s) mice only region 357-383 contains, in part or in whole, a T-cell epitope of the protein.

To determine if the free peptide itself is immunogenic, SJL mice were immunized with 100 ug of the CRM197 peptide 357-383 emulsified in Freund's complete adjuvant. As controls, other groups of mice received 100 ug of CRM197, DT or the CRM197 peptide 306-334. The response of cells obtained from these mice were studied by lymphocyte proliferation assay. As depicted in figure 2, cells from CRM197-primed mice responded to the CRM197 peptide 357-383 as expected.

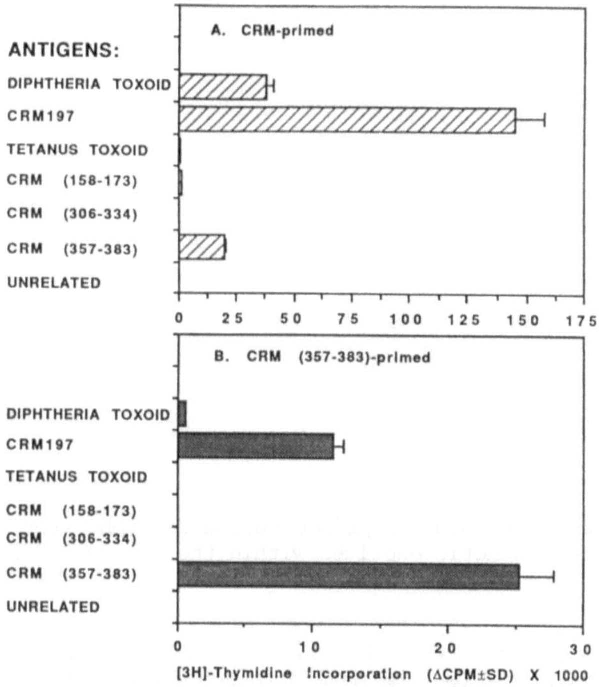

Figure 2. Responses of protein or peptide-primed murine T-lymphocytes to in vitro challenge with the proteins, CRM197 or diphtheria toxoid, or to the CRM197 peptide 357-383.

A similar result was also obtained with cells from DT-primed mice (data not shown). Interestingly, cells obtained from CRM(357-383)-primed mice responded very well in vitro to the CRM197 peptide 357-383 and cross-reacted with the whole protein, CRM197, but did not recognize DT. Clearly, the CRM197 peptide 357-383 is immunogenic in its free form (i.e. not coupled to a carrier protein) since a substantial anti-peptide T-cell response was observed.

Having determined that region 357-383 of CRM197 indeed contained a T-site which was also immunogenic as a free peptide, it was of interest to precisely determine the minimum sequence necessary to activate protein-primed murine T-lymphocytes. In order to map the T-site, a set of nested peptides was synthesized which maintained a fixed C-terminal at glycine 383 and varied the N-terminal from alanine 357 to asparagine 373. The capacity of these peptides to stimulate CRM197 or DT-primed T-cells was compared by assaying T-cell proliferation. As shown in

Figure 3, the inclusive subset of CRM197 peptides 357-383 through 370-383 were nearly equivalent in their capacity to stimulate CRM197-primed lymph node cells. In contrast, the CRM197 peptides 371-383 and 372-383 showed substantial decreases in the capacity to stimulate T-cells which correlated with their decreasing size. The shortest of the CRM197 peptides, 373-383, elicited no significant level of T-cell activity.

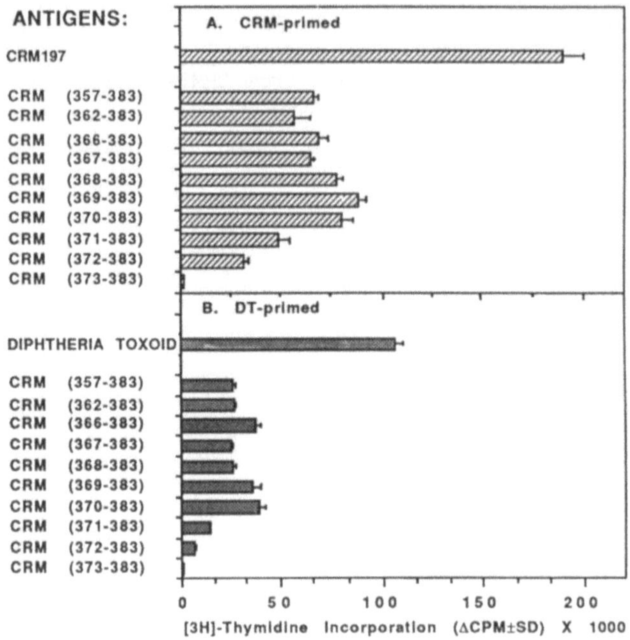

Figure 3. Responses of protein-primed murine T-lymphocytes to a nested set of synthetic peptides within the region 357-383 of CRM197.

A similar pattern of peptide reactivity was also apparent with the DT-primed cells. Thus, the data obtained from <u>in vitro</u> challenge with this set of peptides of both the CRM197 and DT-primed cells strongly suggest that the N-terminal residue of the T-site maps to valine 370 for the polyclonal T-cell response of SJL mice.

In order to determine whether or not the region 357-383 of CRM197 could provide T-helper function, PRP-peptide conjugates were prepared by reductive amination. The peptide conjugates were used to immunize outbred Swiss Webster mice. Each animal received intramuscularly 5 ug of peptide conjugate on days 0 and 14. The conjugates were administered in phosphate buffered saline without an adjuvant. Sera were collected at various intervals and pooled within a group. Specific antibody to PRP was determined by Farr assay and are shown in Figure 4. Control mice immunized with PRP alone or with PRP coupled to the CRM197 peptide 158-173 produced no detectable antibody response (data not shown). In contrast, those mice immunized with either PRP-CRM(357-383) or PRP-CRM (366-383) peptide conjugates produced rising antibody titers which exceed 1.5 ug antibody/ml. Thus, the data suggest that both of the CRM197 peptides could act as carrier molecules and provide the necessary T-site which enabled immune recognition of the PRP to occur.

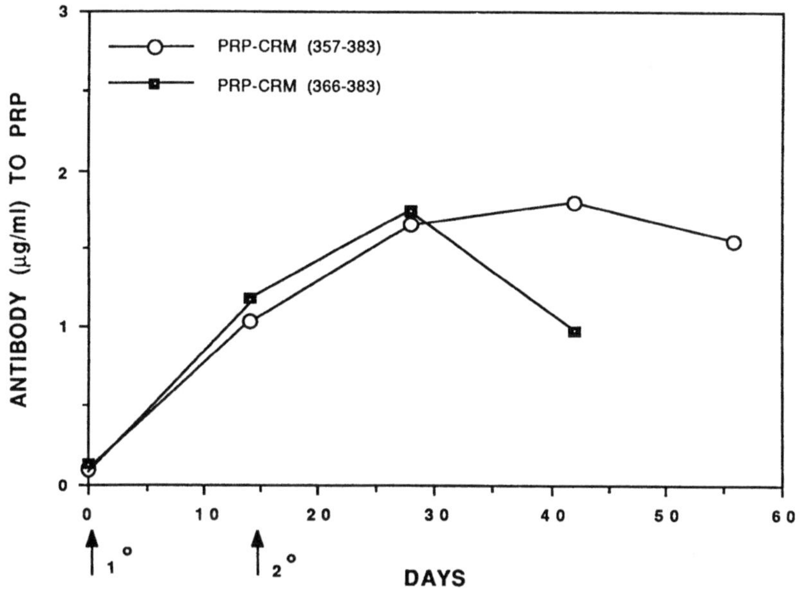

Figure 4. Antibody responses to PRP of Swiss Webster mice immunized with PRP-CRM(357-383) or PRP-CRM(366-383) peptide conjugates. No detectable antibody response was elicited in control mice immunized with PRP alone or with PRP conjugated to CRM(158-173).

In conclusion, a series of synthetic peptides which represented selected regions of CRM197 were examined for their capacity to be recognized as T-cell epitopes. Of these, region 357-383 was found to be recognized by both CRM197 and DT-primed T-cells. Furthermore, this region was effective in priming for an anti-peptide T-cell response. Close inspection of this region with a nested set of peptides, resulted in the mapping of the N-terminal of the T site to residue 370. Significantly, this region when coupled to PRP was able to function as a carrier molecule as demonstrated by the production of an antibody response to PRP.

REFERENCES

Anderson, P., Peter, G., Johnston, R.B.J., Wetterlow, L. H., and Smith, D. H., (1972), Immunization of humans with polyribophosphate, the capsular antigen of Hemophilus influenzae, type b, J. Clin. Invest., 51:39-44.

Anderson, P., Pichichero, M., Edwards, K., Porch, C. R., and Insel, R., (1987), Priming and induction of Haemophilus influenzae type b capsular antibodies in early infancy by Dpo2o, an oligosaccharide-protein conjugate vaccine, J. Ped., 111:644-650.

Anderson, P., Pichichero, M. E., and Insel, R. A., (1985), Immunogens consisting of oligosaccharides from the capsule of Haemophilus influenzae type b coupled to diphtheria toxoid or the toxin protein CRM197, J. Clin. Invest., 76:52-59.

Bixler, G.S., and Pillai, S., (1989), The cellular basis of the immune response to conjugate vaccines, In: "Contributions to

Microbiology and Immunology," J. M. Cruse ed., S. Karger, Switzerland, (in press).

Cates, K. L., Marsh, K. H., and Granoff, D. M., (1985), Serum opsonic activity after immunization of adults with Haemophilus influenzae type b-diphtheria toxoid conjugate vaccine, Infect. Immun., 48:183-189.

Chou, P. Y., and Fasman, G. D., (1974), Predicition of protein conformation, Biochem., 13:222-224.

Chu, C. Y., Schneerson, R., Robbins, J. B., and Rastogi, S. C., (1983), Further studies on the immunogenicity of Haemophilus influenzae type b and pneumococcal type 6A polysaccharide-protein conjugate, Inf. Immun., 40:245-256.

DeLisi, C., and Berzofsky, J. A., (1985), T-cell antigenic sites tend to be amphipathic structures, Proc. Natl. Acad. Sci., 82:7048-7052.

Greenberg, D. P., Ward, J. I., Burkart, K., Christenson, P. D., Guravitz, L., and Marcy, S. M., (1987), Factors influencing immunogenicity and safety of two Haemophilus influenzae type b polysaccharide vaccines in children 18 and 24 months of age, Pediatr. Infect. Dis. J., 6:660-665.

Hopp, T. P., and Wood, K. R., (1983), A computer program for predicting protein antigenic determinants, Mol. Immunol., 20:483-489.

Insel, R. A., and Anderson, P. W., (1986), Oligosaccharide-protein conjugate vaccines induce and prime for oligoclonal IgG antibody responses to the Haemophilus influenzae b capsular polysaccharide in human infants, J. Exp. Med., 163:262-269.

Kyte, J., and Doolittle, R. F., (1982), A simple method for displaying the hydropathic character of a protein, J. Mol. Biol., 157:105-132.

Peltola, H., Kayhty, H., Sivonen, A., and Makela, H., (1977), Haemophilus influenzae type b capsular polysaccharide vaccine in children: A double-blind field study of 100,000 vaccines 3 months to 5 years of age in Finland, Pediatrics, 60:730-737.

Smith, D. H., Peter, G., Ingram, D. L., Harding, A. L., and Anderson, P., (1973), Responses of children immunized with the capsular polysaccharide of Haemophilus influenzae type b, Pediatrics, 52:637-645.

THE USE OF MOUSE CLASS II MHC PEPTIDES TO PRODUCE SITE-SPECIFIC MONOCLONAL ANTIBODIES

Christopher J. Krco, Thomas G. Beito, Roger G. Little, Jay Zeller, Daniel M. McCormick and Chella S. David

Departments of Immunology and Biochemistry
Mayo Medical School, Mayo Clinic
Rochester, MN 55905

INTRODUCTION

We have attempted to produce site-specific monoclonal antibodies against defined regions on mouse class II (Ia) molecules. These studies were designed to produce monoclonal antibodies against defined sequences of Ia molecules which may have applications in the design of novel experimental, functional and therapeutic investigations into the biochemical role of Ia molecules in immunity and disease states (Rothbard and Taylor, 1988). We describe here our initial results using peptides derived from the polymorphic regions 43-61 of the alpha chain subunit and polymorphic regions 57-74/78 of beta chain subunit of I-A molecules from three alleles.

MATERIALS AND METHODS

Each peptide was assembled by solid-phase methods on p-methylbenzhydrylamine resin with N-t-Boc L-amino acid benzyl side-chain derivatives using an automated peptide synthesizer (McCormick and Atassi, 1985). Synthesis was performed under conditions and appropriate recycling that achieved a coupling efficiency of greater than 98.6% for each residue as measured by a quantitative ninhydrin reaction for free NH-terminal groups. Following synthesis, each peptide was removed from the resin support by acidolysis and lyophilized. All peptides were purified by preparative reversed-phase HPLC (Vydac C_{18} TP-preparative column, 1.0 x 25 cm) using a gradient system of 0.1% TFA/water (v/v) and 0.1% TFA containing 80% acetonitrile (v/v). After purification, the sequence of each synthetic peptide was confirmed by gas-phase microsequencing on a protein sequenator.

Mice were inoculated with 50 μg of Ia peptide-KLH conjugate emulsified in complete Freund's adjuvant in the hind right footpad (Krco et al., 1985). Fourteen days later mice were inoculated in the left hind footpad with 50 μg of conjugate in adjuvant. Seven days after the second inoculation (day 21 after the initial inoculation) mice were bled via retro-orbital sinus and sera individually collected and stored. The cycle of boosting every fourteen days and bleeding at day seven post-injection was repeated at different subcutaneous sites until specific anti-peptide antibody was detected.

For the isolation of antibody secreting hybridomas, immune mice were inoculated with 50 μg of free peptide intravenously. Three days later the spleens were removed and single cell suspensions prepared. Lymphocytes were mixed 10:1 with FO myeloma cells and centrifuged. The cell pellets were resuspended in 1 ml of 50% polyethylene glycol 1540 and the cells were allowed to fuse for 90 sec. The cells were diluted in culture medium and centrifuged. The cell pellets were resuspended in Iscove's medium supplemented with 10% calf serum and placed into microtiter wells (100 μl/well). Twenty-four hours later 100 μl of HAT supplemented medium was added to the wells. Every three to four days thereafter the medium was replaced with 100 μl of fresh HAT medium. When cells within the wells approached 75% confluency the supernates were tested for the presence of specific anti-peptide antibody by ELISA. Cells from wells exhibiting specific antibody activity were harvested and subjected to two rounds of limiting dilution cloning. Ascites were produced in vivo from the cloned hybridomas and the monoclonal antibodies were purified by protein A affinity chromatography.

Anti-peptide monoclonal antibody activity was measured by ELISA. Ten μg/ml solution of Ia peptide in 0.01 M carbonate buffer (pH 9.6) was added (100 μl) to Immunolon II (Dynatech) micro-ELISA wells overnight at 37°C. The wells were then washed using 0.01 M phosphate buffer and blocked using 200 μl of a 1% BSA solution. Following a 2 hr incubation at 37°C, the wells were washed again using 0.01 M phosphate buffer. Next, 100 μl of antibody solution was added to the wells. Following a 90 min incubation at 37°C the wells were washed again and 100 μl of a goat anti-mouse immunoglobulin beta-galactosidase conjugate was added to the wells. Following a 30 min incubation at room temperature the wells were washed and 100 μl of a 1 mg/ml solution of PNPG indicator solution was added to the wells. Color development was determined spectrophotometrically at 414 nm.

Antisera was tested for anti-peptide antibody by radioimmunoassay. The RIA's were conducted essentially the same as ELISA with the exception that I-Staph. protein A was used to measure specifically bound antibody.

Table 1. Sequences of Peptides Synthesized

I-A Subunit	Allele	
Alpha	b	K E T V W M L P E F G Q L A S F D P Q
	d	K K T V W R L P E F G Q L I L F E P Q
	q	K E T V W M L P E F G Q L T S F D P Q
Beta	b	D A E Y W N S Q P E I L E R T R A E L D T V
	d	D A E Y W N S Q P E I L E R T R A E V D T A
	q	D A E Y W N S Q P E I L E R T R A E

RESULTS

Three alpha subunit peptides comprising residues 54-61 from the b, d, and q alleles of the I-A subregion of the mouse MHC were synthesized. This series of peptides differ from each other at five residues. In addition, three beta subunit peptides comprising residues 57-78 (b and d alleles) and 57-74 (q allele) were synthesized. In contrast to the alpha peptide series, the beta subunit series of peptides are identical at residues 57-74 but differ at 2 residues between 75-78. All six peptides (Table 1) were conjugated to KLH. Each peptide conjugate was emulsified in complete Freund's adjuvant and injected into ten A.TH mice. In addition, five A.TH mice were immunized with KLH emulsified in complete Freund's adjuvant.

Table 2. Reactivity of Individual Mouse Serum Samples Against Specific Ia Peptide Immunogenes.

Mouse/Sample	cpm ± SEM					
	A_α^q(43-61)	A_α^d(43-61)	A_β^q(57-74)	A_β^b(57-78)	A_β^d(57-78)	KLH
#1	179± 8	9453± 100	115± 28	450± 33	4332±126	·
#2	105± 15	28733± 23	291± 13	456± 12	231± 17	·
#3	2239	17779± 61	115± 25	415± 35	482± 72	·
#4	3013	33239±1054	131± 15	184± 7	4637± 57	·
#5	525	24614± 207	167± 5	254± 3	224± 34	·
#6	1187±139	10402± 91	182± 19	338± 30	168± 6	·
#7	1103± 33	31930± 120	175± 62	295± 69	105± 3	·
#8	471± 71	30077± 170	163± 40	630± 5	288± 3	·
#9	1478± 23	11548± 48	133± 3	436± 48	277± 27	·
#10	313± 11	14929± 145	213±127	135± 4	109± 12	·
NMS	113± 12	303± 6	146± 1	198± 42	181± 27	340± 12
αKLH	108± 7	341± 51	165± 11	142± 19	144± 50	38689± 19

Day 35 sera (1:100)

Individual sera was collected and stored. Radioimmunoassays were conducted on individual sera (Table 2). It was observed that each of the six peptides had different degrees of immunogenicity. Peptide A_α^d (43-61) was the most immunogenic peptide in A.TH mice with 100% (10/10) of the mice producing specific antibody (cpm's of 9500-33,000). Peptide A_α^q (43-61) was less immunogenic in that only 50% (5/10) of the mice producing specific antibodies (cpms of 1100-3000). Peptide A_β^q (57-78) was marginally immunogenic since only 20% (20/10) of the mice elicited a specific antibody response (cpm's of 4300-4600). The two remaining beta peptides A_β^b (57-78) and A_β^q (57-74) were not immunogenic in A.TH mice. None of the sera reacted with KLH (cpm's of less than 350) nor did normal mouse serum react any of the peptides(cpm's of less than 350).

Table 3. Cross-Reactivity of Pooled Anti-A_α^d (43-61) Sera with Class II MHC Peptides

Peptide	Reciprocal Antiserum Dilution					
	200	400	800	1600	3200	6400
KLH	33776±282	26722±379	16911±154	9289± 92	4664± 74	2284± 47
A_α^d(43-61)	11843±629	5943±187	1307±142	1307± 71	635± 6	410± 23
A_α^q(43-61)	5139±227	2906± 58	1267± 26	630± 14	361± 21	198± 2
A_α^b(43-61)	5159±325	2617±124	1136± 46	593± 40	295± 9	196± 4
A_β^q(57-78)	130± 29	82± 21	59± 14	17± 13	39± 8	44± 12
A_β^q(57-74)	388± 33	266± 1	153± 34	122± 39	75± 2	63± 2
A_β^b(57-78)	193± 40	83± 1	58± 17	36± 2	29± 6	37± 7

Day 35 sera (1:100)

Anti-A_α^d (43-61) sera were pooled and assessed for cross-reactivity against all six peptides (Table 3). Anti-A_α^d (43-61) sera reacted strongly with A_α^d peptide (cpm of 11,800) at a 1:100 dilution and had an end point of 1:1600 (cpm of 1300). As expected as predicted by sequence homology (Table 1), anti-A_α^d (43-61) sera exhibited 43% cross-reactivity with A_α^b and

A^q_α peptides (1:100) dilution and no cross-reaction with any of the beta peptides (3% cross-reaction).

Following several fusions, a total of five monoclonal antibodies specific for A^d_α (43-61) were obtained (Table 4). All five anti-A^d_α monoclonal antibodies (IaP-1,2,3,7,8) are specific for A^d_α and do not cross react with b or q alpha allele peptides. In addition, following immunization of B10.RUB1 mice with A^b_β (57-78)-KLH conjugate two anti-A monoclonal antibodies (IaP11,12) were obtained.

Table 4. Specificity of Protein A Purified Monoclonal Antibodies

	O.D. (414 nm)						
		A_α (43-61)			A_β (57-74/78)		
Monoclonal Antibody	KLH	b	d	q	b	d	q
IaP-1	.049	.051	.343	.046	.036	.003	.034
IaP-2	.039	.045	.297	.044	.031	.026	.026
IaP-3	.026	.040	.364	.053	.024	.019	.020
IaP-7	.080	.124	.174	.115	.036	.045	.047
IaP-8	.046	.026	.352	.025	.022	.020	.026
IaP-11	.224	.180	.173	.190	.682	.136	.109
IaP-12	.298	.213	.185	.184	.904	.353	.131
IaP-5	.518	.077	.099	.068	.055	.052	.054

1 μg/well peptide and 5 μg/well Protein A purified antibody

DISCUSSION

Our results can be summarized as follows. One, we have been able to elicit antibody production against defined regions on Ia molecules in mice. Two, three appears to be a genetic control in the ability of strains of mice to elicit anti-peptide antibodies. This strain variability may be a reflection of varying degrees of homology between donor peptide sequence and host Ia sequence such that a greater degree of homology correlates with decreases immunoreactivity. As such, this situation resembles a peptide induced "alloreactivity". Three, monoclonal anti-peptide antibodies can be isolated from immune mice. The monoclonal antibodies isolated to date are peptide specific and do not cross react in binding assays with other allelic peptides. This situation is in contrast to the 50% cross reactions observed using pooled anti-peptide sera. The ability to isolate allele- and site-specific anti-Ia peptide monoclonal antibodies offers the promise of highly selective in vivo therapeutic modalities in the treatment of autoimmune diseases. In particular such allele-specific anti-class II monoclonal antibodies may be a novel way of treating autoimmunity in the heterozygous human population.

ACKNOWLEDGEMENTS

This work was supported by NIH contract, AI-72650, and in part by NIH grant, HD-09140.

REFERENCES

Krco, C.J., Gorzynski, T.J. and Beito, T. (1985) Characterization of a monoclonal antibody that recognizes a determinant unique to human haemoglobin β-chain. J. Immunogenetics 12:197-211.

McCormick, D.J. and Atassi, M.Z. (1985) Antigenic structure of human hemoglobin: Delineation of the antigenic sites (Site 2) within region 41-65 of the alpha chain by immunochemistry of synthetic peptides. J. Protein Chemistry 4:171-184.

Rothbard, J.B. and Taylor, W.R. (1988) A sequence pattern common to the T cell epitopes. EMBO Journal 7:93-100.

EPITOPE MIMICRY BY ANTI-IDIOTYPE SEQUENCES IN REVERSE ORIENTATION

Dennis W. Metzger[1], Clayton W. Naeve[2], and Victor H. Van Cleave[1]

Departments of [1]Immunology and [2]Virology and Molecular Biology, St. Jude Children's Hospital, Memphis, TN 38101
[1]Department of Microbiology and Immunology, University of Tennessee, Memphis, TN 38163

Antigen mimicry by anti-idiotypic internal images offers one approach to vaccine development and immune modulation. However, the rational design of internal image vaccines and an understanding of idiotypic interactions will depend upon a precise understanding of the rules which govern antigen recognition and of the structural basis for internal image expression. Previous studies using the reovirus hemagglutinin and a synthetic peptide (GAT) identified determinants in antibody internal image variable (V) regions that appeared to match continuous sequences within the respective antigens (Ollier et al., 1985; Bruck et al., 1986).

We have used a rabbit immunoglobulin (Ig) epitope, i.e., the V_Ha1 allotype, to examine the nature of antigen mimicry in a well-defined protein system (Metzger et al., 1985; Van Cleave et al., 1986) and have produced two mouse monoclonal antibodies (mAbs) that bear a1-like internal images. These mAbs, termed 6B12 and 61B8, a) are reactive with all homologous and heterologous anti-a1 Ab but not with normal Ig; b) efficiently inhibit the binding of rabbit Ig to anti-a1 Ab; and 3) elicit the production of anti-allotype Ab when injected into normal mice. In addition, when directly examined by immunoelectron microscopy, the V region terminus of each mAb binds anti-a1 Fab fragments at approximately 180° angles (Van Cleave et al., 1986).

Sequencing of the 61B8 and 6B12 hybridoma V_L mRNAs by primer extension revealed only 65% homology between them and no apparent similarities with rabbit allotype residues. However, the mAb V_H region sequences demonstrated greater than 95% identity at the DNA level, even though the hybridomas were derived from separate mice and were of different isotypes (Van Cleave et al., 1988). Interestingly, a reverse homology to the a1 antigen was observed in the mAb V_H region, with a paired glutamic acid (position 56) and threonine (position 57) in complementarity-determining region 2 (CDR2) of both 6B12 and 61B8. Not only is this pairing unique among all known mouse sequences (Dildrop,

This work was supported by grants AI18880 and CA21765 from NIH, and by ALSAC.

1984), but these are the same amino acids in V_H framework region 3 (FR3) of rabbit Ig which correlate with a1 allotype expression (Mage et al., 1984; Kabat et al., 1987). Further homology to rabbit Ig was noted among another three amino acids, also quite rare, flanking positions 56 and 57 (Fig. 1). The ability of the reversed sequence to mimic an a1-like determinant was tested directly by producing synthetic peptides corresponding to the putative antigenic regions of rabbit Ig and the mAb internal images, respectively. Although the homologous residues were presented in opposite orientations, both peptides completely inhibited at similar concentrations the binding of rabbit Ig to an anti-a1 mAb (Metzger et al., 1984). Conservative substitutions in the peptide sequence identified the paired Glu 56 and Thr57 as being critical for expression of the a1 epitope (Van Cleave et al., 1988).

<---------FR 3---------->

Rabbit a1	NH_2 –	Ser	Pro	Thr	Thr	Glu	Asp	Thr	Ala	Thr	Tyr	– COOH
mAb H chains (6B12 and 61B8)	COOH –	Ser	Tyr	Ala	Thr	Glu	Asp	Ile	Asn	Pro	Tyr	– NH_2
Occurrence		<2			83	<2	<2				22	

<---------CDR 2--------->

Fig. 1 Region of observed homology between positions 81-90 of rabbit a1 FR3 and positions 52-60 of the mAb V_H CDR2 (numbering system according to Kabat et al., 1987). Occurence refers to the observed frequency of that residue among sequenced members of the mouse J558 H chain family (i.e., <2 = less than 2%). Note that the rabbit a1 sequence is shown N→C while the mAb sequence is shown C→N. (From Van Cleave, et al., 1988).

To determine whether the conserved residues of rabbit a1 FR3 could bear structural homology to the same conserved, though reversed, residues in the internal image CDR2, we have now used the molecular modelling system MENDYL to predict the Ig structures based on sequence homology with available myeloma protein crystallographic data (Fig. 2), an approach used successfully by others to predict unknown Ig structures (Chothia et al., 1986). The framework regions of myeloma proteins have a high degree of sequence and structural homology which can be used to advantage in modelling studies (Chothia et al., 1985; de la Paz et al., 1986; Chothia et al., 1986). Of the known Fab structures J539, MCPC603, NEW, and KOL, rabbit a1 is most homologous to the sequence of J539. We therefore generated an a1 FR3 model using the coordinates of J539 as the initial structure, the backbone atom coordinates were retained, and 5 of 10 J539 side chains in FR3 were replaced with rabbit a1 residues.

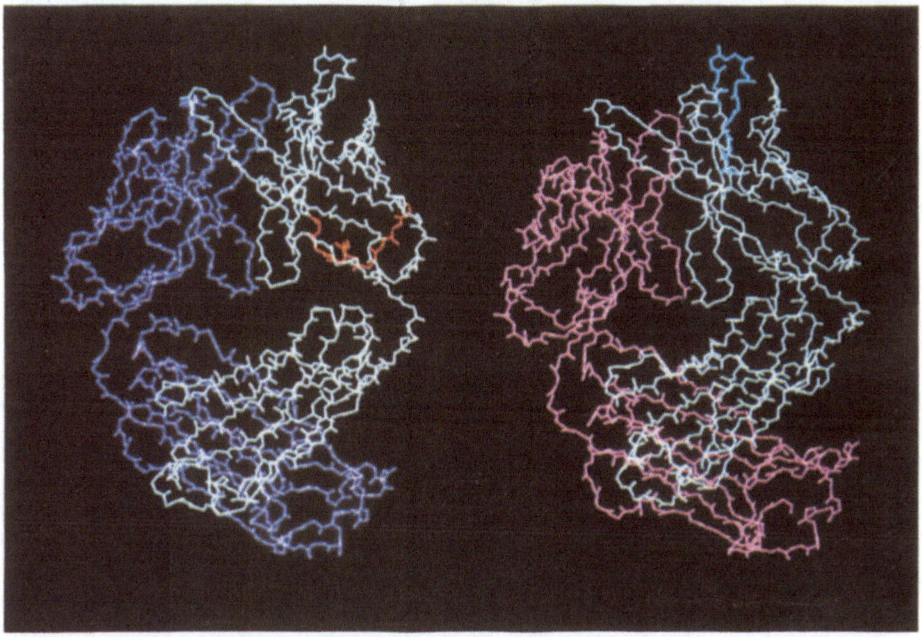

Fig. 2 Modelling the conformation of residues associated with FR3 rabbit a1 allotype and CDR2 a1-like internal images. The macromolecular modelling system MENDYL (Tripos Assoc., Inc., St. Louis, MO) was used to build, minimize, and view the structures interactively. The calculations were performed on a CompactVaxII and displayed on an Evans and Sutherland PS340. Left: the location of the a1 allotope in FR3 (red) on the H chain of Fab J539 (H chain in white, L chain in purple). Right: the location of the internal image CDR2 (blue) on the H chain of Fab J539 (H chain in white, L chain in magenta).

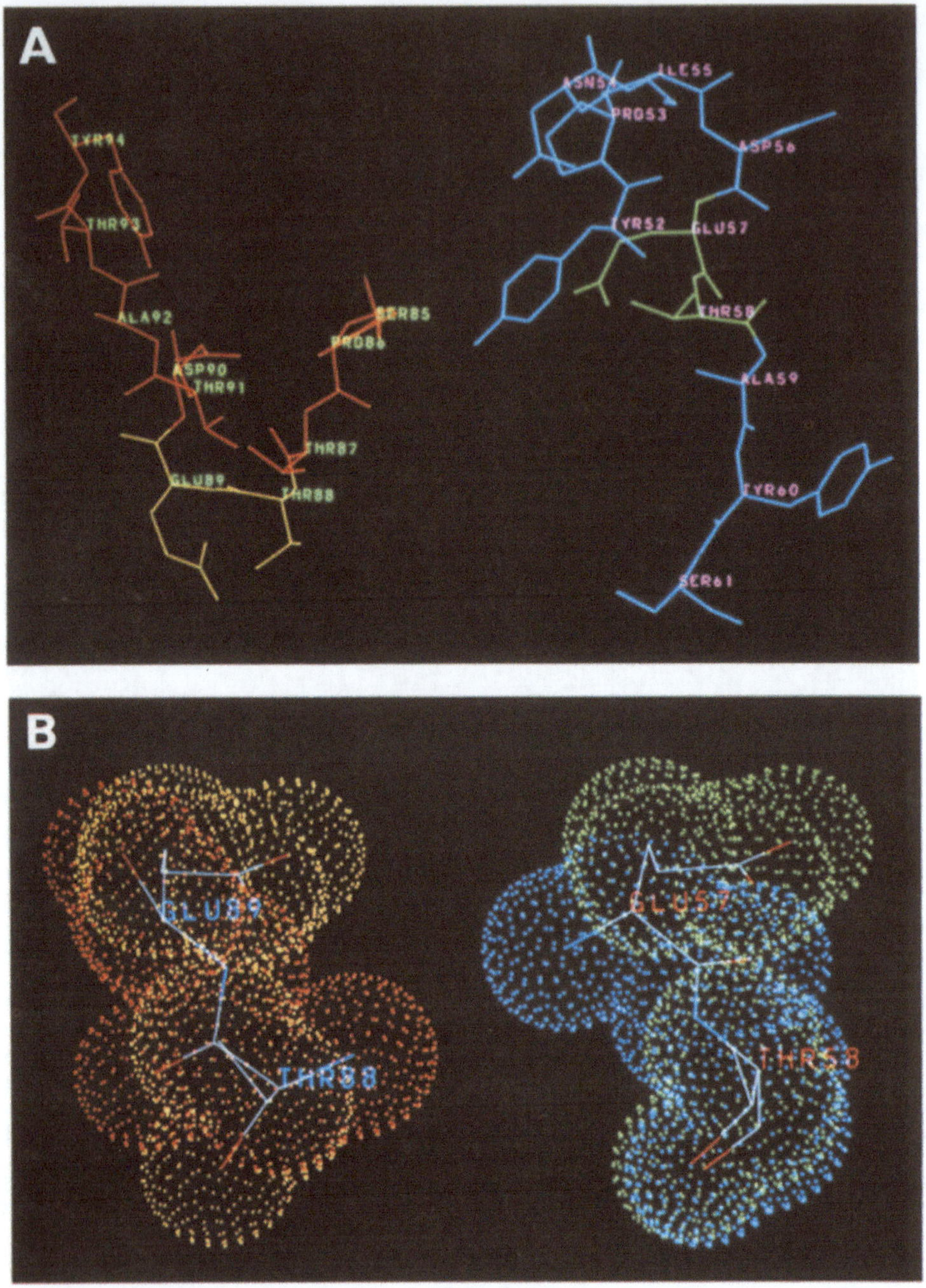

Fig. 3 A) Left: the predicted a1 allotope structure from Ser85 to Tyr94 [Kabat numbering system=Ser81 to Tyr89] (red) with the critical residues Thr88 and Glu89 shown in yellow. Right: the predicted internal image CDR2 structure from Tyr52 to Ser61 [Kabat numbering system=Tyr52 to Ser60] (blue) with the critical residues Glu57 and Thr58 shown in green. B) Left: a1 residues Thr88 and Glu89 color coded by atom type with backbone atoms in red and side chains in yellow. Right: CDR2 residues Glu57 and Thr58 color coded by atom type with backbone atoms in blue and side chain atoms in green.

With regard to CDRs, one might expect these to be difficult to model given their variability in length and sequence; however, 3/6 CDR loops (H1, H2, and L2) show remarkably conserved structure among the myeloma proteins NEW, MCPC603 and KOL (de la Paz et al., 1986). Since the conserved a1 residues in the internal images reside in H2 (a subset of CDR2 residues), we used the coordinates of this region from J539 as starting geometry, replacing 8 of 10 J539 side chains with the corresponding internal image sequence. The structures were then subjected to energy minimization to relax strained bond lengths and angles and to eliminate bad van der Waals contacts (Fig. 3a). Finally, the side chains of the critical residues on the internal image model (Glu57 and Thr58 by the Brookhaven numbering system) were interactively manipulated in an attempt to mimic rabbit a1 (Fig. 3b).

This exercise showed that, although the predicted overall structures for a1 FR3 and CDR2 are much different, the critical Thr and Glu side chain placements are stereochemically feasible and can be almost superimposable in either context (yellow and green atoms in Fig. 3b). Furthermore, the conserved Tyr52, Asp56, and Ser61 in the internal image CDR2 are unlikely to participate in the antigenic site but are probably structurally important: Tyr94 is not surface accessible and is conserved in FR3 of all known Fab structures as is Asp90; Ser85 is also present in KOL and NEW. These results suggest that the backbone atoms (red and blue atoms in Fig. 3b) are not recognized as part of the internal image.

In contrast to previous studies in which internal image epitopes were present in the correct orientation (Ollier et al., 1985; Bruck et al., 1986), the results presented here suggest that the ability to simulate the structural motif of an antigen is dependent upon the configuration of amino acid side chains rather than the orientation of the carbon backbone. With the potential for recognition of protein sequences in reverse orientation, the effective repertoire of the immune system is thus greatly increased. Further studies to define the structural basis for internal image epitopes should lead to novel approaches for vaccine design.

REFERENCES

Bruck, C., Co, M.S., Slaoui, M., Gaulton, G.N., Smith, T., Fields, B.N., Mullins, J.I., and Greene, M.I., 1986, Nucleic acid sequence of an internal image-bearing monoclonal anti-idiotype and its comparison to the sequence of the external antigen, Proc. Natl. Acad. Sci. USA 83:6578.

Chothia, C., Novotny, J., Bruccoleri, R., and Karpus, M., 1985, Domain association in immunoglobulin molecules: the packing of variable domains, J. Mol. Biol. 186:651.

Chothia, C., Lesk, A.M., Levitt, M., Amit, A.G., Mariuzza, R.A., Phillips, S.E.V., and Poljak, R.J., 1986, The predicted structure of immunoglobulin D1.3 and its comparison with the crystal structure, Science 233:755.

de la Paz, P., Sutton, B.J., Darsley, M.J., and Rees, A.R., 1986, Modelling of the combining sites of three anti-lysozyme monoclonal antibodies and of the complex between one of the antibodies and its epitope, EMBO J. 5:415.

Dildrop, R., 1984, A new classification of mouse V_H sequences, Immunol. Today 5:85.

Kabat, E.A., Wu, T.T., Reid-Miller, M., Perry, H.M., and Gottesman, K.S., 1987, "Sequences of Proteins of Immunological Interest", U.S. Department of Health and Human Services, PHS, NIH.

Mage, R.G., Bernstein, K.E., McCartney-Francis, N., Alexander, C.B., Young-Cooper, G.O., Padlan, E.A., and Cohen, G.H., 1984, The structural and genetic basis for expression of normal and latent V_Ha allotypes of the rabbit, Mol. Immunol. 21:1067.

Metzger, D.W., 1984, A mouse monoclonal antibody against rabbit V_H allotype shares the predominant idiotype with a rabbit antibody of the same specificity, Eur. J. Immunol. 14:304.

Metzger, D.W., 1985, The nature of antiidiotype molecules induced by antiallotype. Presence of both latent allotype and allotypic internal images, J. Exp. Med. 162:35.

Ollier, P., Rocca-Serra, J., Somme, G., Theze, J., and Fougereau, M., 1985, The idiotypic network and the internal image: possible regulation of a germ-line network by paucigene encoded Ab2 (anti-idiotypic) antibodies in the GAT system, EMBO J. 4:3681.

Van Cleave, V.H., Murti, K.G., and Metzger, D.W., 1986, Mouse monoclonal antibodies induced by anti-allotype antibody display internal images of the rabbit V_Ha1 allotype: direct visualization by immunoelectron microscopy, Eur. J. Immunol. 16:701.

Van Cleave, V.H., Naeve, C.W., and Metzger, D.W., 1988, Do antibodies recognize amino acid side chains of protein antigens independently of the carbon backbone?, J. Exp. Med. 167:1841.

VACCINE-CONTAINING BIODEGRADABLE MICROSPHERES SPECIFICALLY ENTER THE GUT-ASSOCIATED LYMPHOID TISSUE FOLLOWING ORAL ADMINISTRATION AND INDUCE A DISSEMINATED MUCOSAL IMMUNE RESPONSE

John H. Eldridge[1], Jonathan A. Meulbroek[1], Jay K. Staas[2], Thomas R. Tice[2], and Richard M. Gilley[2]

[1]Department of Microbiology, University of Alabama at Birmingham, Birmingham, AL 35294
[2]Controlled Release Division, Southern Research Institute, Birmingham, AL 35255-5305

ABSTRACT

Biodegradable and biocompatible microspheres have been investigated for their usefulness as a vaccine delivery system for both parenteral and enteral immunization. Microspheres composed of poly(DL-lactide-co-glycolide) which contained a toxoid vaccine of Staphylococcal enterotoxin B were found to strongly potentiate the circulating anti-toxin antibody response following intraperitoneal injection. Following oral administration, microspheres < 10 μm in diameter were specifically taken up into the Peyer's patches of the gut-associated lymphoid tissue, where those ≥ 5 μm remained fixed for an extended period. Microspheres < 5 μm were disseminated within macrophages to the mesenteric lymph nodes, blood circulation and spleen. Oral immunization with enterotoxoid-containing microspheres induced circulating toxin-specific antibodies and a concurrent secretory IgA anti-toxin response in saliva, gut wash fluids and bronchial-alveolar wash (BAW) fluids. In contrast, soluble enterotoxoid was completely ineffective as an oral immunogen.

INTRODUCTION

Currently, extensive research is being conducted in an effort to identify and isolate appropriate and safe vaccine antigens for the immunization of man against a wide variety of infectious agents (Chanock and Lerner, 1984). However, the overriding concern for safety dictates that highly purified carbohydrates and proteins be employed, and these are often poorly immunogenic. In this regard, currently available DNA procedures have proven to be powerful tools in the study of bacterial and viral proteins, and many genes encoding antigens recognized by protective antibodies have been cloned and expressed. Due to their poor immunogenicity, however, very few effective immunogens have been produced through DNA technology. Among the approaches used to enhance vaccine immunogenicity are conjugation to carriers (Arnon et al., 1980) and adjuvants (Warren et al., 1986). Alum is the single adjuvant currently approved for human use and its use is questioned for booster immunizations.

The need for effective vaccination procedures is particularly acute with respect to organisms which produce their pathophysiologic effects through acute infections localized to the gastrointestinal, pulmonary, nasopharangeal and genitourinary surfaces. These areas are bathed in mucus which contains immunoglobulins consisting almost exclusively of secretory IgA (sIgA) (Hanson 1961; Tomasi and Zigelbaum, 1963; Tomasi et al., 1965). This antibody is derived from large numbers of IgA-producing plasma cells which infiltrate the lamina propria regions underlying these mucosal membranes (Brandtzaeg and Baklien, 1976; Brandtzaeg, 1984), and is specifically transported to the lumenal surface through the action of secretory component (Solari, 1985; Mestecky, 1987; Brandtzaeg, 1985). However, parenteral immunization regimens are almost universally ineffective at the induction of sIgA responses. Secretory immunity is most often achieved through the direct immunization of mucosally-associated lymphoid tissues, of which the largest mass is represented by the Peyer's patches (PP) of the gastrointestinal tract.

The finding that Peyer's patches (PP), which are discrete lymphoid follicles along the gastrointestinal tract, possess IgA precursor B cells which can repopulate the lamina propria regions of the gastrointestinal and upper respiratory tracts and differentiate into mature IgA-synthesizing plasma cells, suggested that the induction of IgA responses is not necessarily a local phenomenon (Craig and Cebra, 1971). The PP contain a large subpopulation of B cells that are committed to IgA synthesis, all categories of regulatory T cells, and functional accessory cells, i.e., macrophages and dendritic cells (Kiyono et al., 1982; Spalding et al., 1983). In addition, the PP are covered by a unique epithelium which contains microfold cells (M cells) with highly developed pinocytotic channels that allow sampling of antigens from the gut lumen, and transport of the antigens to the cells in the underlying dome region, with subsequent stimulation in the B cell (follicular) and T cell (parafollicular) zones (Bockman and Cooper, 1973).

Heremans and Bazin and co-workers measured the development of IgA responses in mice orally immunized with soluble (Crabbe et al., 1969) or particulate (Bazin et al., 1970) antigens. A sequential appearance of antigen-specific IgA plasma cells was observed first in the mesenteric lymph nodes, later in the spleen, and finally in the lamina propria of the gastrointestinal tract. Subsequent studies have shown that the oral administration of antigens leads to the production of sIgA antibodies in the gut and also in secretions distant to the gut, e.g., in BAW, colostrum, milk, saliva and tears (Crabbe et al., 1969; Mestecky et al., 1978; Michalek et al., 1976; Montgomery et al., 1974). Coupled with the findings of Craig and Cebra (1971), these results suggest that the PP are an enriched source of precursor IgA B cells, which subsequent to antigen exposure, follow a circular migrational pathway and account for the expression of sIgA at distant mucosal surfaces. This circular traffic pattern provides a common mucosal immune system by continually shuttling sensitized B cells to mucosal sites where they synthesize antibody to gut-encountered environmental antigens and potential pathogens (Hanson et al., 1978).

Extensive studies have demonstrated the feasibility of oral immunization to induce this common mucosal immune system (Mestecky et al., 1978; Bienenstock and Befus, 1980; Mestecky, et al., 1980), but with rare exception the large doses required to achieve adequate local concentrations in the PP have made this approach impractical for purified vaccine antigens. Among the strategies explored to overcome this problem are incorporation of the antigen into liposomes and the use of genetically engineered bacteria which colonize the PP and express plasmid-encoded vaccine antigens (reviewed by Mestecky, 1987). However, practical considerations make both of these approaches unlikely for wide application. Antigen-loaded liposomes are difficult to reproducibly prepare and have a limited shelf life. In the case of a modified bacterial vector the immune response to the carrier microorganism limits its usefulness to a single immunization. We report here on our studies with biodegradable microspheres consisting of lactide/glycolide copolymers, a vaccine delivery system which results in substantially enhanced immunity without the need for adjuvants and which is effective following either parenteral or oral administration.

MATERIALS AND METHODS

Staphylococcal Enterotoxin B Vaccine

A formalinized vaccine of staphylococcal enterotoxin B was prepared as described by Warren et al. (1973). One gram of enterotoxin was dissolved in 0.1 M sodium phosphate buffer, pH 7.5, to 2 mg/ml. Reagent-grade 36% formaldehyde containing 12% methanol diluted (1:6) in the same phosphate buffer was added to the enterotoxin solution. The solution was placed in a slowly-shaking 37°C controlled environment incubator-shaker and the pH was checked daily and maintained at 7.5 ± 0.1 units. At the end of the 30 day incubation period the toxoid was concentrated and washed into borate buffered saline (BBS) using a pressure filtration cell (Amicon), and sterilized by filtration. Conversion of the enterotoxin into enterotoxoid was confirmed by the absence of weight loss in 3 to 3.5 kg rabbits injected intramuscularly with 1 mg of toxoid-converted material.

Immunizations

Groups of BALB/c mice were orally administered either free or microencapsulated enterotoxoid in 0.5 ml of 8 parts tap water : 2 parts 7.5% sodium bicarbonate via a gastric tube. Systemic immunizations were carried out through the intraperitoneal injection of free or microencapsulated enterotoxoid in PBS.

Collection of Biologic Fluids

Blood was collected from the retroorbital plexus in calibrated, heparinized capillary pipettes and the plasma harvested following centrifugation. Pools of 100 µl of plasma per mouse from groups of five mice were heat-inactivated and stored at -70°C until assayed. Saliva and gut wash fluids were collected from groups of mice as described by Elson et al. (1984). In brief, the mice were administered four doses (0.5 ml) of lavage solution, isoosmotic with mouse gut fluid, at 15 min intervals via a gastric tube. Fifteen minutes after the last dose the mice were anesthetized and administered 0.1 mg pilocarpine by intraperitoneal injection. The intestinal contents were discharged over a 10 to 20 min period and collected in 5 ml of a solution of 0.1 mg/ml soybean trypsin-inhibitor (Sigma) in 50 mM EDTA. Concurrent with the intestinal discharge, a large volume of saliva is secreted and 200 µl was collected per mouse. The secretion samples were clarified by centrifugation and sodium azide, phenylmethylsulfonyl fluoride, and fetal calf serum were added as preservative, protease inhibitor and alternate substrate for protease activity, respectively. All samples were stored at -70°C until assayed.

Radioimmunometric Assays of Toxin-Specific Antibodies

Radioimmunometric assays were performed in Immulon strips (Dynatech) coated with toxin at 1 µg/ml in BBS, pH 8.4, overnight at 4°C. Control strips were uncoated, but all were blocked for 2 hr at 25°C with 1% bovine serum albumin (Sigma) in BBS, which was also used as the diluent for all samples and ^{125}I-labeled reagents. Various 2-fold dilutions of test samples were added to washed triplicate wells and incubated for 6 hr at 25°C. After washing, 100,000 cpm of ^{125}I-labeled affinity-purified IgG goat anti-mouse μ, α, or γ chain-specific antibody (Southern Biotechnology Associates) was added per well and incubated overnight at 4°C. Following removal of unbound antibodies by washing, the bound ^{125}I-antibodies were detected in a gamma spectrometer (Beckman). The results are presented as the reciprocal of the sample dilution producing a signal > 3-fold that of the group-matched prebleed at the same dilution (end point titration).

Microsphere Tissue Penetration Studies

Mice were administered a single dose of 20 mg of Coumarin-

6 loaded microspheres suspended in 0.5 ml tap water using a blunt-tipped feeding needle inserted into the stomach. The mice were sacrificed at various times and 3 representative PP from the small intestine (one from the midgut and one each from within 1 cm of the stomach and appendix), the first mesenteric lymph node proximal to the appendix, and the spleen were excised, mounted in OCT compound (Miles), and snap frozen in liquid nitrogen. Tissues were cut into 4 to 6 µm serial sections and all sections were viewed in a fluorescence microscope (Leitz) to quantitate microspheres/tissue. The size was determined using a calibrated eyepiece micrometer (Leitz) and its location noted (the PP dome region was defined as the area being within 200 µm of dome epithelium). Some internalized microspheres were divided in the sectioning process, but individual sections contained a limited number of microspheres and multiple viewings of sequential sections allowed divided microspheres to be counted as one and the diameter to be accurately determined.

RESULTS

Several studies have shown that parenteral immunization with antigens trapped within or incorporated into the lipid bilayer of liposomes results in an enhanced antibody response (Allison and Gregoriadis, 1974). This activity appears, at least in part, to be due to their function as a depot for the antigen (Kramp et al., 1982). In addition, the increased uptake by macrophages has been sited as a mechanism resulting in more efficient presentation to T cells (Dailey et al., 1977). We investigated the immunopotentiating activity of an analogous delivery system which does not suffer from the inherent stability problems of liposomes: biocompatible and biodegradable microspheres. These microspheres were formed by first polymerizing equal molar parts of DL-lactide and glycolide (Fig. 1), and then purifying the copolymer (DL-PLG). The copolymer was then dissolved so as to incorporate the antigen, a toxoid vaccine of Staphylococcal enterotoxin B, into small spherical droplets which were hardened to produce microspheres with the vaccine homogeneously dispersed into the copolymer matrix. DL-PLG copolymers are biocompatible esters that biodegrade *in vivo* into lactic and glycolic acids (Fig. 2). The mechanism of degradation is by hydrolysis of the ester linkages, and the degradation rate is primarily determined by the ratio of lactide to glycolide present in the copolymer (Miller et al., 1977). Fifty µm microspheres composed of 50:50 (wt%:wt%) DL-PLG biodegrade to completion in approximately 6 wk after intramuscular injection, and under the same conditions an 85:15 ratio requires > 25 wk. The vaccine containing microspheres examined in the experiments reported here contained 0.23 wt% enterotoxoid.

Groups of BALB/c mice were immunized by intraperitoneal (IP) injection with various doses of free or microencapsulated enterotoxoid on three occasions separated by 30 d, and plasma samples were obtained at 10 d intervals. Plasma enterotoxin-

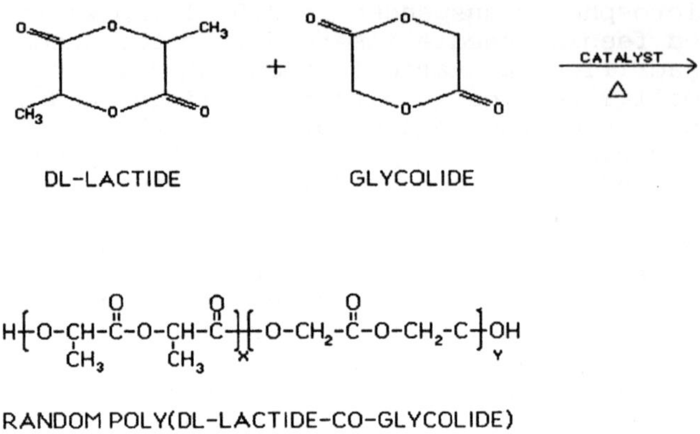

Figure 1. Reaction of DL-lactide and glycolide to yield a random poly(DL-lactide-co-glycolide) copolymer.

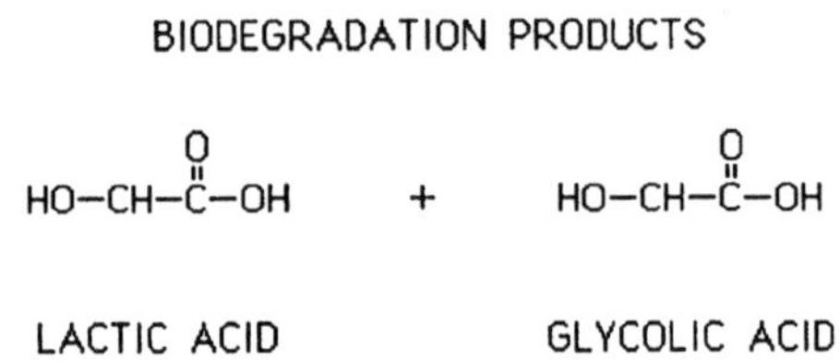

Figure 2. Biocompatible degradation products formed by hydrolysis of poly(DL-lactide-co-glycolide) copolymers *in vivo*.

specific antibodies were quantitated by end-point titration in isotype-specific radioimmunometric assays (Table 1). Primary immunization with the optimal dose of free enterotoxoid, 25 µg, elicited a characteristically poor response which was detected only in the IgM isotype. In contrast, 25 µg of microencapsulated enterotoxoid induced not only higher levels of toxin-specific IgM, but a vigorous IgG response. Microencapsulation allowed 4 times as much enterotoxoid to be administered without causing high zone paralysis, and this higher dose resulted in a further heightening of immunity. This immunopotentiating effect was even more pronounced following secondary and tertiary immunizations (Table 1). Secondary IgG anti-enterotoxin was detected at 512-fold higher dilution in the plasma of mice which received 50 µg of microencapsulated enterotoxoid than in those receiving an optimal regimen of free material. In fact, tertiary immunization with free enterotoxoid was required to raise an anti-enterotoxin response equivalent to that obtained with a single microencapsulated dose of 100 µg. Similar adjuvant activity has been observed for common laboratory protein antigens such as haptenated keyhole limpet hemocyanin and whole virus such as influenza.

Table 1. Plasma IgM and IgG anti-enterotoxin levels on day 20 following primary, secondary, and tertiary parenteral immunization with various doses of soluble or microencapsulated staphylococcal enterotoxoid.

Enterotoxoid Dose (mg) Per Immunization	Form	Plasma Anti-Enterotoxin Titer On Day 20 following Parenteral Immunization: Primary		Secondary		Tertiary	
		IgM	IgG	IgM	IgG	IgM	IgG
100	Microspheres	1,280	10,240	320	163,840	640	655,360
50	Microspheres	1,280	5,120	640	163,840	1,280	327,680
25	Microspheres	640	2,560	640	81,920	640	163,840
50	Soluble	< 20	< 20	80	< 20	640	640
25	Soluble	160	< 20	160	32	80	10,240
12.5	Soluble	< 20	< 20	80	80	40	1,280

Based on their probable ability to deliver high local concentrations of antigen and the protection which the wall matrix would be expected to provide against the low pH of the stomach and the proteolytic enzymes of the gut, we speculated that microspheres would represent a suitable vehicle for the delivery of vaccine antigens to the PP. The uptake of microcapsules from the gut lumen by the phagocytic M cells which overlie the PP, and their progression into the underlying lymphoreticulum, was investigated by orally-administering a suspension of 85:15 DL-PLG microspheres containing the fluorochrome coumarin-6 to mice. At various times, three representative PP, the spleen, and the first mesenteric lymph node (MLN) draining the appendix were excised, frozen, and serially sectioned. Fluorescence microscopic observation revealed that internalized microspheres of various sizes were present in the PP at 24 hr post administration and at all times tested through 35 d (Table 2). At no time were microspheres of any size observed to penetrate at other sites than the PP. The total number of microspheres observed in the PP increased through 4 d and then decreased over the following 31 d to 15% of the peak number. At 1, 2 and 4 d the proportion of < 2 µm (45-47%), 2 to 5 µm (31-35%) and > 5 µm (18-23%) diameter particles remained constant and reflected the size distribution of the input dose. Concurrent with the decrease in total microsphere numbers beginning at 7 d, a progressive shift in the size distribution was observed such that the small (< 2 µm) and medium (2-5 µm) sizes ceased to predominate and the large (> 5 µm) microspheres became the numerically greatest species observed. In addition, the microspheres observed deep in the lymphoreticulum were exclusively ≤ 5 µm in diameter, and it was this fraction which progressively extravasated from the PP (Table 2).

Table 2. Penetration of coumarin-6 85:15 DL-PLG microspheres into and through the Peyer's patches following oral administration

Time (days)	Total number observed	Proportion of diameter (%) Small <2 µm	Medium 2-5 µm	Large > 5 µm	Proportion at Location (%) Dome	Deep
1	296	47	35	18	92	8
2	325	45	32	23	83	17
4	352	46	31	23	76	24
7	196	21	29	41	88	11
14	148	16	29	55	98	2
21	91	7	27	66	98	2
28	63	5	24	71	100	0
35	52	6	19	79	97	3

The peak number of microspheres observed in the MLN occurred at 7 d when their numbers were falling in the PP (Table 3). The size distribution clearly showed that microspheres of > 5 µm in diameter did not enter this tissue. At early time points the majority of particles in the MLN were located in the subcapsular sinus, but with time they progressed into and through the cortex and medullary regions. Similar examination of the spleens showed that microspheres were not detectable in this organ until 4 d post administration, the peak number occurred at 14 d, and no microspheres > 5 µm in diameter were detected (data not shown). Several microsphere formulations with different wall materials have been tested and all have been found to enter the PP and migrate through the body with similar kinetics and restrictions imposed by size. In addition, immunohistochemical studies have shown that the < 5 µm microspheres which extravasate the PP do so within phagocytic cells.

These data suggested that the quality of the immune response induced by orally administering a microencapsulated vaccine could be controlled by the size of the particles. Microspheres < 5 µm in diameter would extravasate the PP within macrophages to stimulate systemic immunity, while those 5 to 10 mm in diameter would stimulate mucosal immunity through the release of antigen in the PP. We addressed these possibilities in experiments in which groups of mice were orally-administered 100 µg of enterotoxoid in solution or as a suspension within 1 to 10 mm 50:50 DL-PLG microspheres, on 3 occasions separated by 30 d. Examination of the plasma endpoint titers of IgM and IgG anti-enterotoxoid antibodies showed that the mice receiving microencapsulated vaccine exhibited a rise in specific antibodies with each oral immunization, while the soluble enterotoxoid was ineffective (Table 4).

Table 3. Migration of coumarin-6 85:15 DL-PLG microspheres into and through the mesenteric lymph nodes following oral administration

Time	Total number	Proportion of diameter (%)			Proportion at Location (%)	
		Small	Medium	Large		
(days)	observed	<2 μm	2-5 μm	> 5 μm	Dome	Deep
1	8	50	50	0	100	0
2	83	76	24	0	95	5
4	97	73	27	0	73	27
7	120	67	32	0	64	36
14	54	83	17	0	9	91
21	20	75	25	0	5	95
28	15	67	32	0	0	100
35	9	44	56	0	0	100

Table 4. Plasma IgM and IgG anti-enterotoxin levels on day 20 following primary, secondary, and tertiary oral immunization with soluble or microencapsulated (50:50 DL-PLG) staphylococcal enterotoxoid

Enterotoxoid Dose (mg) Per		Plasma Anti-toxin Titer on Day 20 Following Oral Immunization					
		Primary		Secondary		Tertiary	
Immunization	Form	IgM	IgG	IgM	IgG	IgM	IgG
100	Microspheres	80	1,280	320	5,120	1,280	40,960
100	Soluble	<20	<20	80	<20	640	<20

Table 5. Enterotoxin-specific IgA antibodies in the saliva and gut fluids of mice on days 10 and 20 after tertiary oral immunization with soluble or microencapsulated enterotoxoid

Enterotoxoid Dose (mg) Per		IgA Anti-Enterotoxin Titer Following Tertiary Oral Immunization			
		Day 10		Day 20	
Immunization	Form	Saliva	Gut Wash	Saliva	Gut Wash
100	Microspheres	1,280	1,024	640	256
100	Soluble	40	< 8	10	< 8

Enterotoxin-specific sIgA antibodies were compared in saliva and gut wash samples obtained on days 10 and 20 following tertiary oral immunization with either encapsulated or unencapsulated enterotoxoid (Table 5). In contrast to the inability of the free enterotoxoid to evoke a response when administered orally, the ingestion of an equal amount of the vaccine incorporated into microspheres resulted in a substantial sIgA anti-enterotoxin response in both saliva and gut secretions. It should be pointed out that the gut secretions are diluted into a total of 5 ml per mouse during collection. Although it is impossible to determine the exact dilution factor this imposes, the IgA concentration is at minimum 10-fold higher in the mucus which bathes the gut, and this has not been taken into account in the measurements presented here.

DISCUSSION

Results obtained in this study illustrate the potential of biocompatible and biodegradable microspheres as a delivery system for vaccine antigens. Their utility as noninflammatory injectable reservoirs for the uniform or pulsatile longterm release of other bioactive agents is also under active investigation (Tice and Cowsar, 1984; Redding, et al., 1984). Although the immunopotentiating activity of the 1 to 10 µm microspheres seen in these studies may be due in part to a depot effect on antigen release, the lower adjuvancy we have observed with microspheres of > 10 µm indicates that this is not the only mechanism. The observation that appropriately-sized microspheres are rapidly phagocytized by macrophages suggests that efficient uptake by antigen-presenting accessory cells may play an important role in immune enhancement. In a similar manner, the effectiveness of microspheres as an oral vaccine delivery vehicle is due not only to the protection provided by the wall material during transit through the gut, but also to the efficient targeted delivery of a high concentration of antigen into the PP. In summary, antigen containing microspheres represent a vaccine delivery system with immunopotentiating activity which has the potential for wide application for both parenteral and oral immunization.

ACKNOWLEDGEMENTS

The authors are indebted to Drs. Jerry McGhee and Jiri Mestecky for their help and encouragement and to Ms. Sandra Roberts for the preparation of this manuscript. This work was supported by USPHS grants AI 24772, AI 21774 and Contract DAMD17-86-6162 from the U.S. Army Medical Research Acquisition Activity.

REFERENCES

Allison, A. C., Gregoriadis, G., 1974, Liposomes as immunological adjuvants, Nature 252:252.

Arnon, R., Sela, M., Parant, M., Chedid, L., 1980, Antiviral response elicited by a completely synthetic antigen with built in adjuvance, Proc. Natl. Acad. Sci. USA 77:6769.
Bazin, H., Levi, G., Doria, G., 1970, Predominant contribution of IgA antibody-forming cells to an immune response detected in extraintestinal lymphoid tissues of germ-free mice exposed to antigen by the oral route, J. Immunol. 105:1049.
Bienenstock, J., Befus, A. D., 1980, Mucosal immunology, Immunology 41:249.
Bockman, D. E., Cooper, M. D., 1973, Pinocytosis by epithelium associated with lymphoid follicles in the bursa of Fabricius, appendix and Peyer's patches. An electron microscope study, Am. J. Anat. 136:455.
Brandtzaeg, P., 1985, Role of J chain and secretory component in receptor-mediated glandular and hepatic transport of immunoglobulins in man, Scandanavian J. of Immunol. 22:111.
Brandtzaeg, P., 1984, Immune functions of human nasal mucosa and tonsils in health and disease, in: "Immunology of the Lung and Upper Respiratory Tract," J. Bienenstock, ed., McGraw-Hill, New York.
Brandtzaeg, P., Baklien, K., 1976, Immunohistochemical studies of the formation and epithelial transport of immunoglobulins in normal and diseases human intestinal mucosa, Scand. J. Gastroenterol. 11(Suppl. 36):1.
Chanock, R. M., Lerner, R. A., 1984, "Proceedings of the Cold Spring Harbor Conference: Modern Approaches to Vaccines," Cold Spring Harbor Laboratory.
Crabbe, P. A., Nash, D. R., Bazin, H., Eyssen, H., Heremans, J. F., 1969, Antibodies of the IgA type in intestinal plasma cells of germfree mice after oral or parenteral immunization with ferritin, J. Exp. Med., 130:723.
Craig, S. W., Cebra, J. J., 1971, Peyer's patches: an enriched source of precursors for IgA-producing immunocytes in the rabbit, J. Exp. Med., 134:188.
Dailey, M. O., Post, W., Hunter, R. L., 1977, Induction of cell-mediated immunity to chemically modified antigens in guinea pigs. II. The interaction between lipid conjugated antigens, macrophages, and T lymphocytes, J. Immunol., 118:963.
Elson, C. O., Ealding, W., Lefkowitz, J., 1984, A lavage technique allowing repeated measurement of IgA in mouse intestinal secretions, J. Immunol. Meth., 67:101.
Hanson, L. A., 1981, Comparative studies of the immune globulins of human milk and of blood serum, Int. Arch. Allergy Appl. Immunol., 18:241.
Hanson, L. A., Ahlstedt, S., Carlsson, B., Kaijser, B., Larsson, P., Mattsby-Baltzer, I., Akerlund Sohl, A., Svanborg-Eden, C., Svennerholm, A. M., 1978, Secretory IgA antibodies to enterobacterial virulence antigens: their induction and possible relevance, Adv. Exp. Med. Biol., 1007:165.
Kramp, W. J., Six, H. R., Kasel, J. A., 1982, Post-immunization clearance of liposome-entrapped adenovirus type 5 hexon, Proc. Soc. Exp. Biol. Med. 169:135.
Kiyono, H., McGhee, J. R., Wannemuehler, M. J., Frangakis, M. V., Spalding, D. M., Michalek, S. M., Koopman, W. J., 1982, *In vitro* immune response to a T cell-dependent antigen by

cultures of dissociated murine Peyer's patch, Proc. Natl. Acad. Sci. (USA) 79:596.

Mestecky, J., 1987, The common mucosal immune system and current strategies for induction of immune responses in external secretions, J. Clin. Immunol. 7:265.

Mestecky, J., McGhee, J. R., Arnold, R. R., Michalek, S. M., Prince, S. J., Babb, J. L., 1978, Selective induction of an immune response in human external secretions by ingestion of bacterial antigen, J. Clin. Invest. 61:731.

Mestecky, J., McGhee, J. R., Crago, S. S., Jackson, S., Kilian, M., Kiyono, H., Babb, J. L., Michalek, S. M., 1980, Molecular-cellular interactions in the secretory IgA response, J. of the Reticuloendothelial Society 28:455.

Michalek, S. M., McGhee, J. R., Mestecky, J., Arnold, R. R., Bozzo, L., 1976, Ingestion of *Streptococcus mutans* induces secretory IgA and caries immunity, Science 192:1238.

Miller, R. A., Brady, J. M., Cutwright, D. E., 1977, Degradation rates of oral resorbable implants (polylactates and polyglycolates): rate modification with changes in PLA/PGA copolymer ratios, J. Biomed. Mater. Res. 11:711.

Montgomery, P. C., Rosner, B. R., Cohn, J., 1974, The secretory antibody response. Anti-DNP antibodies induced by dinitrophenylated type III pneumococcus, Immunol. Commun. 3:143.

Redding, T. W., Schally, A. V., Tice, T. R., Meyers, W. E., 1984, Long acting delivery systems for peptides: inhibition of rat prostate tumors by controlled release of D-Trp^{6}-LH-RH from injectable microcapsules, Proc. Natl. Acad. Sci. USA 81:5845.

Solari, R., Kraehenbuhl, J. -P., 1985, The biosynthesis of secretory component and its role in the transepithelial transport of IgA dimer, Immunol. Today 6:17.

Spalding, D. M., Koopman, W. J., Eldridge, J. H., McGhee, J. R., Steinman, R. M., 1983, Accessory cells in murine Peyer's patch. I. Identification and enrichment of a functional dendritic cell, J. Exp. Med. 157:1646.

Tice, T. R., Cowsar, D. R., 1984, Biodegradable controlled release parenteral systems, Pharmacology Technology Journal 8:26.

Tomasi, T. B., Tam, E. M., Solomon, A., Prendergast, R. A., 1965, Characteristics of an immune system common to certain external secretions, J. Exp. Med. 121:101.

Tomasi, T. B., Zigelbaum, S., 1963, The selective occurrence of γ_1A globulin in certain body fluids, J. Clin. Invest. 42:1552.

Warren, H. S., Vogel, F. R., Chedid, L. A., 1986, Current status of immunological adjuvants, Annu. Rev. Immunol. 4:369.

Warren, J. R., Spero, L., Metzger, J. F., 1973, Antigenicity of formaldehyde-inactivated staphylococcal enterotoxoid B, J. Immunol. 111:885.

POTENTIATION OF CELL-MEDIATED IMMUNE RESPONSES DIRECTED AGAINST P. FALCIPARUM SPOROZOITE PEPTIDE VACCINE BY IMMUNOMODULATORS

Donna M. Russo, John S. Sundy, Henry C. Maguire and William P. Weidanz

Department of Microbiology and Immunology
Hahnemann University, Philadelphia, PA

INTRODUCTION

The development of genetically engineered subunit vaccines has renewed and intensified efforts to immunize third world populations against currently uncontrolled diseases including malaria; however, the success of such vaccines will depend in large part upon efforts to enhance their immunogenicity. It is likely that subunit vaccines will require potent adjuvants in order to be effective. While considerable work has been aimed at enhancing antibody responses to peptide antigens, little is know regarding the potentiation of antigen specific cell-mediated immune responses. We have utilized a delayed type hypersensitivity (DTH) model to investigate cell-mediated immune responses, including those parameters which affect the induction, magnitude and duration of responsiveness. We examined the potency of R32tet32, a subunit vaccine directed against the sporozoite stage of Plasmodium falciparum, to induce and elicit DTH responses in mice and found that these responses were potentiated significantly through the use of selected lymphokines and immunomodulators.

MATERIALS AND METHODS

Animals. Male C57BL/6 mice were purchased from Jackson Laboratories (Bar Harbor, ME) and housed in our AALAC (American Association for Accreditation of Laboratory Animal Care) approved animal facility. All animals were used between 7 and 12 weeks of age. Experimental groups consisted of 4 to 6 animals.

Antigens. The vaccine candidate, R32tet32, $[(NANP)_{15}NVDP]_2$tet1-32 is a recombinant protein comprised of 32 tandem tetrapeptide repeats derived from the repetitive region of the circumsporozoite protein of Plasmodium falciparum fused to the first 32 amino acids of the tetracycline gene read out of frame (Young et al., 1985). This antigen was generously provided by Dr. James Young, Smith Kline and French Laboratories. Previous work conducted by several investigators (Good et al., 1986; Del Guidici et al., 1986; Togna et al., 1986) has demonstrated that R32tet32 contains two T-cell epitopes, one located within the parasite derived repetitive region and another located on the tet tail. Recognition of the parasite derived T-epitope is restricted to mice expressing I-A^b class II molecules.

Immunizations. C57BL/6 (H-2b) mice were clipped and then sensitized intradermally (i.d.) with 75 µg of R32tet32 emulsified in complete Freund's adjuvant (CFA) (GIBCO Laboratories, Grand Island, NY)) or administered in pyrogen-free sterile saline (Abbott) without adjuvant. A total of 0.2 ml of antigen was divided equally and administered bilaterally into the dorsal flanks. At the time of sensitization, certain groups also received specific immunomodulating agents administered i.d. directly adjacent to the site of antigen sensitization. Other groups were injected with these agents systemically (i.v. or i.p.). Five days later, the animals were challenged in test pinnae with 20 ug of $R32tet_{32}$ in 10 ul of saline. An equal volume of sterile saline was injected into the control ear. Ear thickness measurements were taken immediately prior to challenge and at several time points during the 72 hr period post-challenge with a dial thickness gauge (Schesinger, Dobbs Ferry, N.Y.) calibrated in .01 mm increments. Ear swelling associated with antigen-specific DTH was calculated as follows:

48 hr measurement minus baseline measurement = ear value

test ear value minus control ear value = DTH

Immunomodulators; description and bioactivity assays. Human recombinant IL-2 was kindly provided by E. I. Du Pont De Nemours & Company, Glenolden, PA. Prior to use *in vivo*, this reagent was assayed for biological activity *in vitro* by testing its ability to induce proliferation of an IL-2 dependent cell line (Gillis *et al.*, 1980). Recombinant human tumor necrosis factor-alpha (TNF) and recombinant murine interferon-gamma (IFN) were generously provided by Genentech Inc., (San Francisco, CA). The bioactivity of TNF was verified by injecting 10 ug i.v. into mice concomitantly with a sublethal dose (100 ug) of lipopolysaccharide W (LPS) *E. coli* 0127:B8 (Difco Laboratories, Detroit, MI). Animals receiving both TNF and LPS died within 24 hr, while those injected with LPS alone survived. Cyclophosphamide (CY) was purchased from Sigma Chemical Co. (St. Louis, MD).

RESULTS

Immunopotentiation of DTH with Cyclophosphamide and IFN. The effect of two different immunomodulators on the induction of DTH with R32tet32 was examined in C57BL/6 mice. All animals were sensitized with R32tet32 administered i.d. without adjuvant. Groups of mice were pretreated with 90 mg/kg of CY two days prior to sensitization, or received 2 X 10^4 units of recombinant murine IFN administered locally at the time of sensitization. One group received CY before sensitization in addition to IFN at the time of sensitization. The non-specific or toxic effects of these agents were assessed by administering them to groups of mice immunized i.d. with saline. Challenge with R32tet32 demonstrated that the magnitude of the induced responses was significantly increased by the addition of either CY or IFN. Most notable, however, was the observation that these two agents administered together induced DTH responses significantly greater than responses obtained with CY or IFN administration alone.

IL-2 Augments the Induction of DTH to R32tet32. At the time of sensitization with R32tet32, certain groups of mice also received 5 X 10^4 units of human recombinant IL-2 administered i.d. Control animals also received IL-2, but were sensitized with saline. Animals which received IL-2 and antigen developed significantly greater DTH responses upon challenge with R32tet32, as compared to the responses of animals which received antigen alone. Control animals sensitized with IL-2 and saline did not respond to antigenic challenge.

Tumor Necrosis Factor Failed to Potentiate DTH. Groups of mice, sentized with R32tet32, also received various concentrations of TNF. Ten, 20 and 25 µg of TNF corresponding to units of 5 X 10^5 to 1.5 X 10^6 were admin-

Table 1. Enhancement of DTH

Immunomodulator[a]	Dose	Route	Enhancement[b]
IFN	$2X10^4$ units	i.d.	3.4 X
Cy[c]	90 mg/Kg	i.p.	3.5 X
IFN +Cy	as above	as above	5.0 X
IL-2	$5X10^4$ units	i.d. (flank)	2.8 X
IL-2[d]	50 units	i.d. (pinnae)	2.2 X
TNF	$5X10^5$-$1.5X10^6$ units	i.d.	1 X[e]
CFA	(V/V)	i.d.	4-6 X

[a]Agents were administered at the time of antigen sensitization unless otherwise specified.
[b]Defined as ear swelling greater than that observed when animals were sensitized with antigen in the absence of immunomodulator.
[c]Cy was administered two days prior to antigen sensitization.
[d]IL-2 was injected into test pinnae eight hours after antigenic challenge.
[e]Not significant.

istered i.d. TNF did not enhance the magnitude or affect the duration of DTH responses in immunized mice at the concentrations tested.

Efferent Potentiation With IL-2. IL-2 is known to have potent effects on the proliferation of antigen reactive T-cells *in vivo*. We examined its ability to enhance or expand antigen reactive T-cell proliferative events *in vivo* by testing the effect of local as well as systemic administration of IL-2 on the efferent limb of DTH responsiveness. All mice were immunized and challenged with R32tet32. Eight hours post-challenge, one group of animals received a pulse of IL-2 (50 units) at the site of challenge. Control animals were sensitized and challenge with R32tet32 but pulsed with saline. Toxicity controls were sensitized with saline, challenged with R32tet32 and pulsed with IL-2. Ear measurements were taken over a 72 hr period. The local administration of IL-2 administered at this time resulted in increased responses that were apparent as early as 24 hr post challenge and were maintained at significant levels for 72 hr after challenge in contrast to control responses which reached peak intensity at 24 hr and declined sharply thereafter. This alteration of responsiveness was not due to nonspecific effects of IL-2 on the vasculature because toxicity controls challenged with the same concentration of IL-2 did not demonstrate this pattern of reactivity.

DISCUSSION

Subunit vaccines, both those produced by recombinant DNA technology or by peptide synthesis will in most instances require administration with potent adjuvants as well as delivery systems which will optimize immunogenicity. CFA, the most potent adjuvant known currently is not acceptable for use in humans or animals due to the induction of intense inflammatory reactions in addition to conversion of the recipients immune status to tuberculin positive. Currently, only alum conjugates are licensed for use in humans, but while antibody titers are frequently boosted in immunized individuals, in no instance has the use of aluminum salts promoted the induction of cellular immune responses. Therefore, the definition of suitable agents which will potentiate the appropriate cell-mediated immune responses to peptide antigens will be of critical importance to the success of subunit vaccines.

The data presented in this paper demonstrate that DTH responses induced by a single T-epitope located on a subunit vaccine can be potentiated dramatically by concurrent administration with specific immunomodulators. The vaccine candidate, R32tet32, encodes an $H\text{-}2^b$ restricted T-epitope within the parasite derived repeat region, which upon administration without adjuvant primes for meager DTH responses at best and frequently results in nonresponsiveness. This can be overcome if the antigen is emulsified in CFA in which case responses of considerable magnitude and long duration develop consistently. We have shown that responses on this same order of magnitude can be induced by other adjuvants which when employed locally cause minimal toxicity.

IFN proved to be a potent adjuvant when administered locally with R32-tet32. Subsequently elicited antigen-specific DTH responses were routinely increased by 3 or 4 fold. This phenomenon was dose dependent and correlated with the route of delivery as well (not shown). Intravenous administration of IFN was effective in enhancing immune responses but required higher doses than when administered locally. Other immunomodulators including CY and IL-2 also enhanced the induction of DTH, but the most striking potentiation of DTH occurred when pretreatment with CY was used in association with local IFN administration at the time of antigen sensitization. The administration of these two agents together resulted in the development of DTH responses which were significantly greater than responses obtained through the use of CY or IFN alone. In addition, the duration of the DTH response elicited was considerably lengthened (not shown). These observations suggest the possibility of stacking immunoenhancing effects. Methods by which cell-mediated immune responses can be amplified optimally should arise from an understanding of the mechanisms which induce and potentiate these responses.

These data demonstrate that cell-mediated immune responses can be potentiated through the use of selected immunomodulators. Clearly, further studies into the mechanisms of immunopotentiation are warranted. Advancements in this area may lead to measures which will improve the efficacy of subunit vaccines.

ACKNOWLEDGEMENTS

We thank Dr. James Young for generously providing us with R32tet32. This work was supported by UNDP/World Bank/WHO Special Programme for Research and Training in Tropical Diseases and the Gustave and Louise Pfeiffer Research Foundation.

REFERENCES

Del Guidici, G., Cooper, J. A., Merino, J., Verdini, A. S., Pessi, A., Togna, A. R., Engers, H. D., Corradin, G. and Lambert, P. H. 1986. The antibody response in mice to carrier-free synthetic polymers of Plasmodium falciparum circumsporozoite repetitive epitope is I-A restricted: possible implications for malaria vaccines. J. Immunol. 137:2952.

Gillis, S., Scheid, M., Watson, J. 1980. Biochemical and biological characterization of lymphocyte regulatory molecules. III. The isolation and phenotypic characterization of interleukin 2-producing T-cell lymphomas. J. Immunol. 125:2570.

Good, M. F., Berzofsky, J. A., Maloy, W. L., Hayashi, Y., Fujii, N., Hockmeyer, W. T., Miller., L. H. 1986. Genetic control of the immune response in mice to a Plasmodium falciparum sporozoite vaccine.

Widespread nonresponsiveness to a single malaria T epitope in highly repetitive vaccine. J. Exp. Med. 164:655.
Togna, A. R., Del Giudici, G., Verdini, A. S., Bonelli, F., Pessi, A., Engers, H. D., & Corradin, G. 1986. Synthetic P. falciparum circumsporozoite peptides elicit heterogenous L3T4+Tcell proliferative responses in H-2^b mice. J. Immunol. 137:2956.
Young, F. F., Hockmeyer, W. T., Gross, M., Ballou, W. R., Schneider, I., Hollingdale, M. R. Beaudoin, R. L., Hollingdale, M. R.,Miller L. H., Diggs, C. L., Rosenberg, M. 1985. Expression of Plasmodium falciparum circumsporozoite proteins in Escherichia coli for potential use in a human malaria vaccine. Science. 228:958.

DETOXIFICATION OF PERTUSSIS TOXIN BY SITE-DIRECTED MUTAGENESIS

S. Cockle, S. Loosmore, K. Radika, G. Zealey,
H. Boux, K. Phillips and M. Klein

Connaught Research Institute
Willowdale, Ontario, Canada M2R 3T4

INTRODUCTION

Pertussis toxin (PT) is a major virulence factor of Bordetella pertussis and also an important protective antigen in vaccines against whooping cough. Unfortunately, the chemical treatments normally used to inactivate the toxin can seriously reduce its immunogenicity; moreover, there are concerns that residual traces of active toxin might possibly account for the rare occurrence of vaccination side-effects. Following recent achievements in cloning and sequencing the TOX operon encoding PT (Locht & Keith, 1986; Nicosia et al., 1986), the genetic engineering of a PT analogue lacking toxicity but retaining full immunogenicity has become a major goal in the development of improved pertussis vaccines.

PT is a non-covalent hexameric protein of molecular weight 105,000 with two functional moieties. The catalytic A protomer, or subunit S1 (molecular weight 26,000), is an ADP-ribosyltransferase that inactivates GTP-binding regulatory proteins involved in the control of intracellular adenylate cyclase and phospholipase C. The B oligomer, comprising subunits S2 to S5, with S4 occurring twice, mediates interactions of the toxin with target cells. Since many of the biological effects of PT are due to its ADP-ribosyltransferase (ADPR) activity, we undertook to mutate functionally important amino acids in S1 to produce inactive variants. Our approach has been to recombine the mutated S1 gene with the remainder of the TOX operon and achieve secretion of a fully-assembled toxin analogue from a suitable Tox$^-$ Bordètella host. This strategy takes into account that the holotoxin is required for maximum immunogenicity, and that in due course it will also be desirable to inhibit the cell-binding function of PT by appropriate changes to the B oligomer.

This communication describes the identification of a major active-site residue in S1, the construction of recombinant Bordetella strains incorporating mutations to this and other S1 residues, and the production and characterization of a series of immunogenic PT analogues with greatly reduced toxicity.

CHOICE OF SITES FOR MUTAGENESIS

Photocrosslinking of NAD to S1. Mixtures of reduced purified S1 and [carbonyl-^{14}C]NAD or [adenine-^{14}C]NAD were UV-irradiated for up to 3h at 0°C to induce photochemical crosslinking. Radiolabel was efficiently incorporated into S1 from the nicotinamide moiety (0.75 mol/mol) but not from the adenine moiety (0.1 mol/mol), while the NAD glycohydrolase activity was specifically and completely abolished. The nicotinamide-labelled protein was digested with trypsin and the resulting fragments were separated by reverse-phase HPLC, giving only two major radioactive peptides. The purified peptides were partially sequenced, and in each case the first 15 residues unambiguously matched residues 118-132 of mature S1 (Figure 1). Radioactivity was associated only with a modified amino acid corresponding to Glu129, which is therefore at or near the nicotinamide active site. Similar functional glutamates have previously been discovered in diphtheria toxin and P. aeruginosa exotoxin A (Carroll & Collier, 1984; 1987). This amino acid and the adjacent Tyr130, which is conserved in diphtheria toxin, were therefore selected for extensive mutagenesis.

Other Sites. Sequence comparison suggests that in fact Glu106 of S1 more closely resembles the active-site glutamate of diphtheria toxin, so substitutions were also tested at this position. The amino acid sequences of S1, cholera toxin and E. coli heat labile toxin show two longer homology regions in which the highly conserved arginine residues may have functional importance (Figure 1). Thus Arg9, Arg13 and Arg58 were also targets for mutagenesis. Subunit S1 is activated by reduction of its single disulfide bond between Cys41 and Cys201, suggesting a critical role for one or other of these residues: both were mutated.

1 11 21
Asp Asp Pro Pro Ala Thr Val Tyr ARG Tyr Asp Ser ARG Pro Pro Glu Asp Val Phe Gln Asn Gly Phe
31 41
Thr Ala Trp Gly Asn Asn Asp Asn Val Leu Glu His Leu Thr Gly Arg Ser CYS Gln Val Gly Ser Ser
51 61
Asn Ser Ala Phe Val Ser Thr Ser Ser Ser Arg ARG Tyr Thr Glu Val Tyr Leu Glu His Arg Met Gln
71 81 91
Glu Ala Val Glu Ala Glu Arg Ala Gly Arg Gly Thr Gly His Phe Ile Gly Tyr Ile Tyr Glu Val Arg
101 111
Ala Asp Asn Asn Phe Tyr Gly Ala Ala Ser Ser Tyr Phe Glu Tyr Val Asp Thr Tyr Gly Asp Asn Ala
121 131
Gly Arg *Ile Leu Ala Gly Ala Leu Ala Thr Tyr Gln Ser GlU TYR Leu Ala* His Arg Arg Ile Pro Pro
141 151 161
Glu Asn Ile Arg Arg Val Thr Arg Val Tyr His Asn Gly Ile Thr Gly Glu Thr Thr Thr Thr Glu Tyr
171 181
Ser Asn Ala Arg Tyr Val Ser Gln Gln Thr Arg Ala Asn Pro Asn Pro Tyr Thr Ser Arg Arg Ser Val
191 201
Ala Ser Ile Val Gly Thr Leu Val Arg Met Ala Pro Val Val Gly Ala CYS Met Ala Arg Gln Ala Glu
211 221
Ser Ser Glu Ala Met Ala Ala Trp Ser Glu Arg Ala Gly Glu Ala Met Val Leu Val Tyr Tyr Glu Ser
231
Ile Ala Tyr Ser Phe

Fig.1. Amino acid sequence of pertussis toxin subunit S1 derived from the nucleotide sequence of the TOX gene from B. Pertussis 10536. The segment corresponding to the radiolabelled S1 fragments is italicized. Sequence homologies between PT, cholera toxin and E. coli heat-labile toxin are underlined. Residues selected for site-directed mutagenesis are in upper-case type.

CONSTRUCTION OF RECOMBINANT BORDETELLA STRAINS

Site-Directed Mutagenesis. The TOX gene is a polycistron comprising a promoter and structural genes encoding the five subunits arranged in tandem, though in different reading frames. The S1 gene was excised, subcloned into M13 and subjected to site-directed mutagenesis by the phosphorothioate method, with single or multiple base changes or deletions. Twenty-six primary mutations were generated, and several others were derived by constructing crosses among these.

Transformation of Tox⁻ Bordetella. Mutant S1 genes were recombined with the native B oligomer genes to yield new polycistrons (TOX*), which were incorporated into a broad-host-range plasmid for delivery into Tox⁻ Bordetella hosts by bacterial conjugation or electroporation. Alternatively, TOX^* genes were integrated into the host chromosome by conjugative transfer of a "suicide" plasmid. This integration was either random or directed through homologous recombination utilizing the flanking regions of the operon. Either route allowed secretion of PT analogues into the growth medium at levels up to about 3 μg/ml, depending on the mutant. To create a production strain of B. pertussis with the endogenous TOX gene replaced by an appropriate TOX^* variant, the TOX operon was first deleted by homologous recombination with a Kan^r gene sandwiched between the normal TOX flanking sequences, then a TOX^* gene was reintegrated at the same locus. These recombinant strains also secreted fully-assembled PT analogues into the medium.

Purification of PT Analogues. Selected mutant toxins were purified from culture supernatants by fetuin-agarose affinity chromatography and/or hydroxyapatite chromatography and shown to contain the normal complement of five protomers by both SDS-gel electrophoresis and reverse-phase HPLC (data not shown).

BIOLOGICAL PROPERTIES OF PT MUTANTS

CHO Cell Clustering Assay. Culture supernatants were assayed for toxin by enzyme-linked immunoassay (EIA) and screened for toxicity using Chinese hamster ovary (CHO) cells, which adopt a characteristic clustered growth pattern when exposed to PT. This response depends on binding of PT to target cells and also correlates with ADPR activity. Results on 26 mutants and wild-type PT from recombinant Bordetella are summarized in Table 1. Mutations at Glu106 had no effect on activity. Otherwise, residual toxicity was 1% or less in all but two examples, and as low as 0.1% in several cases. Thus the seven residues Glu129, Tyr130, Arg9, Arg13, Arg58, Cys41 and Cys201 all appear to have significant functional or structural roles in subunit S1.

ADP-Ribosyltransferase Activity. The enzymic activity of seventeen purified or semi-purified mutants was directly determined as the ability to catalyse transfer of radiolabelled ADP-ribose from [adenylate-^{32}P]NAD to purified bovine retinal transducin (Watkins et al., 1985). As before, the proteins were quantified by EIA. These results are also included in Table 1, and exhibit good agreement with data from the CHO cell clustering assay.

Antigenic Analysis. All PT mutants were screened by EIA for the presence of an S1 epitope recognized by the mouse monoclonal antibody PS21, which inhibits PT-induced CHO cell clustering and protects mice passively against intracerebral challenge with B. pertussis. This MAb therefore appears to define an immunodominant protective epitope which it

Table 1. Characterization of Recombinant Pertussis Toxin Analogues

PT Analogue	Residual CHO Cell Toxicity (%)[a]	Residual ADPR Activity (%)[a]	S1 Epitope[b]
Glu129 (wt)	100	100	+++++
Del129	0.1		++
Gly129	0.1	0.3	++++
Gln129	0.1	0.1	+/-
Asp129	0.7	0.8	++
Asn129	0.1	0.3	++
Lys129	0.3	0.2	-
Arg129	0.1		-
His129	0.2		-
Pro129	0.2		+
Del130	0.2	0.1	-
Phe130	12.0		++++
Gly129 Ala130	0.2	0.6	-
Gly129 Phe130	1.0		++++
Gly106	100		+++++
Gln106	100		+++++
Del9	0.2		-
Glu9	0.1	0.2	+/-
Lys9	0.1		++++
His9	0.2	0.1	+++
Del13	0.3		-
Glu13	5.0		++++
Glu9 Glu13	1.0	0.9	-
Glu58	0.7	0.6	+++
Ala41	0.3	0.3	-
Ala201	0.2	0.1	-
Ala41 Ala201	0.2	0.1	-

[a]Results are expressed relative to native PT from B. Pertussis as 100%.
[b]The degree of expression of the S1 epitope ranges from 5+ (native PT) to no reaction.

may be important to retain in a subsequent pertussis vaccine. Several mutations at Arg9, Arg58 and Glu129 yielded genetically detoxified holotoxins that were still strongly reactive with PS21 (Table 1). However, other changes at these sites and at Arg13, Cys41 and Cys201 resulted in loss of this recognition. The epitope defined by PS21 is therefore influenced by amino acid residues in different parts of the sequence.

Toxicity and Immunogenicity in Mice. Three purified Glu129 mutants were tested in mice for acute toxicity and histamine sensitization activity (Table 2). While lethal doses were not in fact reached in many cases, it is clear that all exhibited markedly decreased toxicity relative to native PT. Moreover, mice immunized with these mutants were protected at non-toxic doses against intracerebral challenge with virulent B. pertussis (Table 2). Correspondingly, antisera from immunized mice were found to contain high levels of PT-specific antibodies and to inhibit PT-induced CHO cell clustering.

Table 2. Physiological Properties of Selected PT Analogues

PT Analogue	Acute Toxicity LD_{50} (μg)	Histamine Sensitization LD_{50} (μg)[a]	Mouse Protection ED_{50} (μg)[b]
Glu129 (wt)	2	0.2	2
Gly129	>5	3	2
Gln129	>16	>3	16
Asn129	>5	3	2

[a]Mice were challenged with histamine acid phosphate (1mg/10g body weight) 5 days after injection with PT or PT analogue.
[b]Mice were subjected to intracerebral challenge with 20,000 B. Pertussis 18323 cells 14 days post-immunization.

DISCUSSION

A series of pertussis holotoxin analogues has been generated in engineered Bordetella strains by mutation of the catalytic S1 subunit at seven critical sites, including Glu129, which was found by photocross-linking to be near the active site for NAD hydrolysis. Many of these mutants were highly detoxified according to CHO cell clustering and ADPR assays, and retained an immunodominant protective S1 epitope. Three affinity-purified analogues were immunogenic and protective in mice, yet exhibited low toxicity. Other workers have also located functional residues in S1 by mutation of the S1 gene expressed in E. coli (Barbieri & Cortina, 1988; Burnette et al., 1988; Pizza et al., 1988). However, subunit S1 alone has weak immunogenicity, whereas our approach can easily be extended to other subunits without resorting to in vitro assembly of the holotoxin. We therefore feel our results represent an important advance towards the development of a genetically inactivated form of PT for inclusion in a new generation of whooping cough vaccines.

REFERENCES

Barbieri, J.T. and Cortina, G., 1988, ADP-ribosyltransferase mutations in the catalytic S-1 subunit of pertussis toxin, Infect. Immun., 56:1934-1941.

Burnette, W.N., Cieplak, W., Mar, V.L., Kaljot, K.T., Sato, H. and Keith, J.M., 1988, Pertussis toxin S1 mutant with reduced enzyme activity and a conserved protective epitope, Science, 242:72-74.

Carroll, S.F. and Collier, R.J., 1984, NAD binding site of diphtheria toxin: identification of a residue within the nicotinamide subsite by photochemical modification with NAD, Proc. Natl. Acad. Sci. USA, 81: 3307-3311.

Carroll, S.F. and Collier, R.J., 1987, Active site of Pseudomonas aeruginosa exotoxin A. Glutamic acid 553 is photolabelled by NAD and shows functional homology with glutamic acid 148 of diphtheria toxin, J. Biol. Chem., 262:8707-8711.

Locht, C. and Keith, J., 1986, Pertussis toxin gene: nucleotide sequence and genetic organization, Science, 237:1258-1264.

Nicosia, A., Perugini, M., Franzini, C., Casagli, M.C., Borri, M.G., Antoni, G., Almoni, M., Neri, P., Ratti, G. and Rappuoli, R., 1986, Cloning and sequencing of the pertussis toxin genes: operon structure and gene duplication, Proc. Natl. Acad. Sci. USA, 83:4631-4635.

Pizza, M., Bartolini, A., Prugnola, A., Silvestor, S. and Rappuoli, R., 1988, Subunit S1 of pertussis toxin: mapping of the regions essential for ADP-ribosyltransferase activity, Proc. Natl. Acad. Sci. USA, 85:7521-7525.

Watkins, P.A., Burns, D.L., Kanaho, Y., Liu, T.-Y., Hewlett, E.L. and Moss, J., 1985, ADP-ribosylation of transducin by pertussis toxin, J. Biol. Chem., 260:13478-13482.

FILAMENTOUS FUSION PHAGE CLONING VECTORS FOR THE STUDY OF EPITOPES AND DESIGN OF VACCINES

Stephen F. Parmley[1] and George P. Smith[2]

Department of Immunology and Infectious Diseases
Research Institute, Palo Alto Medical Foundation
Palo Alto, CA 94301[1] and Division of Biological
Sciences, Tucker Hall, University of Missouri
Columbia, MO 65211[2]

ABSTRACT

Foreign sequences can be inserted within the minor coat protein, pIII, of filamentous phage, creating a fusion protein that is incorporated into the virion; the virions, which we call "fusion phage," display the foreign amino acids encoded in the insert on their surface. Phage bearing specific antigenic determinants from a target gene can be purified in infectious form from a vast excess of phage bearing other determinants by affinity to antibody directed against the gene product. Fusion phage can be used as a source of antigen as a carrier-hapten conjugate for obtaining immunological reagents in rabbits, and for B epitope mapping. By generating a fusion phage library expressing virtually all possible short, amino acid sequences, it may be possible to study epitopes on immunologically important proteins without the use of synthetic peptides and without ever having cloned the genes.

INTRODUCTION

There are two main strategies for mapping B epitopes on immunologically important proteins when the protein sequence is known and/or the gene has been cloned. First, many short peptides are synthesized covering the entire protein sequence and tested individually for reactivity to antibody against the protein and/or the ability to elicit antibody when used as an immunogen. Second, epitopes can be mapped from a cloned gene by fusing gene fragments in expression vectors, such as λgt11, and testing the reactivity of the resulting fusion proteins to polyclonal and/or monoclonal antibody. Once an epitope has been mapped, it can be further characterized by deletion or substitution of amino acids, in the case of synthetic peptides, or base pairs, in the case of cloned genes. Both methods are arduous undertakings and are limited in the number of different peptides or fusion proteins that can be feasibly tested. The filamentous bacteriophage cloning system reviewed here should provide a much more rapid and easier method for characterizing epitopes. In addition, with the development of the "epitope library" proposed below, it should be possible to access virtually all possible short amino acid sequences to determine their reactivity to antibody. Before outlining the proposal for creating an "epitope library," the virtues of the cloning system will be reviewed.

The vector is based on filamentous bacteriophage, such as M13, f1, and fd, which infect *E. coli* cells displaying F pili. Filamentous phage are rod-shaped nucleoprotein particles composed of single-stranded circular DNA surrounded by a helical array of the major coat protein, pVIII. There are several minor coat proteins, including pIII, which is located at one tip of the virion (Pratt *et al.*, 1969) and is encoded on gene III. The amino-terminal domain of pIII is exposed on the surface of the virion and binds to the F pilus to initiate infection of host cells (Armstrong *et al.*, 1981); the carboxyl-terminal domain is buried in the capsid and is necessary for morphogenesis of phage particles (Crissman and Smith, 1984). When foreign DNA fragments are spliced into gene III, a fusion protein is created that is incorporated in the virion; we call these recombinant virions "fusion phage." These virions retain phage function and display the foreign epitope(s) on their surface (Smith, 1985; de la Cruz *et al.*, 1988; Parmley and Smith, 1988). Fusion phage have been used to map B epitopes and as immunogens (de la Cruz *et al.*, 1988). Phage bearing specific antigenic determinants from a target gene can be purified in infectious form from a vast excess of phage bearing other determinants by affinity to antibody directed against the gene product (Parmley and Smith, 1988). Thus, the vector is "antibody-selectable" because antibody is used to directly *select* clones of interest, unlike conventional expression vectors, where antibody is used to *screen* for clones.

The most recently reported fusion phage vectors, fUSE1 and fUSE2, have cloning sites located 2-3 amino acids downstream of the signal peptide of pIII, so that the majority of the pIII sequence is uninterrupted by inserts (Parmley and Smith, 1988). Several inserts, ranging from 20-335 base pairs have been tested and found to be stable. The foreign amino acids did not impair phage particle production or infectivity and were accessable to antibody. There is little worry, therefore, that other inserts will be masked by the folding of downstream pIII sequence.

An extremely sensitive affinity-purification procedure was developed for isolating fusion phage bearing determinants recognized by antibody. The procedure exploits the super-strong biotin-streptavidin bond and allows target fusion phage to be purified over 10^5-fold in a single round. Target phage can be enriched over background phage 10^9-fold with an additional round of purification. The method is simple, rapid, and requires only minute amounts of antibody (Parmley and Smith, 1988). The sensitivity should allow isolation of extremely rare clones from large libraries.

One limitation of the system in its current state is the difficulty in generating large libraries with the fusion phage vector. Strategies to increase the library size are being explored with the hope of applying the method to genomic libraries, well-represented cDNA libraries, and, as discussed below, novel procedures for studying epitopes. The most promising strategy to increase the library size would employ a bacteriophage λ shuttle vector, so that it can be packaged into λ phage particles *in vitro* and introduced into host cells by infection. Once inside the cell, the fusion phage genome could be excised from the composite genome *in vivo* by the phage P1 *cre-lox*P site-specific recombination system (Sternberg and Hamilton, 1981; Sternberg *et al.*, 1986) as illustrated in Figure 1. This method would allow production of fusion phage libraries of 10^9 clones and, as mentioned above, the affinity-purification procedure allows isolation of clones from backgrounds as high as 10^9, thus it should be possible to isolate virtually any clone--that expresses determinants recognized by an antibody--from libraries of this size.

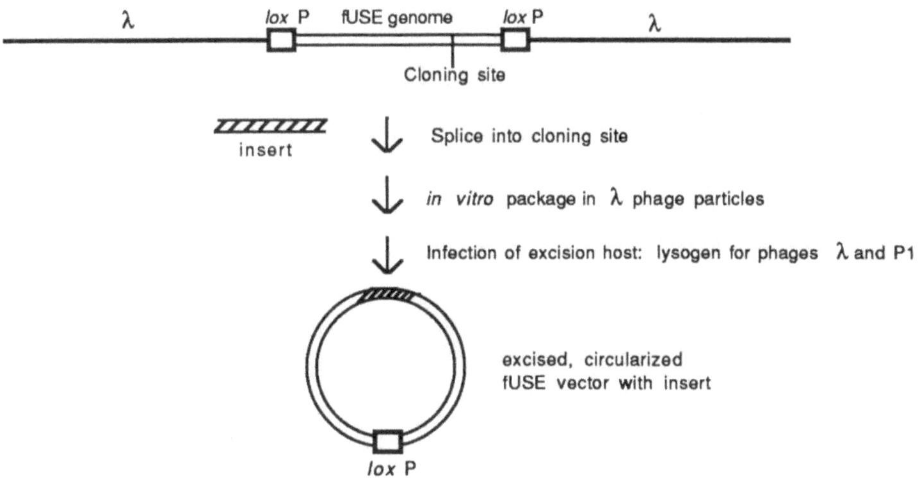

Fig. 1. The proposed λ shuttle vector for introducing the fusion phage genome into host cells by *in vitro* packaging and infection. The fusion phage genome will be incorporated into the λ genome and will be flanked by *lox*P sites. Inserts will be spliced into the fusion phage cloning site, the composite vector will be packaged into λ particles *in vitro*, and infected into host cells which are stable λ and P1 lysogens. The P1 excision system (Sternberg and Hamilton, 1981; Sternberg *et al.*, 1986) will circularize the fusion phage genome and the λ genome will be repressed.

"EPITOPE LIBRARY"

The most novel application of the method would be the development of an "epitope library"; a library containing 10^9 independent clones bearing 30 codons of random coding sequence. This library should contain epitopes recognized by virtually any anti-protein antibody, since (for example) almost every 6-amino-acid epitope would be present many times in various contexts. Non-contiguous epitopes would probably also be represented in the library, since these epitopes can be mimicked by contiguous ones (Geysen *et al.*, 1986). In addition, antibodies may be able to recognize the molecular environments created by amino acid side chains independently from the orientation of the carbon backbone (Van Cleave *et al.*, 1988).

As mentioned above, the fusion phage system is "antibody-selectable," so it should be possible to use functionally interesting antibodies to purify many different clones expressing recognizable epitopes from the "epitope library." The amino acid sequences of the inserts would be determined at the DNA level by directly sequencing single-stranded viral DNA from each clone. In effect, the antibodies themselves are allowed to select the epitopes worthy of investigation.

There may be enough information obtained from sequencing the inserts from many clones purified with a interesting monospecific or monoclonal antibody (say, one that protects against a disease) to deduce the epitope recognized by the antibody--all without having to clone DNA from the target organism. If an immunologically important antigen has been sequenced or cloned, it should be possible to locate contiguous epitopes by comparing the sequences of inserts to protein sequence, eliminating the need to clone segments of the gene or synthesize peptides.

Likewise, non-contiguous epitopes could be mapped to the known three-dimensional structure of the protein. Antibody-reactive fusion phage clones may be useful as immunological reagents themselves, or the information obtained from sequencing many inserts could be used to design vaccines. In all the envisioned applications, relatively arduous, expensive experimental manipulations (cloning genes, synthesizing and characterizing peptides, etc.) would be replaced with affinity-purification from an all-purpose epitope library (presumably commercially available), DNA sequencing, and data processing. The latter two clearly will become increasingly easy and informative in the coming years.

REFERENCES

Armstrong, J., Perham, R.N., and Walker, J.E. (1981), Domain structure of bacteriophage fd adsorption protein, FEBS Lett. 135, 167-172.

Crissman, J.W. and Smith, G.P. (1984), Gene-III protein of filamentous phages: evidence for a carboxyl-terminal domain with a role in morphogenesis, Virology 132, 445-455.

de la Cruz, V.F., Lal, A.A., and McCutchan, T.F. (1988), Immunogenicity and epitope mapping of foreign sequences via genetically engineered filamentous phage, J. Biol. Chem. 263, 4318-4322.

Geysen, H.M., Rodda, S.M., and Mason, T.J. (1986), A priori delineation of a peptide which mimicks a discontinuous antigenic determinant, Mol. Immunol. 23, 709-715.

Parmley, S.F. and Smith, G.P. (1988), Antibody-selectable filamentous fd phage vectors: affinity purification of target genes, Gene 73, 305-318.

Pratt, D., Tzagoloff, H., and Beaudoin, J. (1969), Conditional lethal mutants of the small filamentous coliphage M13. II. two genes for coat proteins, Virology 39, 42-53.

Smith, G.P. (1985), Filamentous fusion phage: novel expression vectors that display cloned antigens on the virion surface, Science 228, 1315-1317.

Sternberg, N. and Hamilton, D. (1981), Bacteriophage P1 site-specific recombination. I. recombination between *lox*P sites, J. Mol. Biol. 150, 467-486.

Sternberg, N., Sauer, B., Hoess, R., and Abremski, K. (1986), Bacteriophage P1 *cre* gene and its regulatory region, J. Mol. Biol. 187, 197-212.

Van Cleave, V.H., Naeve, C.W., and Metzger, D.W. (1988), Do antibodies recognize amino acid side chains of protein antigens independently of the carbon backbone?, J. Exp. Med. 167, 1841-1848.

RECOMBINANT-DERIVED MODIFIED-LIVE HERPESVIRUS VACCINES

Saul Kit

Division of Biochemical Virology
Baylor College of Medicine
Houston, TX 77030 USA

INTRODUCTION

Swine pseudorabies (PR; Aujezky's disease [AD]) and infectious bovine rhinotracheitis (IBR) are important veterinary diseases, each of which cost American farmers over 30 million dollars annually. Killed and modified-live virus (MLV) vaccines have been used in the United States and in other countries for over 25 years to control AD and IBR, but these conventional vaccines have had numerous shortcomings with regard to safety and efficacy. First, despite the intensive use of the conventional vaccines, AD- and IBR morbidity and mortality have increased. For example, about 10% of the sixty million swine in the United States are now infected with PRV compared with about 1% in 1972. Second, animals vaccinated with conventional killed or MLV vaccines remained susceptible to infection at a later time with virulent field strains, and often shed virulent virus, thereby spreading infection in the herd. Third, the intramuscular (IM) administration of MLV IBR vaccines has been hampered by the hazards of vaccine-induced abortions. As an alternative, temperature-sensitive IBR vaccines have been used, but these specialized vaccines could only be administered by the inconvenient and laborious intranasal (IN) route. Virus replication in the deep tissues of the body to amplify the immune response was, by definition, restricted. Fourth, some of the conventional AD vaccines, though safe for swine, were unsafe for other animal species. Incidents in which lambs and chickens were killed through accidental vaccination with conventional AD vaccines fueled suspicions about the safety of all MLV vaccines and underlined the need for greater attenuation. Given the lack of confidence in conventional MLV vaccines, drastic test and slaughter programs were mandated to control AD in England and Denmark, and similar programs have been considered in regions of the United States. Fifth, conventional MLV vaccines are inherently unstable and can revert to virulence. Finally, vaccinated animals develop antibodies to the vaccine virus, and, until now, the vaccine-induced antibodies were indistinguishable from field virus-induced antibodies. Therefore, all seropositive animals were considered by government authorities to be potential carriers of latent virulent viruses, and the movement of seropositive animals across state lines or national boundaries was inhibited or forbidden. These restrictions were, of course, disadvantageous to producers.

It seemed probable to us in 1982 that conventional vaccines could be vastly improved through genetic engineering. Therefore, programs to develop recombinant-derived, species-specific veterinary vaccines were initiated in my laboratory. Only 4 years later, on January 16, 1986, the United States Department of Agriculture (USDA) licensed for manufacture and sale our first generation AD vaccine, "OMNIVAC." On August 30, 1988, the USDA made public an Environmental Assessment and Finding of No Significant Impact for our second generation AD vaccine, "OMNIMARK," thereby authorizing open field safety studies, the final steps in the licensing process. A license for the manufacture and sale of OMNIMARK is expected early in 1989.

This morning, I would like to describe the strategy employed to improve the safety and efficacy of AD vaccines. I will also describe the properties of the IBRV vaccines which were designed in parallel with AD vaccines.* Finally, I will briefly summarize animal testing experiments and the 3-year field experience with the OMNIVAC and OMNIMARK vaccines.

DECREASED VIRULENCE OF THYMIDINE KINASE (TK)-NEGATIVE MUTANTS

The starting material for the development of new bovine IBR vaccine was IBRV Los Angeles (LA), a virulent strain isolated during an outbreak of IBR disease. The starting material for the genetic engineering of new porcine AD vaccines was the Bucharest (BUK) strain of PRV. At least a billion doses of BUK strain vaccines have been used over the last 25 years in Europe and the United States, attesting to the safety of the parental virus and parts derived therefrom.

The original PRV(BUK) vaccine was attenuated by the traditional method of serially passing a PRV isolate 400-800 times in heterologous host cells, i.e., chick embryo fibrolasts (CEF). Adaption to CEF by this procedure frequently results in deletions in the unique-short (U_S) and terminally reiterated (TR_S) regions of the PRV genome. The deletion contributes to attenuation by reducing virulence for pigs and for one-day-old chicks, but, paradoxically, by giving an advantage for growth in vitro in CEF cells (Mettenleiter et al., 1985; 1987b; 1988). The specific BUK strain with which we began our work was a mixture of at least two U_S deletion mutant genotypes. Specifically, BUK clones 1, 5, and 7 exhibited a 3.6 Kbp deletion in the BamHI-7 plus BamHI-12 restriction fragments (also KpnI-I), whereas clone 3 had a larger 4.4 Kbp deletion in these fragments.

Studies by Mettenleiter et al. (1985) have shown that the BamHI-7 fragment of PRV encoded glycoprotein gI. This gI glycoprotein gene was deleted in some, but not all, of the conventional vaccine strains that had been passaged serially in CEF cells. The glycoprotein gI gene was also deleted in our PRV(BUK) clone 3 cells, which exhibited the 4.4 Kbp deletion, but not in the PRV(BUK) clone 5 and clone 7 cells, which had a 3.6 Kbp deletion. The deletion of PRV(BUK) clones 5 and 7 did not extend far enough into the BamHI-7 fragment to remove all gI epitopes. In 1982, we arbitrarily chose to attenuate clones 5 and 7, which exhibited the smaller U_S + TR_S, by engineering a thymidine kinase (TK) gene deletion. Subsequently, the gI-negative mutant, PRV(BUK) clone 3, was also attenuated by TK deletion.

The virulence of herpesviruses is known to be multifactorial (Kit et al., 1985; Lomniczi et al., 1984; 1987a; 1987b; Mettenleiter et al., 1987b). As shown by Field and Wildy (1978), one of the factors highly important for the pathogenicity of human herpes simplex virus (HSV) is the thymidine kinase (TK) gene). Furthermore, herpes simplex virus mutants which failed to express any TK activity replicated to high titers in growing S-phase cells, but replicated poorly in stationary phase tissue culture cells. Field and Wildy (1978) pointed out that herpes simplex viruses are associated with

*Some of the data presented in this paper were taken from the U. S. Patents Nos. 4,514,497, 4,569,840, 4,703,011, 4,609,548, 4,711,850, and 4,753,884.

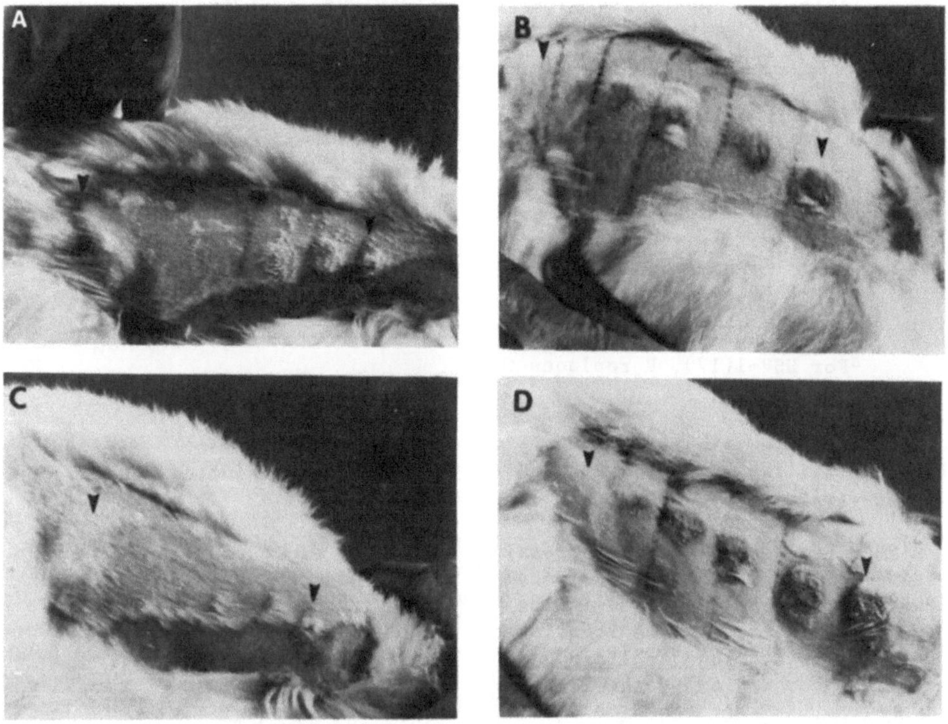

Fig. 1. Vaccination of rabbits with 0.1 ml/rabbit of TK^- vaccinia (1004B) mutant and with TK^+ vaccinia clone 1-1 (Dubbs and Kit, 1964a). The titer of each of the viruses was 1.6 X 10^8 PFU/ml. New Zealand rabbits weighing 6 lbs were vaccinated with serial dilutions (10^8 to 10^3, left and right arrows, respectively) of the viruses. Rabbits vaccinated with TK^- vaccinia (1004B) (A, C) at 10^3 and 10^4 dilution of virus developed small vesicles of about 0.5 to 1.0 cm. By days 6 and 8, considerable healing had occurred. Rabbits vaccinated with parental TK^+ vaccinia had very large lesions by days 6 and 8, about 1.5 to 2.0 cm, except at the 10^7 dilution. Considerable erythema with ulcerated regions 1.0 to 1.5 cm in diameter were observed in rabbits vaccinated with TK^+ vaccinia (Dubbs and Kit, 1981, unpublished experiments).

quiescent, non-dividing neuronal cells, and that TK^- herpes simplex virus mutants might replicate poorly in neuronal cells. This, in part, might contribute to their avirulence. Numerous studies have now confirmed that TK^- human herpesviruses (Klein, 1982; Stanberry et al., 1988), TK^- marmoset herpesviruses, and TK^- bovine and porcine herpesviruses (Kit et al., 1983; 1985; 1986) have greatly reduced virulence and are less likely to recrudesce from latency (Klein, 1982; Stanberry et al., 1985; Tenser et al., 1983).

Regardless of whether the Field and Wildy (1978) hypothesis fully explains the low pathogenicity of TK^- herpesvirus mutants, it is interesting to note that reduced pathogenicity is also exhibited by TK^- poxvirus mutants. Experiments carried out by D. R. Dubbs in my laboratory in 1981 showed that TK^- vaccinia virus mutants were less virulent for rabbits than wild-type TK^+ vaccinia. Figure 1 shows that rabbits vaccinated with 1 to 10 TK^+ vaccinia virus plaque-forming units (PFU) developed large ulcerated lesions which were slow to heal, whereas only small vesicles which healed rapidly developed

Table 1. Sequence Homology of Herpesvirus TK Proteins: Amino Acid No. Downstream for NTP Binding Sites

Virus and Amino Acid No. after NTP	Amino Acid Sequence
IBRV (NTP + 18)	P E P M A Y W R T M F G T
PRV (NTP + 13)	P E P M A Y W R T L F D T
MarHV (NTP + 19)	Q E P M A Y W R S T F A S
HSV-1 and HSV-2 (NTP + 18)[a]	P E P M T Y W Q V L G A S

[a]For HSV-1(17), R replaqes Q.

after vaccination with about 10^5 PFU of a TK$^-$ vaccinia mutant. Similarly, Buller et al. (1985) found that TK$^-$ vaccinia viruses exhibited greatly decreased virulence for mice.

Thus, the strategy for increasing the safety of conventional vaccines by several orders of magnitude entailed engineering PRV and IBRV deletion mutants irreversibly altered in the TK gene. To accomplish this, it was necessary: (i) to map the location of the PRV and IBRV TK genes on the viral DNAs; (ii) to subclone smaller DNA fragments containing the genes, so as to define the boundaries of the TK genes; (iii) to sequence each of the TK genes and, on the basis of the data obtained through sequencing; (iv) to prepare plasmids with deletions (and/or insertions) in the viral TK genes; and (v) to carry out "marker transfer" experiments so that the deletion mutations would be transferred from the plasmids to infectious viral DNA by recombination and yield virus TK deletion mutants.

In order to map the location of the TK gene by "marker rescue" experiments, it was necessary to have stable TK$^-$ virus mutants. These TK$^-$ mutants were selected by treating TK$^+$ PRV and IBRV with bromodeoxyuridine (BrdUrd), iododeoxyuridine (IdUrd), and arabinosylthymine (araT) as described previously for the isolation of vaccinia virus and herpes simplex virus TK$^-$ mutants (Dubbs and Kit, 1964a; 1964b). TK$^-$ host cell mutants were likewise needed for "marker rescue" experiments. For this purpose, TK$^-$ rabbit skin [Rab(BU)] cells were isolated as described for TK$^-$ mouse LM(TK$^-$) cells (Kit et al., 1963; Kit and Qavi, 1983). The mutagen selected TK$^-$ virus used as the immediate parent of OMNIVAC and OMNIMARK was derived from PRV(BUK) clones 5 and 7 and named PRV(BUK-5A). The mutagen induced TK$^-$ IBRV mutant was named IBRV(B8-D53) (Kit et al., 1985; Kit and Qavi, 1983). These mutagen-induced TK$^-$ PRV and IBRV mutants were stable and did not revert to the TK$^+$ phenotype, even after hypoxanthine-aminopterin-thymidine (HAT) selection. By using infectious DNA from the TK$^-$ PRV(BUK-5A) and IBRV(B8-D53) mutants, along with cloned fragments of viral DNA, it was possible to show that the PRV TK gene was located in the BamHI-11 (KpnI-J_L) fragment near the boundary of the BamHI-9 fragment, while the IBRV TK gene was located in the middle of the HindIII-A fragment at about 0.47 units on the physical map of IBRV DNA. The approximate boundaries of the PRV and IBRV TK genes were next established by subcloning smaller restriction fragments and by additional marker rescue experiments with the TK$^-$ virus mutants.

To identify open reading frames of the PRV TK genes, and putative transcriptional and translational start signals, translational stop signals and the polyadenylation signals, DNA sequencing experiments were carried out on a 5 Kbp PRV DNA fragment consisting of about 2.5 Kbp of the BamHI-9 fragment plus 2.5 Kbp of the BamHI-11 fragment immediately downstream from the BamHI-9

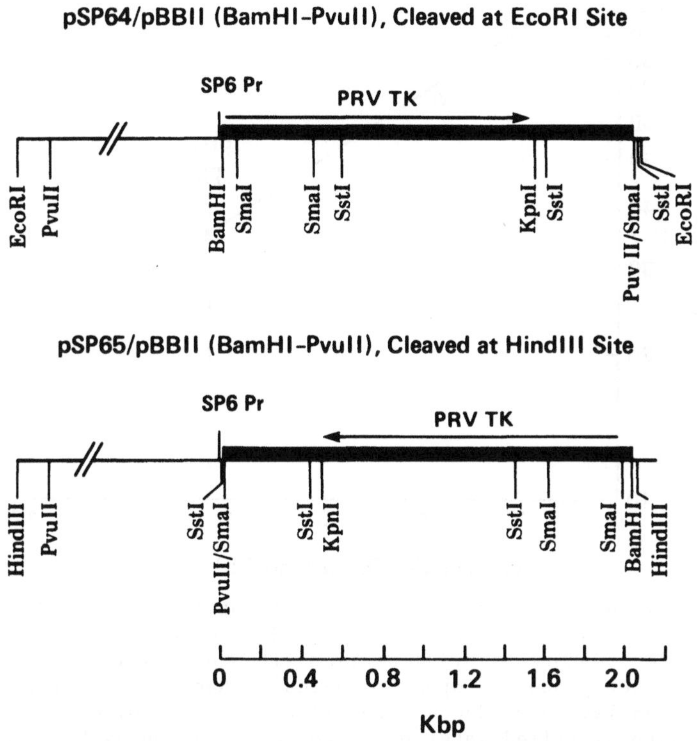

Fig. 2. Restriction nuclease maps of PRV(BUK-5) TK gene subcloned in two orientations in phage pSP64 adjacent to the phage SP6 promoter (Pr). The 2.1 Kbp BamHI-PvuII fragment contains the coding region of the PRV TK gene plus sequences about 1 Kbp downstream from the TK gene coding sequence.

fragment. In the case of the IBRV TK gene, DNA sequencing analyses were carried out on the 2.8 Kbp subfragment of the IBRV HindIII-A fragment.

From the results of the DNA sequencing studies, it was predicted that the PRV and the IBRV TK genes encoded polypeptides of 349 and 358 amino acids, respectively. The predicted sizes of the PRV and IBRV TK polypeptides were 36,830 and 36,950, respectively, somewhat smaller than the HSV-1 TK polypeptide, which contains 376 amino acids and has a molecular weight of about 41,000. However, 45 N-terminal amino acids of the HSV-1 TK polypeptide are not essential for TK activity. If this N-terminal tail of the HSV-1 TK polypeptide were eliminated, the size of the HSV-1 TK polypeptide would be closer to that of the PRV and IBRV TK polypeptides.

The PRV and IBRV TK polypeptides showed significant amino acid sequence homology to the known sequence of the HSV-1 TK polypeptide and to that of other viral and cellular TK polypeptides. For example, amino acid sequences homologous to the putative ATP-binding sequences of HSV-1, HSV-2, marmoset herpesvirus (MarHV), vaccinia, and chick TK polypeptides were found at the amino terminal portion of the PRV and IBRV TK polypeptide sequences (Otsuka and Kit, 1984; Kit, 1985). Table 1 illustrates a particularly notable amino acid sequence homology 13 to 19 amino acids downstream from the putative ATP-binding amino acid sequence of the IBRV, PRV, MarHV, HSV-1, and HSV-2 TK polypeptides.

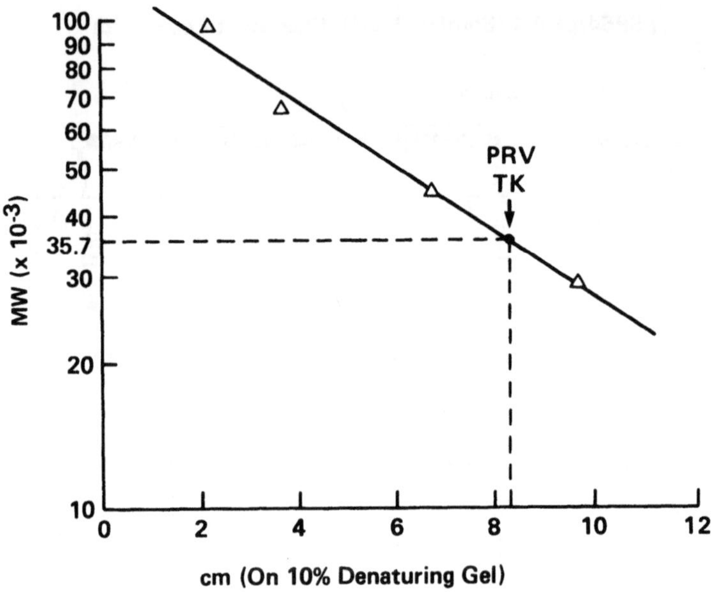

Fig. 3. Molecular weight (mol. wght.) estimation of the PRV TK gene translation product after SDS-polyacrylamide gel electrophoresis by comparison with known mol. wght. marker proteins (open triangles). Rabbit reticulocyte translation products were diluted 1:5, 1:10, or 1:20 with distilled water prior to denaturation by boiling 2 min with buffer D (0.0625 M Tris-HCl, pH 6.8, 0.3% SDS, 5% 2-mercaptoethanol, 10% glycerol, and 0.001% bromophenol blue). The negative slopes from graphs of 100 (log [relative mobility X 100]) versus gel concentration were determined and plotted against the known mol. wght. of protein standards (bovine serum albumin dimer, 132 kD; bovine serum albumin monomer, 66 kD; ovalbumin 45 kD; carbonic anhydrase, 29 kD; and α-lactalbumin, 14.2 kD) to determine the mol. wght. of the PRV TK gene translation product (Sheppard et al., 1986).

Consistent with the Kozak model of translation initiation, the nucleotide sequences at the ATG start signals were GGGATGC and GCCATGG for the PRV and IBRV TK genes. The methionine coding ATG sequences were the first ATG codons downstream from putative transcriptional signals. Putative polyadenylation signals were also found at 21 and 110 nucleotides, respectively, downstream from the TAA stop codons of the PRV and IBRV TK genes.

To further analyze the putative viral TK gene sequences, a 2.1 Kbp BamHI to PvuII subfragment of the PRV BamHI-11 fragment was subcloned in two orientations in phage SP64 and SP65 adjacent to the SP6 promoter (Sheppard et al., 1986) (Fig. 2). The pSP64/pBB11 sequence was then transcribed *in vitro* with the SP6 RNA polymerase plus ribonucleotide triphosphates. The transcript was capped with S-adenosylmethionine, GTP, and guanylyl transferase, and translated with nuclease-treated rabbit reticulocyte lysate and medium containing

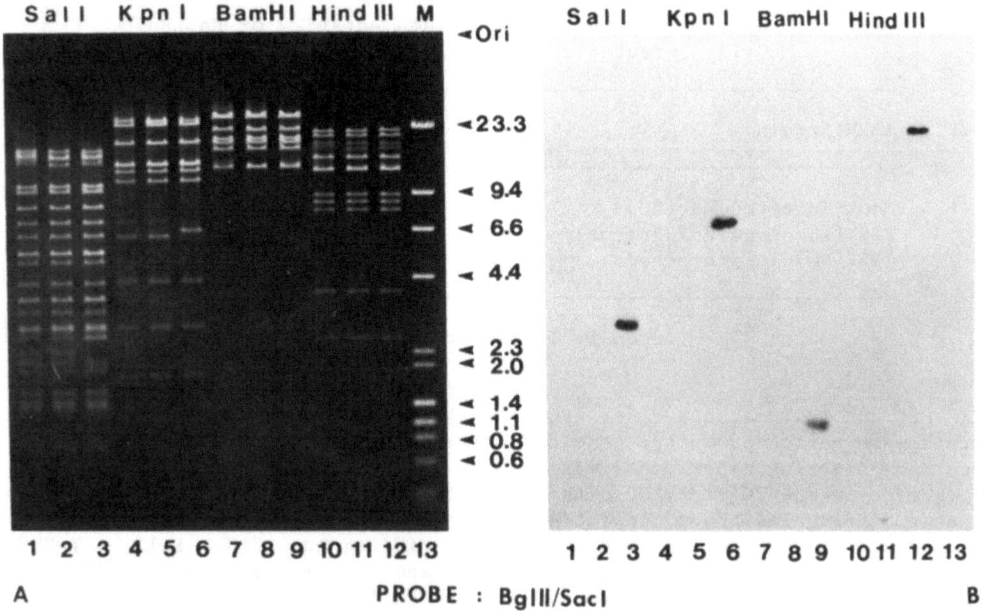

Fig. 4. Southern blot hybridization experiments demonstrating IBRV TK deletion mutants. The DNAs from parental TK^+ IBRV (lanes 3, 6, 9, and 12) and from recombinants TK^- IBRV(NG)dltk clone 1 (lanes 1, 4, 7, and 10) and clone 5 (lanes 2, 5, 8, and 11) were cleaved with SalI, KpnI, BamHI, and HindIII. The fragments were separated by agarose gel electrophoesis, stained with ethidium bromide, and photographed (A). Fragments were then transferred to nitrocellulose filters and hybridized with a ^{32}P-probe consisting of the 400 bp BglII to SacI fragment which had been deleted from the IBRV TK gene. The filters were washed, and exposed to X-ray film (B). Lane M shows the phage lambda HindIII and ϕX174 HaeIII fragments used as mol. wght. (Kbp) markers.

^{35}S-methionine. The translation product was then denatured and analyzed by SDS-PAGE. The results indicated that the mol wght of the translation product was 35,700, as predicted from the amino acid sequence (Fig. 3). In contrast, a translation product was not obtained after transcription of pSP65/pBB11. In similar experiments, a 2 Kbp EcoRI to PstI fragment containing the IBRV TK gene was subcloned in the correct orientation for TK gene transcription adjacent to the SP6 promoter of phage pSP65, transcribed, and translated *in vitro*. Analyses indicated that the translation product was a 37,000 mol wght polypeptide, again, in accordance with predictions.

Based upon predicted restriction endonuclease sites in the viral TK genes, nucleotide sequences of 148 bp and 343 bp, respectively, were deleted from DNA fragments containing the PRV and IBRV TK genes. The PRV TK gene deletion removed an essential portion of the coding sequence and altered the reading frame. In the case of the IBRV TK gene, a 40 bp oligonucleotide sequence (NG) with stop codons in all three reading frames was inserted in place of the deleted 343 bp (BglII to SacI) fragment of the IBRV TK gene so that a deletional/insertion mutant was obtained.

The final steps in the construction of the TK deletion mutants consisted of marker transfer experiments with infectious TK^+ viral DNA and the TK^- deletion plasmids, followed by nucleoside analog selection to isolate candidate

Table 2. Cytosol Thymidine Kinase (TK) Activity of Rab(BU) Cells Infected for 7 Hr with IBRV Strains

IBRV Strain	TK Activity[a]
Mock-infected Rab(BU)	0.04
IBRV(Los Angeles)-infected	1.74
IBRV(NG)dltk-infected	0.03

[a]Picomoles dTMP formed in 10 min at 38° C/μgram protein.

TK^- PRV and IBRV mutants. For proof of structure, Southern blotting experiments were performed on the candidate DNAs (Kit et al., 1985). Some of the Southern blotting experiments carried out to prove that a TK deletion had been made in the TK^- IBRV(NG)dltk mutant are shown in Fig. 4. It may be seen that a ^{32}P-labeled probe consisting of the 343 bp BglII/SacI fragment which had been deleted from the IBRV TK gene, hybridized to 2.8 Kbp SalI, 6.7 Kbp KpnI, 1.1 Kbp BamHI, and 2.1 Kbp HindIII fragments of the parental TK^+ IBRV DNA, but failed to hybridize to any of the DNA fragments of the recombinant TK^- IBRV(NG)dltk clones (Fig. 4B). The sizes of the restriction endonuclease fragments of the TK^- IBRV(NG)dltk clones were also shortened by about 400 bp, as expected (Fig. 4A). Southern blotting experiments likewise demonstrated that a probe consisting of the "NG" oligonucleotide sequence, which was inserted into the TK gene of the deletion mutants, hybridized to the DNA of the TK gene deletion mutants but not to the DNA of the parental virus, whereas a probe consisting of the intact TK gene DNA hybridized to both parental TK^+ IBRV and TK^- IBRV(NG)dltk DNAs.

To verify that the TK^- virus mutants exhibited a TK^- phenotype, TK enzyme assay experiments, thymidine plaque autoradiography, and conventional radioautography on cells growing on glass slides were carried out (Kit et al., 1985). As illustrated by the enzyme experiments on TK^+ and TK^- IBRV viruses shown in Table 2, TK^+ IBRV encoded a TK-inducing activity, but the TK deletion mutant, IBRV(NG)dltk, lacked TK-inducing activity. Similar experiments definitely confirmed that the phenotypes of the PRV TK deletion mutant, PRV(BUK-dl3B) (OMNIVAC), and PRV(dlg92/dltk) (OMNIMARK) were TK^- (Kit et al., 1985; 1987b).

DELETION OF ENVELOPE GLYCPROTEIN GENES TO PRODUCE "MARKER" VACCINES

The second major improvement in the conventional herpesvirus vaccines consisted of the genetic engineering of envelope glycoprotein gene deletions so as to obtain "marker" vaccines. Herpesvirus-infected cells synthesize about seven structural glycoproteins that become embedded in the membranes of infected cells and on the surface of mature enveloped virus particle. Although the role of glycoproteins in the herpesvirus life cycle is not understood, it is known that the viral-induced glycoproteins are major antigens that interact with the host immune system eliciting humoral and cell-mediated immune responses. One or more glycoproteins very likely function in the attachment and penetration of virus particles into cells, in syncytia formation, and in the spread of progeny virus particles from cell to cell, but the mechanisms by which the individual glycoproteins or complexes of glycoproteins carry out these roles has not been elucidated. We reasoned that if a glycoprotein gene nonessential for virus replication could be identified, that glycoprotein gene might be deleted from the virus. Animals vaccinated with the glycoprotein deletion mutant would develop antibodies to all of the

Table 3. Comparison of Herpesvirus Glycoproteins Homologous to HSV-1 gC

Glycoprotein	No. Amino Acids	Polypeptide (kD)	Glycoprotein (kD)	Ref.
HSV-1 gC-1	511	55	130	1[a]
HSV-2 gC-2	479	52	79	1
PRV g92(gIII)	479	51	92	2
IBRV gIII	509	55	95	3
VZV gp66	560	61	-	4
EHV-1 gp13	468	51	90	5

[a]1: Swain et al., 1985; 2: Robbins et al., 1986; 3: Kit et al., unpublished; 4: Davison and Scott, 1986; 5: Allen and Coogle, 1988.

viral glycoproteins except for the deleted glycoprotein. Animals infected with virulent herpesvirus field strains would develop antibodies to all of the viral glycoproteins, including the glycoprotein deleted from the marker vaccine virus. It would be possible through simple serological tests to distinguish vaccinated animals from those infected with virulent viruses from other sources (Kit et al., 1987a; 1987b).

It was known that HSV-1 glycoprotein gC-1, a major component of the viral envelope, acts as a C3b receptor. Also, gC-1 is not essential for HSV-1 infectivity (Frink et al., 1983; Glorioso et al., 1985; Swain et al., 1985). HSV-2 glycoprotein gC-2 is smaller than the HSV-1 gC glycoprotein, but shares extensive amino acid sequence homology with it, and is also nonessential for virus replication (Dowbenko and Lasky, 1984). Robbins et al. (1984) identified a PRV glycoprotein gene, g92(gIII), which mapped at a position on the PRV genome homologous to that of the HSV-1 gC-1 gene. Sequencing studies revealed that the PRV g92(gIII) polypeptide predicted from the nucleic acid sequence was about the same size as the HSV-2 gC-2 polypeptide (Table 3) and shared significant amino acid sequence homology with the HSV-1 and HSV-2 gC polypeptides (Robbins et al., 1986). Experiments to learn whether the HSV-2 gC-2 and PRV g92(gIII) genes, like the HSV-2 gC-1 gene, act as C3b receptors would be worthwhile. Likewise, it was important to investigate whether the PRV g92(gIII) gene and the other HSV-1 gC-1 homologues were required for virus replication. Therefore, starting with OMINVAC[PRV(BUK-dl3b)], we engineered a 1.1 Kbp SalI deletion from the coding region of the PRV g92(gIII) gene (Kit et al., 1987a; 1987b). The resulting deletion mutant, PRV(dltkdlg92) (OMNIMARK), does not synthesize glycoprotein g92(gIII) and does not induce anti-g92(gIII) antibodies in pigs, but replicates to titers of about 5 x 10^7 PFU/ml in swine testicle cells, rabbit skin cells, and cells of several other species, showing that PRV glycoprotein g92(gIII) is not essential for PRV replication. Furthemore, animals vaccinated with PRV(dltkdlg92) (OMNIMARK) were protected from Aujeszky's disease after challenge with virulent PRV strains (Kit et al., 1987b; Kit, 1989; Thacker et al., 1988a).

The gene of IBRV glycoprotein, gIII, which is homologous to the genes of HSV-1 gC and PRV g92(gIII), was also mapped, cloned, and sequenced in Houston. Figure 5 shows that PRV g92(gIII) and IBRV gIII share extensive amino acid sequence homology, particularly at the carboxy-terminal half of the polypeptide. The antigenicity and hydrophobicity characteristics of PRV g92(gIII) and IBRV(gIII) polypeptides are shown in Fig. 6. The sizes of several herpesvirus polypeptides and mature glycoproteins homologous to HSV-1 gC and PRV g92(gIII) are presented in Table 3. These proteins all have the N-

```
MGPLGRAWLIAAIFAWALLSARRGLAEEAEASPSPPPSPSPTETESSAGTTGATPPTPNS    60
                                               :.  .: : :
                                MASLARAMLALLALYAAAIAAAPSTTTA    28

PDATPEDSTPGATTPVGTPEPPSVSEHDPPVTNSTPPPAPPEDGRPGGAGNASRDGRPSG   120
 : ::     : ..       :      ::      : :      :      ::   :
LDTTPNGGGGGNSSEGELSPSP------PPTPAPASPEAGAVSTPPVPPPSVSRRKPPRN    82

GGRPRPPRPSKAPPKERKWMLCEREAVAASYAEPLYVHCGVADNATGGARLELWFHRVGR   180
  : :     ::    ::  .:      :  .. .   :.:     .:   :. : : ::
NNRTR-VHGDKATAHGRKRIVCRERLFSARVGDAVSFGCAV--FPRAGETFEVRFYRRGR   139

FRSTRGDDEAVRNPFPRAPPV--LLFVAQNGSIAYRSA---ELGDNYIFPSPADPRNLPL   235
:::  .: :    :     :   :::   :.:.:   :    . .            . .
FRSPDADPEYFDEPPRPELPRERLLFSSANASLAHADALAPVVVEGERATVANVSGEVSV   199

TVRSLTAATEGVYTWRRDMGTKSQRKVVTVTTHR--APAVSVEPQPALEGAGYAAVCRAA   293
 :    : ::::::::   .  .  .   :.      :   .   : : :  . :::
RVAAADAETEGVYTWRVLSANGTEVRSANVSLLLYSQPEFGLSAPPVLFGEPFRAVCVVR   259

EYYPPRSTRLHWFRNGYPVEARHARDVFTVDDSGLFSRTSVLTLEDA---------TPTA   344
.::: :: :: ::    ::.:         :. :  .: ::. .  :            :
DYYPRRSVRLRWFADEHPVDAAFVTNSTVADELGRRTRVSVVNVTRADVPGLAAADAADA   319

HPPNLRCDVSWFQSANMERRFYAAGTPAVYRPPELRVYFEGGEAVCEARCVPEGRVSLRW   404
  : :::.  :.       ::  :  : :: :  . : :  : :::.. :::     : :
LAPSLRCEAVWYRDSVASQRFSEALRPHVYHPAAVSVRFVEGFAVCDGLCVPPE-ARLAW   378

TVRDGIAPSRTEQTGVCAERPGLVNLRGVRLLSTTDGPVDYTCTATGYPAPLPEFSATAT   464
.     :       : ::: :::.:.:  : ::  ::::::::   : :  :: :  :
SDH---AADTVYHLGACAEHPGLLNVRSARPLSDLDGPVDYTCRLEGLPSQLPVFEDTQR   435

YDASPGLIGSPVLVSVVAVACGLGAVG-LLLVAASCLRRKARARL                  508
:::::. .  ::. : . :  :.: .. .:.. :.:.  .  .
YDASPASVSWPVVSSMIVVIAGIGILAIVLVIMATCVYYRQAGP                   479
Aligned 492, Matches 148, Mismatches 344, Homology 30%
```

Fig. 5. Homologies between predicted amino acid sequence of IBRV gIII and PRV g92(gIII) polypeptides. The IBRV gIII amino acid sequence in single letter code from 1 to 60 begins on the top line. The PRV gIII amino acid sequence from 1 to 28 is shown on the second line. Identical amino acids (:), similar amino acids (.), and gaps introduced by the alignment alogarithm (-) are indicated. The sequences were aligned with the Genepro (Riverside Scientific Enterprises, Seattle, WA) program based on the strategy of Wilbur and Lipman (1983).

terminal signal sequences, multiple Asn-X-Ser/Thr N-glycosylation sites, hydrophilic amino acid sequences downstream from the N-terminal signal sequence, hydrophobic membrane anchor regions, and charged C-terminal (cytoplasmic) tails typical of membrane-spanning glycoproteins (Allen and Coogle, 1988; Dowbenko and Lasky, 1984; McGeoch, 1985; Robbins et al., 1986; Von Heijne, 1986). Antigen epitopes predicted by the method of Hopp and Woods (1981) were present principally in the N-terminal half of the glycoproteins.

To learn whether the IBRV gIII gene was required for IBRV replication, we next prepared an IBRV gIII deletion mutant from IBRV(NG)dltk and designated it IBRV(NG)dltkdlgIII. The IBRV gIII deletion mutant was viable, demonstrating that the IBRV gIII gene, like the homologous HSV-1 gC, HSV-2 gC, and the PRV g92(gIII) genes was not required for virus replication. In addition, the IBRV gIII deletion mutant made available a "marker" IBRV vaccine analogous to the AD vaccine, OMNIMARK.

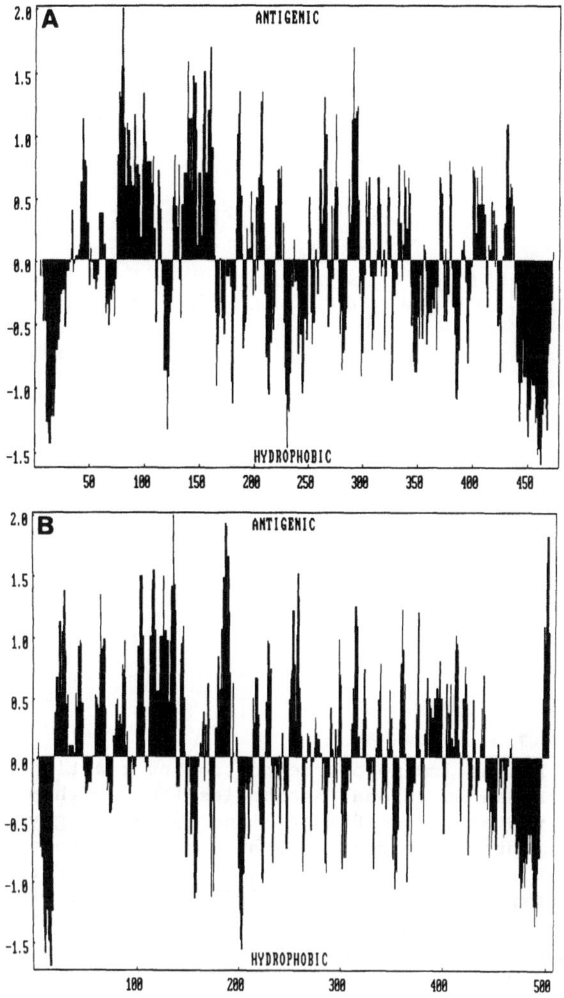

Fig. 6. Comparisons of the antigenicity and hydrophobicity of (A) PRV glycoprotein g92(gIII) and (B) IBRV glycoprotein gIII (Hopp and Woods, 1981).

SEROLOGICAL DIFFERENTIATION OF VACCINATED PIGS FROM THOSE INFECTED WITH VIRULENT PRV STRAINS

Marker vaccines, such as OMNIMARK and IBRV(NG)dltkdlgIII which permit producers to distinguish vaccinated animals from those infected by virulent field strains have potential for use: (i) in vaccination in the face of an outbreak of disease to protect animals and prevent or limit the spread of virulent virus; (ii) for large-scale immunization/eradication programs where test and slaughter procedures are costly or logistically difficult to mount or sustain; (iii) for removing the quarantine stigma from infected herds; and (iv) to reduce the hazards of moving animals carrying latent herpesvirus infections across state and national boundaries. Programs to monitor the serological status of pigs and to cull infected sows from breeding herds in places where AD is prevalent and vaccination is practiced are now receiving

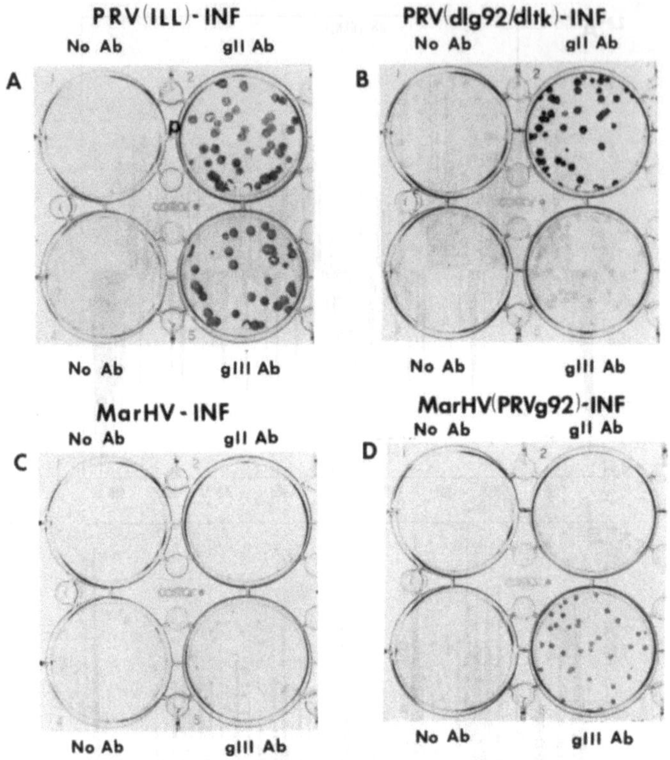

Fig. 7. Plaque immunology demonstrating reaction of mouse monoclonal antibodies (MCA) with pseudorabies virus (PRV) gII and g92(gIII) glycoprotein antigens. MarHV = marmoset herpesvirus; MarHV(PRVg92) = a recombinant MarHV expressing the PRV g92(gIII) protein; PRV(Ill) = a virulent field strain of PRV; PRV(dlg92/dltk) = OMNIMARK vaccine with deletions in tk and g92(gIII) genes. Plaques in Rab-9 (rabbit skin) cells were detected by biotin-rabbit anti-porcine IgG (H+L), peroxidase-conjugated streptavidine, chloronapthal, and hydrogen peroxide.

increased consideration because of the availability of marker vaccines. Commercial diagnostic kits are, of course, needed to accompany the marker vaccines (Kit, 1989). The feasibility of such kits is illustrated by the immunoplaque assay experiments shown in Fig. 7. Plaques formed by PRV(Ill), a typical virulent wild-type strain, and by OMNIMARK[PRV(dlg92/dltk)] react positively with monoclonal antibodies to the major PRV glycoprotein gII, but plaques made by marmoset herpesvirus (MarHV) and the MarHV(PRVg92) recombinant virus, which expressed the PRV g92(gIII) protein, do not. PRV(Ill) plaques also react positively with monoclonal antibodies to PRV g92(gIII), as do plaques made by recombinant MarHV(PRVg92), but plaques made by MarHV and OMNIMARK [PRV(dlg92/dltk)] do not. A sensitive, commercial diagnostic CELISA kit based on competition of field sera with peroxidase-conjugated, anti-PRVgIII monoclonal antibodies has been developed by Fermenta Animal Health, Kansas City, MO for use in conjunction with OMNIMARK.

The development of the OMNIMARK marker vaccine was possible because the gene encoding PRV glycoprotein g92(gIII) is nonessential for PRV replication. Recent investigations have revealed that three additional glycoprotein genes, gI, g63, and gX, are not required for PRV replication in cultured cells (Kit et al., 1987b; Lomniczi et al., 1987b; Mettenleiter et al., 1985; 1987a; Petrovskis et al., 1988). Likewise, the HSV-1 glycoprotein genes homologous to gI, g63, and gX are nonessential for in vitro replication (Longnecker and Roizman, 1987). PRV marker vaccines harboring deletions in the gI and gX glycoprotein genes have been proposed as alternatives to the OMNIMARK, gIII-deletion vaccine and the relative merits of each of the markers have been debated.

Even though it was not known in 1982, the conventional Bartha, NIA-4, and Bucharest PRV vaccines were, in fact, gI deletion mutants. This gI marker is favored by Van Oirschot et al. (1987) as the marker of choice for PRV vaccines to be used in Europe. As indicated earlier, our TK^- PRV(BUK) clone 3 vaccine is also an example of a gI marker vaccine (Kit, 1989). On the other hand, Petrovskis et al. (1988) have favored the use of gX deletion mutants as marker vaccines. They reported that after challenge with virulent viruses, pigs previously vaccinated with a gI^+ gX^- PRV vaccine showed significantly less growth arrest, pyrexia, and shedding of challenge virus than did pigs vaccinated with double gI^- gX^- PRV mutants. Unfortunately, however, Petrovskis et al. (1988) have not compared the response postchallenge of pigs vaccinated with single mutants of gI^- vs gX^-. Furthermore, McGinley and Platt (1988) found that pigs vaccinated with subunit vaccines and later latently infected with field strains often failed to develop detectable gX antibodies. Hence, pigs infected with virulent viruses that could spread in the herd might be missed by using gX as the diagnostic test marker.

One important characteristic of the marker chosen for a vaccine is that the animal should respond immunologically to that chosen protein in virtually all infections, thus allowing the differentiation test to be of maximum value. Of the glycoproteins nonessential for PRV replication, we believe that the g92(gIII) marker chosen for OMNIMARK best fits this criterion. Diagnostic tests utilizing g92(gIII) are the most likely to detect antibodies from all animals exposed to and potentially carriers of disease.

When pigs are infected with wild-type PRV, glycoprotein g92(gIII) is one of the principal targets for antibody formation and for cellular immunity. But when the infecting virus lacks g92(gIII) in the viral envelope, as in the case of OMNIMARK, this lack of g92(gIII) can be compensated by increased utilization of the other viral glycoproteins. Pigs vaccinated with OMNIMARK develop antibodies which neutralize the infectivity of virulent field strains. They exhibit anamnestic responses after challenge and they are fully protected from challenge with virulent virus, showing that g92(gIII) is not essential for the generation of a protective immune response (Kit et al., 1987a; Oliver et al., 1988).

ANIMAL STUDIES

Although tests have not as yet been carried out with the second generation IBRV vaccine, the parental first generation vaccine has been tested in calves and pregnant cows and found to be safe and effective (Kit et al., 1986). Experiments demonstrating the safety and efficacy of OMNIMARK have likewise been carried out. Safety and efficacy experiments in mice, rabbits, hamsters, and guinea pigs were performed by Craig Dees (Fermenta Animal Health). Those in pigs were performed by Bruce Lawhorn (Texas A&M University), Brad Thacker (Michigan State University), and R. W. Worthington (Ministry of Agriculture and Fisheries, New Zealand).

The studies on mice demonstrated that intraperitoneal injection of OMNIMARK at $10^{7.5}$ $TCID_{50}$ per mouse did not kill outbred CFR, nude/nude, or beige mice. Two licensed TK^{+} AD MLV vaccines killed three out of five outbred mice in 7 days after intraperitoneal injection of $10^{4.59}$ and $10^{3.89}$ $TCID_{50}$, respectively (Kit et al., 1987a).

Adverse reactions were not observed after intracranial inoculation of OMNIMARK in pigs at 10^{5} $TCID_{50}$, or after IM inoculation of 3-day-old piglets at $10^{8.1}$ $TCID_{50}$. The inoculation of calves and lambs did not produce clinical signs of disease. OMNIMARK has not been recovered from nasal swabs or from tissues of pigs vaccinated IM, nor was transmission to unvaccinated sentinel pigs detected. OMNIMARK protected pigs from severe intranasal (IN) challenge by virulent U.S.A. field strains or from a virulent New Zealand strain (Oliver et al., 1988; Thacker et al., 1988a). Parenthetically, the first generation virus, OMNIVAC, also protected pigs challenged with a virulent European PRV strain (Van Oirschot and Gielkens, 1987). OMNIMARK vaccination reduced disease symptoms, replication, and the shedding from nasal swabs of the virulent challenge strain. OMNIMARK vaccination also reduced the development of latent infections by virulent challenge virus (Thacker et al., 1988b). This capacity to reduce the development of latency by challenge virus may not be absolute and may depend on the age of the vaccinated pigs, the vaccination dosage, and the dosage of the challenge virus. Nevertheless, the overall effect of OMNIMARK vaccination is to reduce infection of other animals and dissemination of virulent virus in the environment.

Recent experiments conducted in New Zealand provide an example of the efficacy of OMNIMARK. Six pigs were vaccinated IM at 6 weeks and at 8 weeks of age with 2×10^{7} $TCID_{50}$ of OMNIMARK (Oliver et al., 1988). Two nonvaccinates served as contact controls. Nasal and tonsillar swabs were collected daily for 7 days postvaccination. All vaccinates and in-contact nonvaccinates remained healthy following vaccination and gained weight. No pig became febrile. Virus was not isolated from nasal or tonsillar swabs of any of these pigs.

At 10 weeks of age, the above pigs and four age-matched piglets from the same source were challenge-inoculated IN with 2×10^{7} $TCID_{50}$ of a virulent New Zealand strain. Following challenge, vaccinated pigs became febrile but continued to gain weight and all six vaccinates survived. All nonvaccinated pigs died or became moribund within 6 days of challenge. Tonsil and nasal swabs were PRV-positive until the control pigs died or became moribund. Isolations were made from one or more tissues of all nonvaccinates.

With regard to the first generation PRV vaccine, OMNIVAC has been used for almost 3 years. Over 1.5 million doses have been given to pregnant pigs in all stages of gestation and to newborn piglets without adverse effects or complaints from the users to either the U.S.D.A. or the manufacturer. OMNIVAC is user friendly; that is, it fits well into management practices. Piglets can be vaccinated at birth at the same time that their teeth are cut and they are injected with iron. The gene deletions of OMNIVAC and OMNIMARK are stable characteristics of these vaccine viruses with a probability of reversion being essentially zero.

OMNIVAC has been used successfully to eliminate Aujeszky's disease on the Reeves farm in Farmville, North Carolina (Reeves, 1988). The endemic nature of PRV in the Reeves' swine herd was established through a serological survey in September of 1986 which showed that 25 out of 33 breeding pigs had titers of 1:8 to 1:256, and six had titers of 1:4. Nine out of 16 finishing floor pigs, 3 to 6 months of age, exhibited titers of 1:8 to 1:32, and 15 out of 21 growers, 6 weeks to 3 months of age had titers of 1:4 or 1:8. The entire breeding herd and all finishing floor pigs were vaccinated, and a routine vaccination schedule was instituted on pigs 1 to 3 days of age, and again at 10 to 12 weeks of age, after maternal antibodies had declined. Replacement gilts were vaccinated at 6 months of age after demonstrating a seronegative status. A turnover of the breeding herd was instituted and all replacements purchased from seronegative herds were quarantined for 14 days and vaccinated. By 1988, the finishing floor herd had been converted into one in which no positive animals were determined. In this herd, the number of serologically positive pigs has been reduced in the face of a natural challenge. The PRV elimination program had a sound economic basis and the endemic status of the herd was overcome without depopulation and repopulation. The results obtained with OMNIVAC encourage the view that with proper management practices, saturation vaccination, and appropriate testing and culling, OMNIMARK vaccination programs can facilitate elimination of AD in swine herds.

CONCLUDING REMARKS

Finally, I would like to call attention to a problem that deserves further investigation. As already mentioned, herpesviruses encode seven or more glycoproteins and, yet, four of the seven are nonessential for virus replication. Why are so many glycoproteins used to make herpesvirus membranes when only one or two membrane glycoproteins suffice for some of the other viruses (e. g., influenza, toga, or Rift Valley Fever viruses)? What are the roles for the essential and nonessential glycoproteins in the strategy of the herpesvirus genomes? Why is it that one or two glycoprotein genes can be deleted without impairing the virus-induced protective response? Only fragmentary data are presently available to foreshadow the answers to these questions. Possibly, the deletion of a glycoprotein gene with the associated loss of an antigenic target is outweighed for the host by a concommitant loss of immunopathology. PRV glycoprotein mutants may not replicate as well as parental glycoprotein-positive viruses in swine alveolar macrophage, or interfere with alveolar macrophage functions (Iglesias et al., 1988a; 1988b).

We know that herpesviruses can occasionally trigger autoimmune responses through molecular mimicry (Srinivasappa et al., 1986), inhibit the expression of Class I major histocompatability complex antigens (Jennings et al., 1985), and induce suppressor cells (Horohov et al., 1985). Hypotheses to explain how the deletion of particular herpesvirus glycoprotein genes might yield a net gain in host protective response can therefore be formulated. In Houston we now have for comparison a series of herpesvirus mutants with deletions in one or two glycoprotein genes in which all of the mutants are derived from a common parental virus. This system could provide an interesting model with which immunobiologists might sort out some of the puzzling problems of herpesvirus pathobiology.

ACKNOWLEDGMENT

This investigation was aided by a research contract from NovaGene, Inc., Houston, TX.

REFERENCES

Allen, G. P., and Coogle, L. D., 1988, Characterization of an equine herpesvirus type 1 gene encoding a glycoprotein (gp13) with homology to herpes simplex virus glycoprotein C, J. Virol., 62:2850.

Buller, R. M. L., Smith, G. L., Cremer, K., Notkins, A. L., and Moss, B., 1985, Decreased virulence of recombinant vaccinia virus expression vectors is associated with a thymidine kinase-negative phenotype, Nature, 317:813.

Davison, A. J., and Scott, J. E., 1986, The complete DNA sequence of varicella-zoster virus, J. Gen. Virol., 67:1759.
Dowbenko, D. J., and Lasky, L. A., 1984, Extensive homology between the herpes simplex virus type 2 glycoprotein F gene and the herpes simplex virus type 1 glycoprotein gene, J. Virol., 52:154.
Dubbs, D. R., and Kit, S., 1964a, Isolation and properties of vaccinia mutants deficient in thymidine kinase-inducing activity, Virology, 22:214.
Dubbs, D. R., and Kit, S., 1964b, Mutant strains of herpes simplex deficient in thymidine kinase-inducing activity, Virology 22:493.
Field, H. J., and Wildy, P., 1978, The pathogenicity of thymidine kinase deficient mutants of herpes simplex in mice, J. Hyg. (Cambridge), 81:267.
Frink, R. J., Eisenberg. R., Cohen, G., and Wagner, E. K., 1983, Detailed analysis of the portion of the herpes simplex virus type 1 genome encoding glycoprotein C., J. Virol., 45:634.
Glorioso, J., Kees, U., Kumel, G., Kirchner, H., and Krammer, P., 1985, Identification of herpes simplex virus type 1 (HSV-1) glycoprotein gC as the immunodominant antigen for HSV-1-specific memory cytotoxic T lymphocytes, J. Immunol., 135:575.
Hopp, T. P., and Woods, K. R., 1981, Prediction of protein antigenic determinants from amino acid sequences, Proc. Nat. Acad. Sci. USA, 78:3824.
Horohov, D. W., Moore, R. N., and Rouse, B. T., 1985, Production of soluble suppresor factors by herpes simplex virus-stimulated splenocytes from herpes simplex virus-immune mice, J. Virol., 54:798.
Iglesias, G., Pijoan, C., and Molitor, T., 1988a, Replication of pseudorabies virus in swine alveolar macrophages, Chapt. IV, in: "Proc. 10th Int. Cong. Pig. Vet. Soc.," Rio de Janiero, Brazil.
Iglesias, G., Pijoan, C., and Molitor, T., 1988b, Effects of pseudorabies virus infection on alveolar macrophages functions, Chapt. IV, in: "Proc. 10th Int. Cong. Pig Vet. Soc.," Rio de Janiero, Brazil.
Jennings, S. T., Rice, P. L., Kloszewski, E. D., Anderson, R. W., Thompson, D. L., and Tevethia, S. S., 1985, Effect of herpes simplex virus types 1 and 2 on surface expression of class I major histocompatability complex antigens on infected cells, J. Virol., 56:757.
Kit, S., 1985, Thymidine kinase, Microbiol. Sci., 2:369.
Kit, S., 1989, Safety and efficacy of genetically engineered Aujeszky's disease vaccines, in: "Papers presented at the Commis. Europ. Commun. (CEC) Symposium on Vaccination and Control of Aujeszky's Disease," Brussels, Belgium (in press).
Kit, S., Dubbs, D. R., Piekarski, L. J., and Hsu, T. C., 1963, Deletion of thymidine kinase activity from L cells resistant to bromodeoxyuridine, Exp. Cell Res., 31:297.
Kit, S., Kit, M., Bartkoski, M. J., and Dees, C., 1987a, Genetically engineered pseudorabies virus vaccine with deletions in thymidine kinase and glycoprotein genes, in: "Vaccines 87. Modern Approaches to New Vaccines: Prevention of AIDS and Other Viral, Bacterial, and Parasitic Diseases," R. M. Chanock, R. A. Lerner, F. Brown, and H. Ginsberg, eds., Cold Spring Harbor Laboratory.
Kit, S., Kit, M., and McConnell, S., 1986, Intramuscular and intravaginal vaccination of pregnant cows with thymidine kinase-negative, temperature-resistant infectious bovine rhinotracheitis virus (bovine herpesvirus 1), Vaccine, 4:55.
Kit, S., Kit, M., and Pirtle, E. C., 1985, Attenuated properties of thymidine kinase-negative deletion mutant of pseudorabies virus, Am. J. Vet. Res., 46:1359.
Kit, S., and Qavi, H., 1983, Thymidine kinase (TK) induction after infection of TK-deficient rabbit cell mutants with bovine herpesvirus type 1 (BHV-1). Isolation of TK^- BHV-1 mutants, Virology, 130:381.
Kit, S., Qavi, H., Dubbs, D. R., and Otsuka, H., 1983, Attenuated marmoset

herpesvirus isolated from recombinants of virulent marmoset herpesvirus and hybrid plasmids, J. Med. Virol., 12:25.
Kit, S., Sheppard, M., Ichimura, H., and Kit, M., 1987b, Second-generation pseudorabies virus vaccine with deletions in thymidine kinase and glycoprotein genes, Am. J. Vet. Res., 48:780.
Klein, R. J., 1982, The pathogenesis of acute, latent, and recurrent herpes simplex virus infections, Arch. Virol., 72:143.
Lomniczi, B., Kaplan, A. S., and Ben-Porat, T., 1987a, Multiple defects in the genome of pseudorabies virus can affect virulence without detectably affecting replication in cell culture, Virology, 161:181.
Lomniczi, B., Watanabe, S., Ben-Porat, T., and Kaplan, A. S., 1984, Genetic basis of the neurovirulence of pseudorabies virus, J. Virol., 52:198.
Lomniczi, B., Watanabe, S., Ben-Porat, T., and Kaplan, A. S., 1987b, Genome location and identification of functions defective in the Bartha vaccine strain of pseudorabies virus, J. Virol., 61:796.
Longnecker, R., and Roizman, B., 1987, Clustering of genes dispensable for growth in culture in the S component of the HSV-1 genome, Science, 236: 573.
McGeoch, D. J., 1985, On the predictive recognition of signal peptide sequences, Vir. Res., 3:271.
McGinley, M. J., and Platt, K. B., 1988, Studies on the ability of 98-kilodalton pseudorabies virus diagnostic antigen to detect latent infections induced by low-dose exposure to the virus, Am. J. Vet. Res., 49:1489.
Mettenleiter, T. C., Lomniczi, B., Sugg, N., Schrueurs, C., and Ben-Porat, T., 1988, Host cell-specific growth advantage of pseudorabies virus with a deletion in the genome sequences encoding a structural glycoprotein, J. Virol., 62:12.
Mettenleiter, T. C., Lukacs, N., and Rziha, H.-J., 1985, Pseudorabies virus avirulent strains fail to express a major glycoprotein, J. Virol., 56: 307.
Mettenleiter, T. C., Schreurs, C., Thiel, H.-J., and Rziha, H.-J., 1987a, Variability of pseudorabies virus glycoprotein I expression, Virology, 158:141.
Mettenleiter, T. C., Zsak, L., Kaplan, A. S., Ben-Porat, T., and Lomniczi, B., 1987b, Role of a structural glycoprotein of pseudorabies in virus virulence, J. Virol., 61:4030.
Oliver, R. E., Motha, M. X. J., Worthington, R. W., Penrose, M. E., Poole, W. S., 1988, Protection of piglets against a virulent New Zealand isolate of Aujeszky's disease virus by vaccination with a novel genetically-engineered vaccine, New Zeald. Vet. J., 36:100.
Otsuka, H., and Kit, S., 1984, Nucleotide sequence of the marmoset herpesvirus thymidine kinase gene and predicted amino acid sequence of thymidine kinase polypeptide, Virology, 135:316.
Petrovskis, E. A., Wardley, R. C., Berlinski, P. J., Thomsen, D. R., Meyer, A. L., and Post, L. E., 1988, Important role of pseudorabies virus glycoprotein I (GI) in efficacy of live vaccines, in: "Abstracts of 5th Int. Symp. Immunobiol. Proteins and Peptides," Chateau Lake Louise, Alberta, Canada.
Reeves, D. E., 1988, Results of a field study using a genetically altered modified live pseudorabies vaccine in an eradication program, in: "Proc. Annu. Mtg. Am. Assoc. Swine Practitioners," St. Louis, MO.
Robbins, A. K., Watson, R. J., Whealy, M. E., Hays, W. W., and Enquist, L. W., 1986, Characterization of a pseudorabies virus glycoprotein gene with homology to herpes simplex virus type 1 and type 2 glycoprotein C, J. Virol., 58:339.
Robbins, A. K., Weis, J. H., Enquist, L. W., and Watson, R. J., 1984, Construction of E. coli expression plasmid libraries: localization of a pseudorabies virus glycoprotein gene, J. Molec. Appl. Genet., 2:485.
Sheppard, M., Otsuka, H., Kit, M., and Kit, S., 1986, Transcription and in vitro translation of pseudorabies virus (PRV) thymidine kinase (TK) gene, in: "Abstracts of 86th Annu. Mtg. Amer. Soc. Microbiol.," Washington, D. C.

Srinivasappa, J., Saegusa, J., Prabhakar, B. S., Gentry, M. K., Buchmeier, J., Wiktor, T. J., Koprowski, H., Oldstone, M. B. A., and Notkins, A., L., 1986, Molecular mimicry: frequency of reactivity of monoclonal antiviral antibodies with normal tissues, J. Virol., 57:397.

Stanberry, L. R., Bernstein, D. I., Kit, S., and Myers, M. G., 1985, Genital reinfection after recovery from initial genital infection with herpes simplex virus type 2 in guinea pigs, J. Infect. Dis., 153:1055.

Swain, M. A., Peet, R. W., Galloway, D. A., 1985, Characterization of the gene encoding herpes simplex virus type 2 glycoprotein C and comparison with the type 1 counterpart, J. Virol., 53:561.

Tenser, R. B., Ressel, S. J., Fralish, F. A., and Jones, J. C., 1983, The role of pseudorabies virus thymidine kinase expression in trigeminal ganglion infection, J. Gen. Virol., 64:1369.

Thacker, B., Maes, R., Bartkoski, M., and Kolar, J., 1988a, Immunogenicity, efficacy, and safety of a modified-live, thymidine kinase negative, gIII deleted pseudorabies virus vaccine, Chapt. IV, in: "Proc. 10th Int. Cong. Pig Vet. Soc.," Rio de Janiero, Brazil.

Thacker, B., Maes, R., Gonzalez, P., and Han C., 1988b. Development of latency in vaccinated or passibly immune pigs experimentally infected with pseudorabies virus, Chapt. IV, in: "Proc. 10th Int. Cong. Pig Vet. Soc.," Rio de Janiero, Brazil.

Van Oirschot, J. T., and Gielkens, A. L. J., 1987, Vaccines against Aujeszky's disease: comparison of efficacy, DNA fingerprints and antibody response to glycoprotein I, Vet. Quart., 9 (Suppl. 1):37S.

Van Oirschot, J. T., Rziha, H. J.,, Moonen, P. J. L. M., Pol, J. M. A., and Van Zaane, D., 1986, Differentiation of serum antibodies from pigs vaccinated or infected with Aujeszky's disease virus by a competitive enzyme immunoassay, J. Gen. Virol., 67:1179.

Von Heijne, G., 1986, A new method for predicting signal sequence cleavage sites, Nucl. Acids Res., 14:4683.

Wilbur, W. J., and Lipman, D. J., 1983, Rapid similarity searches of nucleic acid and protein data banks, Proc. Nat. Acad. Sci. USA, 80:726.

ADVANCES IN UNDERSTANDING STRUCTURE-FUNCTION RELATIONSHIPS IN HEPATITIS B VIRUS (HBV) ENVELOPE (ENV) PROTEINS

A.R. Neurath and Nathan Strick

The Lindsley F. Kimball Research Institute
of the New York Blood Center
310 East 67th Street
New York, NY 10021

INTRODUCTION

HBV belongs to the group of hepadnaviridae, each member of which has a narrow host-specificity. HBV is infectious for humans and chimpanzees. Duck hepatitis B virus (DHBV) is not infectious for humans. However, transfection of human hepatoma cells with DHBV DNA results in replication of DHBV in these cells (Pugh et al., 1988). Thus, the narrow host range of hepadnaviruses must be determined by the specificity of virus-host cell interactions during early steps of infection. These steps are likely to correspond to virus adsorption, penetration, and uncoating, processes in which viral envelope proteins are involved. In order to fully understand the molecular basis of the virus life cycle and to design on a rational basis preventive and curative measures against hepatitis B, it is necessary to understand structure-function relationships in HBV env proteins.

Knowledge concerning HBV env proteins was summarized in several reviews (Shih, 1985; Neurath and Kent, 1985, 1988; Neurath, in press). This report will briefly recapitulate selected characteristics of HBV env proteins and will deal with new findings concerning HBV-cell receptor interactions and the design of synthetic peptides mimicking epitopes on the S region of HBV env proteins.

PROTEIN COMPONENTS OF THE HBV ENVELOPE

The HBV envelope consists of six distinct but related proteins all coded for by a single gene (Figs. 1 and 2). S- (P25, GP29), M- (GP33, GP36), and L-protein (P39, GP42) arise from translation of distinct mRNA transcripts having identical 3' ends but differing in their 5' ends and from initiation of translation at either 5' end or at internal ATG initiator codons (Price et al., 1988). Differences in glycosylation of the translational products result in six distinct env components. M-protein consists of S-protein + the preS2 sequence and L-protein consists of S-protein + preS2 + preS1 sequences. The exact arrangement of the distinct proteins within the HBV envelope is not known. In addition to mature virions, HBV env proteins occur in subviral particles lacking the viral nucleocapsid core (Fig. 1). These particles usually contain less M- and L-protein than the HBV envelope. In addition to spherical 20-nm particles, less abundant

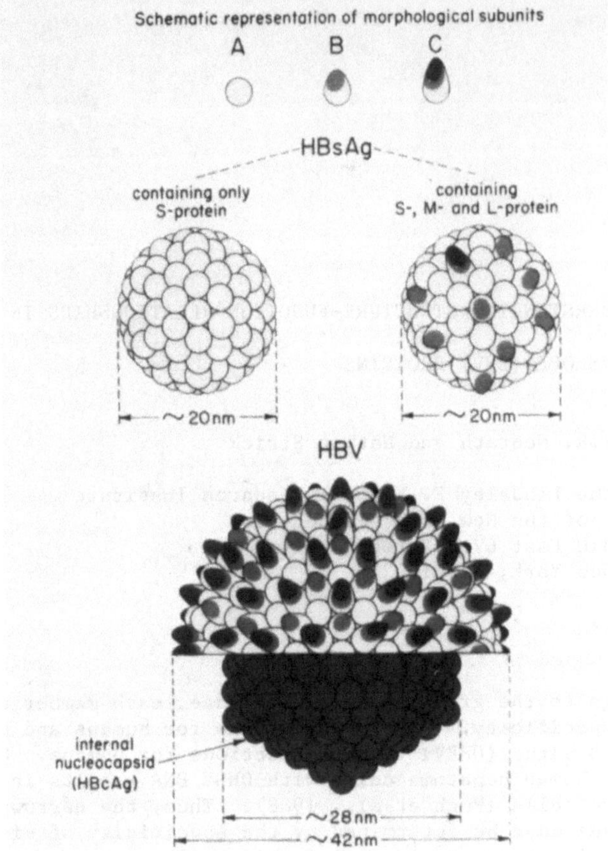

Fig. 1. Schematic model of HBV and of subviral HBsAg particles. Individual structural subunits containing: (A) S-protein only; (B) M-protein; and (C) L-protein are shown on top of the figure. HBV particles contain an internal nucleocapsid shown in the bottom split-open section. S-protein corresponds to white areas, preS2 sequences to gray areas, and preS1 sequences to black areas. For further explanations see Fig. 2.

tubular subviral particles occur in plasma of infected individuals. The tubules resemble in protein composition the HBV envelope more closely than do the 20-nm spherical particles. The recognition of HBV env proteins by anti-peptide antisera indicates that most of the HBV env (L-protein) sequence is exposed, except regions corresponding approximately to amino acid residues 74-93 in the preS sequence and 210-222, 255-283, 340-385, and possibly 386-400 in S-protein (Shih, 1985; Neurath and Kent, 1988; Lerner et al., 1981; Neurath et al., 1986a; Milich et al., 1987). On the other hand, antisera against native HBV env proteins recognize peptides from a more restricted portion of the L-protein sequence (Fig. 2). This can either be due to a limited repertoire of antibodies elicited by the native protein or due to dissimilarities between epitopes on synthetic peptides and the cognate sequences on the native protein. Detailed studies indicate that dominant B cell epitopes are located on the N-terminal half of the preS1 sequence, near the C terminus of the preS1 sequence, and in the N-terminal half of the preS2 sequence. All these epitopes can be mimicked with high fidelity by synthetic peptide analogs (Neurath and Kent, 1988; Neurath, in press; Neurath et al., 1987a). On the other hand, it proved much more difficult to mimick B cell epitopes on the native S-protein

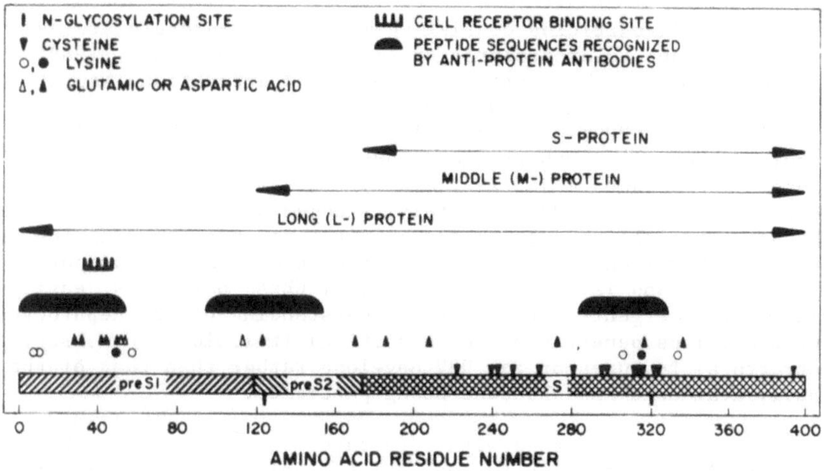

Fig. 2. Schematic representation of HBV env proteins and their relatedness. Closed and open symbols correspond to conserved and variable residues, respectively, in sequences of distinct HBV env protein subtypes.

sequence by synthetic peptides. This can be explained by the presence of 14 cysteine residues in the S-protein sequence, all of which are involved in disulfide bonds (Guerrero et al., 1988). Reduction and alkylation of HBsAg consisting mostly of S-protein drastically alters its antigenicity and immunogenicity (Fig. 3) (Neurath and Strick, 1980), although there are common, nondominant, epitopes shared between the native and denatured forms of HBsAg (Swenson et al., 1988). Linear and cyclic synthetic peptides optimally recognized by anti-S-protein antibodies contain lysine and aspartic acid residues playing a critical role in the group-specific and subtype-specific antigenicity of S-protein (Neurath et al., 1981; Peterson et al., 1984).

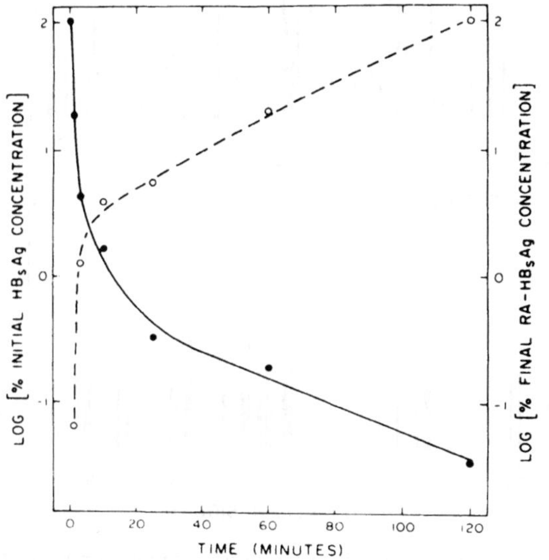

Fig. 3. Kinetics of conversion of native HBsAg by reduction and alkylation into an antigenically distinct form detectable by antibodies to reduced and alkylated HBsAg (RA-HBsAg). Reprinted with permission from Neurath and Strick (1980).

Subviral HBsAg particles consisting mainly of S-protein are extraordinarily resistant to proteolytic enzymes (Guerrero et al., 1988) and maintain their morphological integrity after lipid extraction (Neurath et al., 1978), although they contain about 23% lipids. There are about 100 copies of S-protein (P25 + GP29) per subviral particle, linked by 2-4 intermolecular disulfide bonds; it was proposed that among the 14 cysteine residues of S-protein, 3 residues--295, 297, and 395--are most likely involved (Guerrero et al., 1988). Unfortunately, studies of the structure of subviral HBsAg particles are only partly relevant to understanding the architecture of the virus envelope, which contains a much higher proportion of M- and L-protein, and is, unlike the subviral HBsAg particles, easily dissociated by detergents. For better understanding of HBV-hepatocyte interactions (virus penetration and uncoating) it would be necessary to study structural features of the HBV envelope rather than that of the much more easily available and abundant HBsAg particles.

For all studies concerned with structure-function relationships in HBV env proteins, it is necessary to consider the amino acid variability of the env protein sequence, reflected in the existence of antigenic subtypes of the virus. This variability is more extensive in the preS region as compared with the S region (Fig. 4).

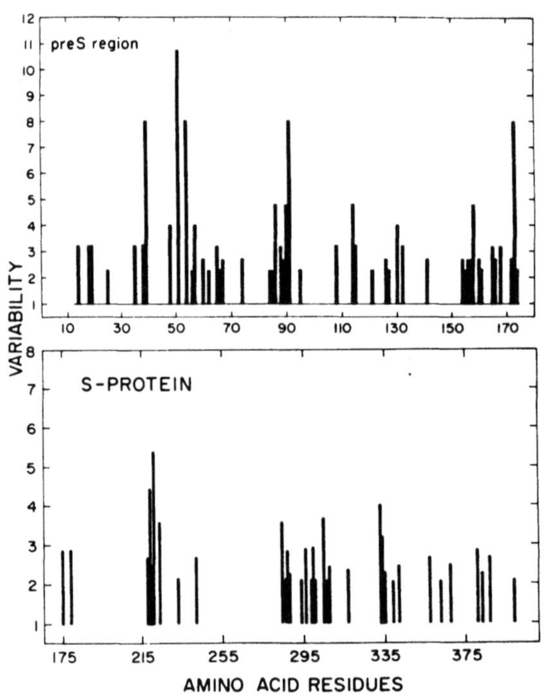

Fig. 4. Variability of amino acid sequences corresponding to preS and S regions of distinct HBV subtypes. Variability analysis was done according to Wu and Kabat (1970). Top part reprinted with permission from Neurath et al. (1987c). Bottom part is derived from sequencing data on 16 distinct HBV isolates (Okamoto et al., 1986). Note the different scales on ordinates.

EFFECT OF SUBTYPE-RELATED AMINO ACID REPLACEMENTS WITHIN THE HBV RECEPTOR BINDING SITE ON THE INTERACTION BETWEEN HUMAN HEPATOMA CELLS AND HBV

The cell receptor binding site on HBV has been localized within residues preS(21-47) of the preS1 sequence (Neurath et al., 1986b). Additional studies revealed that a truncated peptide preS(32-47) also has binding activity, albeit a lower one (Neurath et al., 1988). All synthetic peptides with binding activity were acetylated at the N-terminus. A non-acetylated peptide, preS(12-47), lacked binding activity. This finding may be related to the observation that L-protein is acylated at its N-terminus with myristic acid (Persing et al., 1987) and that myristylation of L-protein is required for viral infectivity (Macrae et al., 1988). The localization of a receptor binding site within the preS1 sequence is in agreement with results indicating that the host range of hepadnaviruses is not determined by the S-protein sequence (Seeger et al., 1987).

There is a limited degree of amino acid variability within the PreS(21-47) sequence, since 4/27 residues are not conserved (Fig. 5). Previously reported studies (Neurath et al., 1986b) defined the receptor binding site using synthetic peptides derived from the adw2 sequence and HBV env protein of the adw (w subspecificity undefined) subtype. As expected, not only HBsAg adw and the synthetic peptide preS(21-47) adw2 but also HBsAg ayw and the synthetic peptide preS(21-47) ayw, each covalently linked to cellulose, adsorbed human hepatoma HepG2 cells to a much greater extent than control cellulose with covalently attached bovine serum albumin (BSA) (Fig. 6A). Antibodies to synthetic peptides from the preS1 sequence from one HBV subtype recognized preS1 sequences of the native HBV env protein of all serological subtypes (Neurath et al., 1987c). Therefore, it was unexpected that the peptide preS(21-47) ayw and the corresponding antipeptide antiserum inhibited the reaction between HBsAg adw and HepG2 cells much less efficiently than analogous reagents corresponding to the adw2 subtype (compare Fig. 6B with column 2 in Fig. 6A). In a reverse experiment, significant inhibition of the attachment of HepG2 cells to HBsAg ayw cellulose was observed only with a homologous synthetic peptide but not with preS(21-47) adw2 (compare Fig. 6C with column 3 in Fig. 6A). These results indicate that the attachment of HBV env protein to hepatocytes is influenced by subtype-related amino acid replacements within the receptor binding site of HBV.

COMPARATIVE IMMUNOGENICITIES OF PEPTIDES DERIVED FROM THE PRES1, PRES2, AND S REGIONS OF THE HBV ENV PROTEIN

To serve as useful reagents for structure-function relationship studies in proteins, synthetic peptides must mimic as closely as possible cognate

Subtype	2 1	2	3	4	5	6	7	8	9	3 0	1	2	3	4	5	6	7	8	9	4 0	1	2	3	4	5	6	7
1 adw2	P	L	G	F	F	P	D	H	Q	L	D	P	A	F	G	A	N	S	N	N	P	D	W	D	F	N	P
2 adw	P	L	G	F	L	P	D	H	Q	L	D	P	A	F	G	A	N	S	T	N	P	D	W	D	F	N	P
3 adyw	P	L	G	F	F	P	D	H	Q	L	D	P	A	F	R	A	N	T	N	N	P	D	W	D	F	N	P
4 ayw	P	L	G	F	F	P	D	H	Q	L	D	P	A	F	R	A	N	T	A	N	P	D	W	D	F	N	P
5 ayw	P	L	G	F	F	P	D	H	Q	L	D	P	A	F	R	A	N	T	A	N	P	D	W	D	F	N	P
6 adr	P	L	G	F	F	P	D	H	Q	L	D	P	A	F	G	A	N	S	N	N	P	D	W	D	F	N	P
7 adr	P	L	G	F	F	P	D	H	Q	L	D	P	A	F	G	A	N	S	N	N	P	D	W	D	F	N	P
8 adr	P	L	G	F	F	P	D	H	Q	L	D	P	A	F	G	A	N	S	H	N	P	D	W	D	F	N	P
9 ayr	P	L	G	F	F	P	D	H	Q	L	D	P	A	F	G	A	N	S	N	N	P	D	W	D	F	N	P
	P	L	G	F		P	D	H	Q	L	D	P	A	F		A	N			N	P	D	W	D	F	N	P

Fig. 5. Amino acid variability within the preS(21-47) region of the preS1 sequence carrying the HBV receptor binding site for hepatocytes (Neurath et al., 1986b).

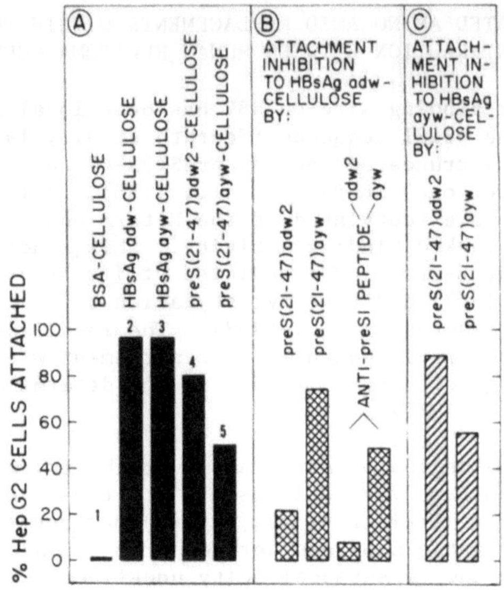

Fig. 6. Influence of HBV subtype-dependent amino acid sequence variations on HBV-HepG2 receptor interactions. For details see Neurath et al. (1986b).

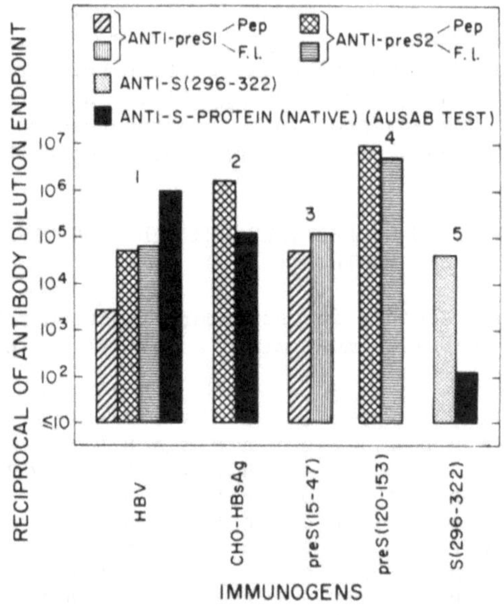

Fig. 7. Comparative immunogenicities of HBV, recombinant HBsAg particles containing S- and M-protein (CHO-HBsAg), and of selected synthetic peptides from the preS1 (3), preS2 (4), and S (5) regions of the HBV env protein. Double-antibody RIA tests were carried out with plates coated with peptides (Pep) or with antigens containing the full-length (F.l.) preS1 or preS2 sequence. AUSAB = test from Abbott Laboratories.

sequences in the native protein. To be highly immunogenic, synthetic peptides have to contain both B and T cell epitopes, each of which should resemble the corresponding epitopes on the native protein. Selected peptides (Neurath and Kent, 1988; Neurath, in press; Neurath et al., 1987a; Milich, 1987) satisfy these requirements. The peptide preS(15-47) (column group 3, Fig. 7) elicited in rabbits higher levels of antibodies recognizing the peptide and the full-length native preS1 sequence than did HBV (column group 1, Fig. 7). Similarly, the peptide preS2(120-153) elicited higher levels of antibodies recognizing the peptide and the full-length preS2 sequence than did HBV or recombinant HBsAg particles containing S-protein and the preS2 sequence (CHO-HBsAg; compare column groups 4, 1, and 2 in Fig. 7). The level of antibodies recognizing the homologous peptide and the native HBV env protein elicited by the latter two synthetic peptides were very similar, indicating a nearly perfect mimicry of B cell epitopes on the native protein by these synthetic peptides (Neurath et al., 1988; Neurath et al., 1987b). On the other hand, one of the best selected peptides from the S region (Zheng et al., 1986) (column group 5, Fig. 7) elicited much lower levels of anti-S antibodies than HBV or CHO-HBsAg (black columns in groups 1 and 2). The level of antibodies recognizing the homologous peptide was about 300 times higher than that of antibodies recognizing native S protein. This suggests a limited cross-reactivity between the synthetic peptide and the cognate sequence on S protein. Unlike the selected peptides from the preS region, the peptide S(296-322) probably does not contain dominant T cell epitopes which are located at a site non-adjacent in the primary sequence (Celis et al., 1988). Additional research is required for designing synthetic analogs mimicking B and T cell epitopes on S-protein.

The specificity of antisera to distinct regions of the HBV env protein can be assessed by immunoblotting. Antibodies elicited by native HBV recognize predominantly M- (GP33 and GP36) and L- (P39 and GP42) proteins, while the most abundant S-protein components P25 and GP29 (lane A, Fig. 8) are barely detectable or undetectable (lane B). This is in agreement with the limited immunological cross-reactivity between native and denatured S-protein (see above). Antisera to peptides derived from preS1 and preS2 sequences recognize L-protein (lane C) and L- and M-protein (lane D),

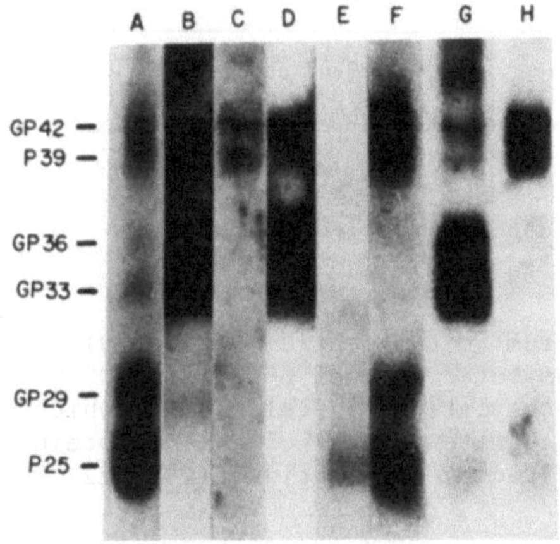

Fig. 8. Separation of HBV env components by polyacrylamide gel electrophoresis and their detection by anti-HBV and by distinct anti-peptide antisera. (A) HBV env components stained by silver; B-H: immunoblots; (B) by anti-HBV; (C) by anti-preS(21-47); (D) by anti-preS(120-145); (E) by anti-S(309-329); (F) by anti-S(296-322); (G) by anti-hybrid-preS(120-153)-S(296-392); (H) by anti-preS(12-47)-S(296-322).

Table 1. Antigenicity and immunogenicity of selected peptides derived from the S protein sequence and of hybrid preS-S-peptides

PEPTIDE IMMUNOGEN	ANTIGENICITY Recognition by rabbit anti-HBs; endpoint dilution	IMMUNOGENICITY Dilution endpoints in double-antibody RIA with: homologous peptide	HBsAg	AUSAB RIA Endpoint dilution
1. cyclic S(313-321)	Not done	1/2,500+	1/500+	N.D.*
2. cyclic S(298-321)	1/5,000		1/2,400+	1/8
3. cyclic S(303-321)	1/5,000		1/1,600+	1/3
4. cyclic S(308-321)	1/5,000	1/9,600++	1/6,400++	1/10
5. cyclic S(303-321)		1/12,800++	1/2,400++	N.D.*
6. cyclic S(298-321)		1/6,400++	1/1,600++	N.D.*
7. S(296-322) Cys→Ala at 298,311-313	1/240	1/3,200++	1/1,600++	N.D.*
8. cyclic S(296-322)**	1/2,000	1/36,000++	1/7,800++	1/120
9. preS(120-153)-S(296-322)**		1/140,000++	1/4600++	1/30
10. preS(12-47)-S(296-322)**		1/22,500++	1/3200++	1/50
11. preS(120-153)-S(308-321)		<1/40++	1/100++	N.D.*
12. preS(12-47)-S(308-321)		1/40++	1/90++	N.D.*
13. S(296-322)** dimeric or polymeric	1/100 (1/800 with human anti-HBs)	1/38,000++	1/12,800++	1/5,000

+Immunogen = peptide linked to NH_2-dextran, adjuvant = alum

++Immunogen = free peptide, complete + incomplete Freund's adjuvant

*N.D. = not detectable

**With a Gly-Gly-Tyr extension at the C-terminus

respectively. Antiserum to the peptide S(309-329) recognizes predominantly P25 but not the glycosylated form of S-protein (GP29) (lane E). This can be explained by N-glycosylation at residue 320, which may prevent the access of antibodies to the cognate segment of the S-protein sequence. On the other hand, antibodies elicited by the peptide S(296-322) recognized both P25 and GP29 (lane F).

In attempts to increase the immunogenicity of S(296-322), this peptide was covalently linked to peptides from the preS1 and the preS2 region, known to contain T cell epitopes, preS(12-47) and preS(120-153), respectively. The hybrid peptides elicited lower levels of antibodies with anti-S specificity than unconjugated S(296-322) (lanes G, H; Table 1, lines 8-10). The antisera contained predominantly antibodies with anti-preS2 and anti-preS1 specificities.

It is of interest to note that antibodies to peptides derived from S-protein recognized both native (for example, column group 5, Fig. 7) and denatured (lanes E, F, Fig. 8) S-protein. In this respect, the repertoire of antibodies elicited by the peptide differs noticeably from the repertoire of antibodies elicited by the native HBV env protein in which antibodies recognizing denatured S-protein represent a minor, frequently undetectable specificity (compare the black column in group 1, Fig. 7 with lane B, Fig. 8).

ATTEMPTS TO MIMIC DISCONTINUOUS EPITOPES ON S-PROTEIN

Research with peptides from the preS region of the HBV env protein contributed significantly to elucidating structure-function relationships in HBV env proteins and demonstrated the potential of selected synthetic peptides from this region as components of hepatitis B vaccines (Neurath and Kent, 1985, 1988; Neurath, in press; Neurath et al., 1987a). To keep the progress in studies on HBV env proteins in balance, it appears at this time necessary to continue efforts in designing synthetic analogs mimicking B cell epitopes, and possibly other functionally important regions of S-protein. Such studies are difficult, since very little is known about the structure of native S-protein and no data concerning the correct disulfide pairing in this protein are available.

In the first series of experiments peptides 1-4 listed in Table 1 were tested for antigenicity and immunogenicity. The peptides had cysteine residues at positions 313 and 321. Other cysteine residues from the sequence of the S-protein (subtype adw2) were replaced by alanines (for the sequence of S-protein see Table 2). All these peptides had homoserine lactone at the C termini preceded by two glycine residues. The peptides were oxidized with potassium ferricyanide to form disulfide bonds leading to the generation of a cyclic peptide. The peptides were linked to an amino derivative of dextran 2000 prepared as described by Inman (1975) using glutaraldehyde as cross-linking agent. The resulting complexes were adsorbed on alum and used for immunization. The antibody responses to native S-protein (HBsAg) were low compared with antipeptide responses (Table 1). After five doses of these immunogens, the rabbits were hyperimmunized with free peptides in combination with complete Freund's adjuvant. The hyperimmunization resulted in an increased antibody response to S-protein as measured by double antibody RIA tests which detect predominantly low avidity IgG antibodies (Neurath et al., 1982). Very few if any antibodies were detected by a commercial test for measuring anti-S-protein antibodies (AUSAB RIA). A longer peptide (No. 7, Table 1) was also unsatisfactory. Somewhat better results were obtained with peptide No. 8 having cysteine

Table 2. Amino acid sequence of peptides derived from the S(279-335) region of HBsAg S-protein (subtype adw2)

Peptide 1 S(279-294)	P V <u>C</u> P L I P G S T T T S T G P
Peptide 2 S(295-310)	<u>C</u> K T <u>C</u> T T P A Q G N S M F P S
Peptide 3 S(311-320)	<u>C</u> <u>C</u> <u>C</u> T K P T D G N
Peptide 4 S(321-335)	<u>C</u> T <u>C</u> I P I P S S W A F A K Y

residues at positions 311 and 312 replaced by aminobutyric acid and having tyrosine at the C terminus preceded by a Gly-Gly-spacer (Zheng et al., 1986). In attempts to increase the immunogenicity of the latter peptide, it was covalently linked to peptides derived form the preS region, preS(120-153) and preS(12-47), which have both B and T cell epitopes. The linking of the peptides was accomplished by mixing the reduced peptide S(296-322) with the respective peptides from the preS sequence both having Gly-Gly-Cys residues at the C-terminus. These C-terminal Cys residues had been derivatized by pyridine-2-thione by treatment with 2-pyridyl disulfide. The hybrid peptides did not elicit a higher anti-S response in comparison with cyclic S(296-322) (compare lines 8-10 in Table 1). Similarly, linking of a shorter peptide S(308-321) to peptides from the preS sequence resulted in diminished rather than increased antibody responses to the S region (compare lines 4, 11, and 12 in Table 1). The most promising immunogen in this series resulted from dimerization (polymerization?) induced by mixing reduced S(296-322) with an equivalent amount of the same peptide the sulfhydryl groups of which were derivatized with pyridine-2-thione (line 13).

The antigenicity and immunogenicity of S-protein depends on the maintenance of up to five intramolecular disulfide bonds (Fig. 2), the pairing of which is unknown. Since B cell epitopes are likely to be located around disulfide bonds, attempts were made to delineate the epitopes by studying the properties of disulfide-containing dipeptides derived from nonadjacent portions of the primary sequence (Atassi, 1978). Peptides 1-4 (Table 2) were synthesized. The properties of individual peptides as well as of di- and tripeptides generated by disulfide bond formation between the individual peptides were studied. Three of the peptides contain two or more cysteine residues. Therefore, di- or tripeptides generated by disulfide bond formation cannot represent uniform products and remain undefined. However, based on results obtained, additional peptides containing a single cysteine residue can be prepared, so that it will be possible to study the properties of better defined di- or tripeptides. Disulfide bonds between the peptides were generated using the same approach described for the preparation of hybrid preS-S peptides (Table 1). The single peptides as well as all possible combinations of dipeptides were poorly recognized by anti-HBs antibodies in double antibody RIA tests (Table 3). A product generated by disulfide bond formation between peptides 2, 3, and 4 was optimally recognized by anti-HBs (Table 3). The same tripeptide elicited the highest level of anti-S-protein antibodies measured by double antibody RIA (Table 4) but only a low level of antibodies measurable by the AUSAB test (endpoint = 1:10), and was thus inferior as compared with some of the peptides listed in Table 2.

In a complementary set of experiments, peptides derived from the sequence S(295-325) (Table 2) were synthesized, each peptide having five out of seven cysteine residues replaced by serines at distinct positions. Each of the peptides, having two cysteine residues, was oxidized by ferricyanide to form intramolecular disulfide bonds, and tested for immunogenicity using consecutively free peptides and peptides linked to BSA by glutaraldehyde. The results (Fig. 8) demonstrate the importance of disulfide bonds between particular cysteine residues within the S-protein sequence. The best results were obtained with peptides expected to have disulfide bonds between residues 295-312, 298-312, and 313-321. Most of the antisera with relatively high anti-S levels in double antibody RIA tests had low antibody dilution endpoints when the AUSAB test was used to screen for anti-HBs (endpoint dilutions about 1:10), except one of the two antisera against peptide with disulfide bonds in position S(298-312) with an endpoint titer of 1:5000.

Although none of the S-protein analogs described above compare in antigenicity and immunogenicity with peptides derived from the preS region,

Table 3. Antigenicity of peptides 1, 2, 3, and 4 and of di- and tripeptides generated by disulfide bond formation between the distinct peptides

Peptide	Dilution endpoints with: human anti-HBs	rabbit anti-HBs
1	<1:40	<1:40
2	<1:40	<1:40
3	<1:40	<1:40
4	<1:40	<1:40
1+2	<1:40	<1:40
1+3	<1:40	1:40
1+4	<1:40	1:80
2+3	<1:40	<1:40
2+4	<1:40	<1:40
3+4	<1:40	1:100
1+2+3	1:500	<1:40
1+3+4	1:500	1:100
2+3+4	1:2500	1:500

Table 4. Elicitation of anti-HBs (anti-S-protein) by tripeptides generated by disulfide bond formation between peptides 1, 2, 3, and 4

Peptide	Dilution endpoints of anti-HBs antibodies measured by double-antibody RIA
1+2+3	1:300*
1+3+4	1:335*
2+3+4	1:7,200*

*Geometric mean dilution endpoints corresponding to 2 rabbit antisera per peptide.

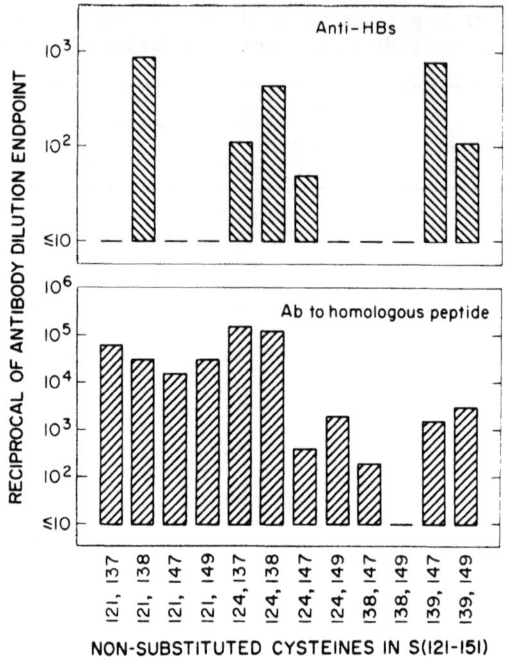

Fig. 9. Immunogenicity of a series of peptides corresponding to the S-protein sequence S(295-325) having 5 out of 7 cysteine residues replaced by serines. Free peptides and subsequently peptides linked to bovine serum albumin were used to immunize rabbits. The antisera were tested for antibodies recognizing the homologous peptide (bottom) and HBsAg (top). To conform with numbering of amino acid residues in the HBV env protein, add 174 to numbers on the abscissa.

it is expected that further modifications of selected S-peptide immunogens and their combination with peptides corresponding to S-protein T cell epitopes will lead to the design of improved analogs and to better understanding of structure-function relationships in the HBV env protein.

ACKNOWLEDGMENTS

Help in preparation of this manuscript was provided by L. Greene, T. Huima, and M. Zucker. Synthetic peptides for the described studies were provided by Dr. S.B.H. Kent, California Institute of Technology; Neosystem Laboratories, Strasbourg, France; and Cambridge Research Biochemicals Ltd., Cambridge, England. The studies were supported in part by a grant from the American Cancer Society.

REFERENCES

Atassi, M.Z., Precise determination of the entire antigenic structure of lysozyme: Molecular features of protein antigenic structures and potential of surface-simulation synthesis - A powerful new concept for protein binding sites, Immunochemistry, 15:909 (1978).
Celis, E., Ou, D., and Otvos, L.Jr., Recognition of hepatitis B surface

antigen by human T lymphocytes: Proliferative and cytotoxic responses to a major antigenic determinant defined by synthetic peptides, J. Immunol., 140:1808 (1988).

Guerrero, E., Gavilanes, F., and Peterson, D.L., Model for the protein arrangement in HBsAg particles based on physical and chemical studies, in: "Viral hepatitis and Liver Disease," A.J. Zuckerman, ed., Alan R. Liss, New York (1988).

Inman, J.K., Thymus-independent antigens: The preparation of covalent, hapten-Ficoll conjugates, J. Immunol., 114:704 (1975).

Lerner, R.A., Green, N., Alexander, H., Liu, F.-T., Sutcliffe, J.G., and Shinnick, T.M., Chemically synthesized peptides predicted from the nucleotide sequence of the hepatitis B virus genome elicit antibodies reactive with the native envelope protein of Dane particles, Proc. Natl. Acad. Sci. USA, 78:3403 (1981).

Macrae, D.R., Persing, D.H., Varmus, H.E., and Ganem, D., Myristylation of the pre S protein of DHBV is required for viral infectivity, in: "Abstracts of Papers presented at the 1988 meeting on Molecular Biology of Hepatitis B Viruses," D. Ganem and W. Gerlich, eds., University of California, San Diego (1988).

Milich, D.R., Genetic and molecular basis for T- and B-cell recognition of hepatitis B viral antigens, Immunol. Rev., 99:71 (1987).

Milich, D.R., McLachlan, A., Moriarty, A., and Thornton, G.B., A single 10-residue pre-S(1) peptide can prime T cell help for antibody production to multiple epitopes within the pre-S(1), pre-S(2), and S regions of HBsAg, J. Immunol., 138:4457 (1987)

Neurath, A.R., Chemical synthesis of hepatitis B vaccines, in: "Immunology of Prophylactic Immunization," A.J. Zuckerman, ed., MTP Press Limited, Lancaster (in press).

Neurath, A.R., and Kent, S.B.H., Antigenic structure of human hepatitis viruses, in: "Immunochemistry of Viruses," M.H.V. Van Regenmortel and A.R. Neurath, eds., Elsevier, Amsterdam (1985).

Neurath, A.R., and Kent, S.B.H., The pre-S region of hepadnavirus envelope proteins, in: "Advances in Virus Research, Vol. 34," K. Maramorosch, F.A. Murphy, and A.J. Shatkin, eds., Academic Press, San Diego (1988).

Neurath, A.R., and Strick, N., Antibodies as immunological probes for studying the denaturation of HBsAg, J. Med. Virol., 6:309 (1980).

Neurath, A.R., Strick, N., and Huang, C.A., Properties of delipidated hepatitis B surface antigen (HBsAg) and preparation of its proteolytic cleavage fragments carrying HBsAg-specific antigenic determinants, Intervirology, 10:265 (1978).

Neurath, A.R., Strick, N., and Oleszko, W.R., Localization of a hepatitis B surface antigen determinant deduced from results of chemical modifications, J. Virol. Meth., 3:115 (1981).

Neurath, A.R., Kent, S.B.H., and Strick, N., Specificity of antibodies elicited by a synthetic peptide having a sequence in common with a fragment of a virus protein, the hepatitis B surface antigen, Proc. Natl. Acad. Sci. USA, 79:7871 (1982).

Neurath, A.R., Kent, S.B.H., Parker, K., Prince, A.M., Strick, N., Brotman, B., and Sproul, P., Antibodies to a synthetic peptide from the preS 120-145 region of the hepatitis B virus envelope are virus-neutralizing, Vaccine, 4:35 (1986a).

Neurath, A.R., Kent, S.B.H., Strick, N., and Parker, K., Identification and chemical synthesis of a host cell receptor binding site on hepatitis B virus, Cell, 46:429 (1986b).

Neurath, A.R., Jameson, B.A., and Huima, T., Hepatitis B virus proteins eliciting protective immunity, Microbiol. Sci., 4:45 (1987a).

Neurath, A.R., Kent, S.B.H., Adamowicz, P, Riottot, M.M., Price, P., Strick, N., Parker, K., Petit, M.A., Budkowska, A., Girard, M., and Pillot, J., Immunological cross-reactivity between preS2 sequences of the hepatitis B virus envelope proteins corresponding to serological subtypes adw2 and ayw, Mol. Immunol., 24:561 (1987b).

Neurath, A.R., Kent, S.B.H., Strick, N., Parker, K., Courouce, A.-M., Riottot, M.M., Petit, M.A., Budkowska, A., Girard,M., and Pillot, J., Antibodies to synthetic peptides from the pre-S1 and pre-S2 regions of one subtype of the hepatitis B virus (HBV) envelope protein recognize all HBV subtypes, Mol. Immunol., 24:975 (1987c).

Neurath, A.R., Kent, S.B.H., Strick, N., and Parker, K., Delineation of contiguous determinants essential for biological functions of the pre-S sequence of the hepatitis B virus envelope protein: Its antigenicity, immunogenicity and cell-receptor recognition, Ann. Inst. Pasteur/Virol., 139:13 (1988).

Okamoto, H., Imai, M., Shimozaki, M., Hoshi, Y., Iizuka, H., Gotanda, T., Tsuda, F., Miyakawa, Y., and Mayumi, M., Nucleotide sequence of a cloned hepatitis B virus genome, subtype ayr: Comparison with genomes of the other three subtypes, J. Gen. Virol., 67:2305 (1986).

Persing, D.H., Varmus, H.E., Ganem, D., The preS1 protein of hepatitis B virus is acylated at its amino terminus with myristic acid, J. Virol., 61:1672 (1987).

Peterson, D.L., Paul, D.A., Lam, J., Tribby, I.I.E., and Achord, D.T., Antigenic structure of hepatitis B surface antigen: Identification of the "d" subtype determinant by chemical modification and use of monoclonal antibodies, J. Immunol., 132:920 (1984).

Price, P.M., Mohamad, A., Zelent, A., Neurath, A.R., and Acs, G., Translational selection in the expression of the hepatitis B virus envelope proteins, DNA, 7:417 (1988).

Pugh, J.C., Yaginuma, K., Koike, K., and Summers, J., Duck hepatitis B virus (DHBV) particles produced by transient expression of DHBV DNA in a human hepatoma cell line are infectious in vitro, J. Virol., 62:3513 (1988).

Seeger, C., Marion, P.L., Ganem, D., and Varmus, H.E., In vitro recombinants of ground squirrel and woodchuck hepatitis viral DNAs produce infectious virus in squirrels, J. Virol., 61:3241 (1987).

Shih, J.W.-K., New technologies, in: "Hepatitis B," R.J. Gerety, ed., Academic Press Inc., Orlando (1985).

Swenson, P.D., Peterson, D.L., and Hu, P.-S., Antigenic analysis of HBsAg with monoclonal antibodies specific for S protein and pre-S2 sequences, in: "Viral Hepatitis and Liver Disease," A.J. Zuckerman, ed., Alan R. Liss, New York (1988).

Wu, T.T., and Kabat, E.A., An analysis of the sequences of the variable regions of Bence Jones proteins and myeloma light chains and their implications for antibody complementarity, J. Exp. Med., 132:211 (1970).

Zheng, J. Hu, P.S., Xu, L.G., Liu, Z.P., and Huang, W.T., Serotypic antigenic structure and immunological activity of the synthetic peptide vaccine for hepatitis-B, Biopolymers, 25:S201 (1986).

ANTIGENIC VARIATION DURING PERSISTENT LENTIVIRUS INFECTIONS AND ITS SIMPLICATIONS FOR VACCINE DEVELOPMENT

R. Montelaro, J. Ball, P. Rwambo and C. Issel

Louisiana State University
Department of Biochemistry, 322 Choppin
Baton Rouge, LA 70803

INTRODUCTION

During the past decade the role of genetic variation during RNA virus replication has gained increased attention as a strategy for viral persistence. Through mechanisms such as genomic point mutations, recombination, and reassortment of multipartite genomes, RNA viruses are capable of modifying their genetic constitution, often resulting in altered biologic (host cell specificity, pathogenicity) or immunologic properties (Holland et al., 1982). The propensity for variation among animal RNA viruses by mutation is usually attributed to the relatively high infidelity of virus-encoded RNA polymerases that replicate the viral genome. Compared to DNA polymerases that display error rates of 10^{-8} to 10^{-11} nucleotide substitutions per site per year, viral RNA polymerases reveal mutation rates of about 10^{-4} to 10^{-5}. Moreover, the genetic variations introduced during viral RNA replication in the cell cytoplasm tend to accumulate, as RNA polymerases lack the editing and repair functions typically associated with nuclear DNA polymerases. Thus, RNA viruses appear to constitute a highly variable system of genetic information in which nature is constantly experimenting with novel genetic constitutions.

Until recently, the retroviruses were, in general, believed to be an exception to the variability observed in other RNA viruses. Retroviruses, contain an RNA genome that is replicated in cells via DNA proviral intermediates synthesized by the virus encoded reverse transcriptase. The involvement of a nuclear DNA intermediate replicated by host cell DNA polymerases suggested a genetic stability for retroviruses that was relatively unique among RNA viruses.
The view of retroviruses as genetically constant was based primarily on studies with members of the oncovirus subfamily, including avian, murine, feline and bovine leukemia and sarcoma viruses. Viral persistence in these viruses was

usually related to integrated provirus in host cell chromosomes and, in certain cases, the disruption of the host immune system. More recent studies with the lentivirus subfamily of retroviruses, however, have revealed completely new aspects of retrovirus persistence mechanisms that allow the virus to persist even in the face of fully competent immune responses. Genomic mutations during lentivirus replication have been estimated to be about 10^{-1} to 10^{-3} nucleotide substitutions per site per year, representing one error per 2,000 nucleotides copied (Hahn et al. 1986; Preston et al., 1988; Roberts et al., 1988). The mutation rate observed for lentivirus reverse transcription is 100 to 1000-fold greater than that observed with other viral RNA polymerases and constitutes an infidelity that is about 10-fold greater than that observed with various oncovirus reverse transcriptases (Roberts et al., 1988).

The prevalence of antigenic variation during persistent lentivirus infections constitutes a major challenge in current infectious disease research. Thus, the purpose of this review is to summarize the nature and role of antigenic variation during animal lentivirus infections and to discuss the implications of this variation on vaccine development.

LENTIVIRUSES - GENERAL PROPERTIES

The lentivirus subfamily of retroviruses currently includes visna/maedi virus of sheep, the caprine arthritis-encephalitis virus (CAEV) of goats, equine infectious anemia virus (EIAV) of horses, and the immunodeficiency viruses of man (HIV), nonhuman primates (SIV), cats (FIV), and cows (BIV). This taxonomic group is based upon similarities in virion morphology, protein serologic properties, genomic organization and nucleotide sequences, host cell specificities, and patterns of pathogenesis.

A general model for a lentivirus virion is presented in Figure 1. Overall the virus has a diameter of approximately 100 nm with a lipid envelope enclosing an oblong core containing two copies of the positive single stranded RNA genome. The viral genome (8,500-10,000 bases) containing three major genes (5'-gag-pol-env-3') encodes the structural proteins of the virus (Stephens et al., 1986; Rushlow et al., 1986). The gag gene encodes the 3-4 internal proteins of the virus including the matrix (MA) protein immediately beneath the lipid bilayer, the major core shell protein (CA), the basic ribonucleoprotein (NC) associated with the RNA genome, and in certain viruses additional low molecular weight proteins of unknown location or function. Each virus particle contains several thousand copies of each of the gag proteins (Parekh et al., 1980). The pol gene encodes the reverse transcriptase (RT) enzyme of which 5-10 copies are associated with the viral RNA in each virion. The env gene encodes the surface proteins of the virus including the highly glycosylated surface glycoprotein (SU) and the sparsely glycosylated hydrophobic transmembrane protein (TM). There are several hundred copies of each env protein per virus particle, perhaps arranged in a symmetrical pattern in the lipid bilayer as trimers of the SU and TM glycoproteins (Ozel, et al., 1988). In addition to the three major genes, lentiviruses also encode several nonstructural proteins via complex splicing mechanisms of gene sequences located between the pol and env genes and at the 3'-end of the env gene. These nonstructural proteins apparently control viral gene expression through complex patterns of positive and negative regulatory interactions.

The predominant humoral immune response during persistent lentiviral infections is directed against the envelope glycoproteins, although minor antibody reactivity can be observed against the core proteins, RT, and certain regulatory proteins. Current diagnostic assays are usually based on detecting antibody to the major core protein, while subunit vaccine trials are primarily focused on the envelope glycoproteins as immunogens.

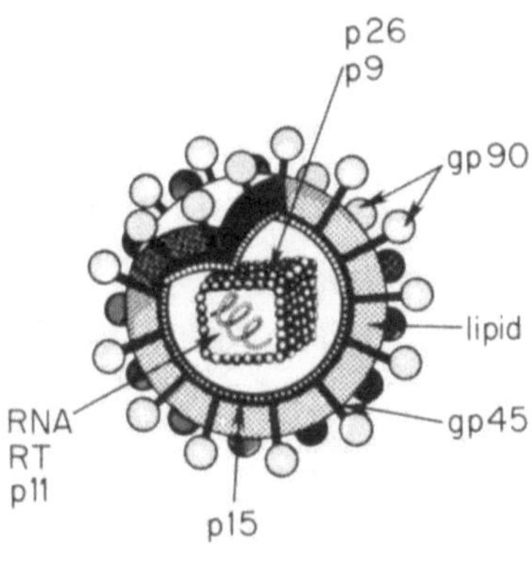

Figure 1. Schematic model of a lentivirus virion with the EIAV protein nomenclature (Parekh et al., 1980) indicated for the major retrovirus structural components (SU surface protein, TM-transmembrane protein, MA-matrix protein, CA-capsid protein, NC-nucleocapsid protein, RT-reverse transcriptase) (Leis et al., 1988).

LENTIVIRUS - REPLICATION AND PERSISTENCE MECHANISMS

A schematic representation of a lentivirus replication cycle is summarized in Figure 2. Lentivirus typically establish cytopathic infections in cells of the immune system.

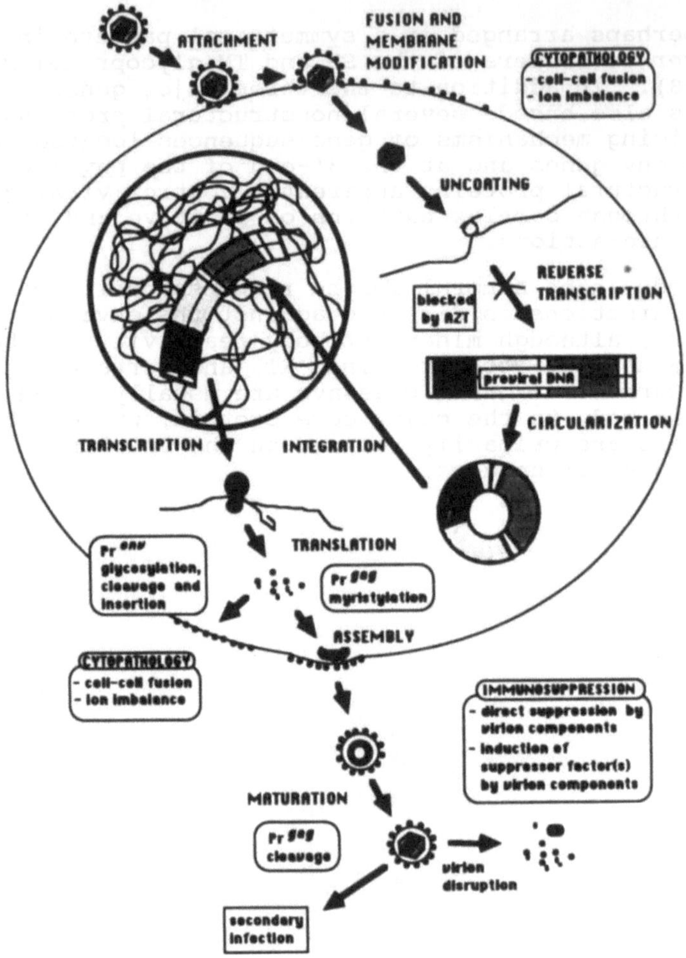

Figure 2. Replication cycle of a lentivirus indicating critical virus-cell interactions (Drawing courtesy of R. Garry, Tulane University Medical Center).

All lentivirus infections involve cells of the monocyte-macrophage lineage as targets, and certain lentivirus infections can also involve T-lymphocytes as host cells. Although lentiviruses induce lytic infections in cells, they are capable of establishing a persistent infection and chronic disease in infected animals, frequently in the presence of competent host immune defenses.

The life cycle of a lentivirus provides certain obvious mechanisms for eluding the immune system of the host. The viral RNA is transcribed by RT into proviral DNA which integrates into the host cell genome and can be eliminated only by death of the cell. The integrated provirus may be expressed immediately or remain latent for years. Thus, infection of replicating bone marrow cells or immune progenitor cells can result in a population of inapparently infected mature immune cells in which viral replication can be induced at anytime by unknown factors. Once virus production is induced, the spread of virus is mediated by free virus or by cell to cell transfer facilitated by the cell membrane fusion properties of the TM glycoprotein. This latter mechanism of transmission has the potential to evade host immune systems.

However, proviral integration in a productive cytolytic infection provides only temporary survival of the viral genome; the virus must replicate and disseminate to other target cells to persist for the life of the animal. Since budding of free virus is accessible to antibody, lentiviruses have evolved several strategies to escape host immunosurvellience systems.

First, epitopes responsible for virus neutralization may not be readily recognized by the host immune system. Hence, the infected animal fails to produce effective neutralizing antibodies, while producing a multitude of non-neutralizing antibodies against several viral proteins. The lack of neutralizing antibodies allows the virus to persist and replicate in the animal, and the virus-binding non-neutralizing antibodies may actually enhance targeting of infectious virus to uninfected immune cells. For example, goats infected naturally or experimentally with CAEV become persistently infected and produce antibodies that bind to all viral proteins, but these antibodies are ineffective in neutralizing the virus (Narayan et al., 1984). In fact, the lack of neutralizing antibodies in infected goats is difficult to overcome, even when strong adjuvants are used together with purified or disrupted CAEV preparations. Other lentivirus infections, including HIV, EIAV, and visna virus, are also characterized by relatively low levels of neutralizing antibodies. It has been postulated that the extensive glycosylation of the envelope glycoprotein may partially or completely sequester neutralizing epitopes from immune surveillence systems.

Second, certain animals infected with lentiviruses produce neutralizing antibodies, but these antibodies display relatively poor neutralization activities. Because these neutralizing antibodies are of relatively low titer and require prolonged incubation with virus to accomplish inactivation, the infecting virus can continue to spread from cell to cell before neutralization occurs. Neutralizing antibody titers are especially weak during the early stages of infection by visna virus and EIAV, but become somewhat stronger during later stages of infection (Narayan et al., 1978; Rwambo et al., 1989).

Third, lentiviruses undergo relatively rapid genomic variations during replication, some of which can alter the structure and antigenic properties of the virus and allow it to circumvent established immune responses. The evolution of antigenic variants can confound immunosurvellience and propagate virus infection even in the presence of fully competent host immune systems. The propensity for antigenic variation during persistent infections has been documented for visna virus, EIAV, HIV, and CAEV using neutralization assays and monoclonal antibody binding assays ((Narayan et al., 1978; Montelaro et al., 1984; Payne et al., 1987; McGuire et al., 1988). Lentivirus isolates recovered from different infected animals may differ in nucleotide and deduced amino acid sequences by as much as 15-20%. In addition, these viral isolates can differ significantly in their antigenic (and biologic) properties. The observed variations are located predominantly in the env gene, with only minor variations observed in the gag and pol sequences. However, it is clear that extensive variation in the envelope glycoproteins is not required to alter antigenic properties. Antigenic variants of lentiviruses recovered from a single infected animal over a period of months or years reveal variations of only 1-2%, representing changes of only 4-5 amino acids in the SU and TM proteins of the virus. In fact, neutralization distinct variants may differ by a single amino acid in the envelope glycoprotein. To date, no two independent isolates of a particular lentivirus appear to be identical; each isolate may be distinguished by genetic sequence or biologic and antigenic properties. Even individual lentivirus isolates from infected animals contain a remarkably large mixture of genotypically, and presumably antigenically, distinct variants that evolve in parallel and coexist during a persistent infection (Saag et al., 1988).

Fourth, extensive replication of lentiviruses can result in immunosuppression, mediated either by specific proteins of the virus or by depletion of essential immune cell populations. Immunosuppression is a hallmark of critical stages of HIV and SIV infections, but is also present to various degrees in other lentivirus infections.

Taken together, the above battery of escape mechanisms constitute a formidable challenge to natural protective systems of the host and to the development of effective vaccines for lentivirus infections. Relative to vaccine development, the extensive propensity for variation among lentiviruses presents the greatest obstacle to producing a lasting broadly protective vaccine. Thus, the nature and role of variation in lentiviral persistence and in determining the immunologic properties of viral proteins has been the subject of intense investigation during the past several years. The results of these studies with HIV will be summarized elsewhere in these proceedings. However, studies of HIV antigens and the significance of variations are limited by the lack of an appropriate experimental infection system in animals. Thus, alternate lentivirus systems provide important models for characterizing lentiviral antigens, for assessing the significance of antigenic variation, and for evaluating different vaccine strategies. Among the available animal lentivirus systems, EIAV has been the most extensively characterized. The remainder of this presentation summarizes the salient features of EIAV infections and immunobiology and reviews their implication for lentivirus vaccine development.

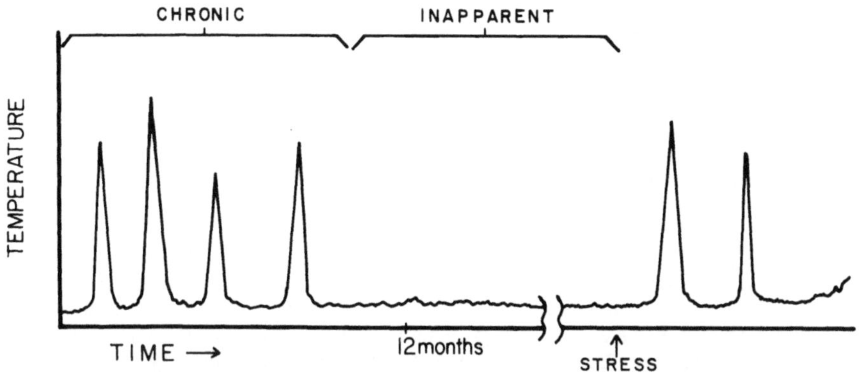

Figure 3. Schematic representation of the clinical response of a horse after infection by EIAV. Clinical episodes of EIA are represented by peaks of fever above normal animal temperatures. The chronic form of disease is characterized by recurrent clinical episodes which occur at unpredictable intervals during the first year after infection. At this time animals usually no longer display any obvious clinical signs, although recrudescence of disease can be induced years later by stress or the administration of immunosuppressants. The clinical response of animals is variable and depends, in part, on host resistance factors, virus virulence factors and environmental factors (eg. weather, work, hauling, etc.).

PROPERTIES OF PERSISTENT EIAV INFECTIONS

Equine infectious anemia (EIA) is one of the first diseases to be assigned a viral ("filterable") etiology in 1904 (Vallee and Carre, 1904) and offers a uniquely dynamic lentivirus infection in which to examine the nature and role of antigenic variation during a persistent infection. Moreover, EIA is the only lentivirus system in which the infected animal routinely brings viral replication under immunologic control, despite rapid and varied changes in the surface antigens of the infecting virus. Thus, an elucidation of the antigenicity of EIAV antigens at the molecular level and their interactions with host immune systems during the course of a persistent infection can provide a critical foundation for designing an effective lentivirus vaccine.

The clinical responses of horses following natural or experimental infections by EIAV can be divided into characteristic stages (Figure 3). Acute EIA, characterized by fever and hemorrhages, is typically associated with the first exposure to virus and appears to correlate with virus replication in and destruction of macrophages. The more classical symptoms of EIA (weight loss, hemolytic anemia, edema, leukopenia, etc.) are observed later during recurring cycles of viremia and illness which appear at irregular intervals separated by weeks or months. This stage of disease is called chronic EIA, and is the hallmark of most EIAV infections. The frequency and severity of these disease episodes decline with time, and the chronic stage of disease usually ends within the first year after infection and an average of 6-8 clinical episodes. The exact progression of disease, however, depends on the infecting strain of EIAV, the individual physical characteristics of the animal, and its environment.

Naturally infected horses seldom die from EIAV infection. Following the chronic stage of disease, most infected horses progress to an inapparent stage which lacks clinical symptoms or and detectable levels of plasma viremia. The animal continues to harbor EIAV, however, as evidenced by the fact that disease can be efficiently transmitted by the experimental transfer of whole blood (100 ml) from inapparently infected to naive horses. In addition, it has been demonstrated that recrudescence of chronic EIA can be induced in some inapparently infected horses by administration of immunosuppressants (eg., dexamethasone) or by stress, even in animals that have been free of febrile episodes for years (Kono et al., 1976). This ability to induce viral replication and clinical disease in inapparent carriers suggests an active immunologic management of viral replication that can be overcome by suppressing or stressing the immune system. The concept of active immunologic management is further supported by the recent observation that horses with combined immunodeficiency (lacking functional B and T lymphocytes and unable to manifest antigen-specific immune responses) are incapable of terminating viremia after experimental EIAV infections, while normal horses infected in parallel effectively cleared detectable virus from their plasma (Perryman et al., 1988).

EIAV ANTIGENS

A variety of immunoassays indicate that the predominant antibody response in EIAV - infected horses is directed against the envelope glycoproteins, gp90 and gp45, and that only minor antibody populations are produced against the core proteins (p26, p15, p11, and p9), and against nonstructural, putative regulatory proteins (Montelaro et al., 1984a and unpublished observations). The nature and kinetics of antigen-specific immune responses are discussed below. In this section, details of the molecular nature of the envelope antigens are described.

The relative contributions of the peptide and oligosaccharide moieties of EIAV envelope glycoproteins have been assayed by measuring the reactivity of immune sera from infected horses against native EIAV glycoproteins and against purified gp90 and gp45 that have been deglycosylated with specific endoglycosidases (Montelaro et al., 1984b and unpublished observations). The results of these studies indicate that the antibody produced during persistent infection is directed predominantly against the peptide portions of the EIAV glycoproteins; only minor antibody reactivity is detectable with the oligosaccharide components of the virus. Moreover, the antibodies appear to be specific for continuous linear rather than conformational glycoprotein epitopes, as the levels of antibody binding are not significantly affected by treatment of the EIAV glycoproteins with detergents, salts, or reducing agents to modify protein conformation.

A panel of monoclonal antibodies has been produced against the prototype strain of EIAV and used to identify specific epitopes of the EIAV glycoproteins (Hussain et al., 1987 and 1988). The results of these studies identify six antibody binding sites (90A-F) which by competitive antibody binding assays appear to define four distinct epitopes (designated 90A, 90B/F, 90C/D and E. Two of these epitopes, 90C/D and E, are involved in virus neutralization by monoclonal antibodies. Two non-overlapping and non-neutralizing epitopes (45A and 45B) are recognized by a limited panel of monoclonal antibodies specific for EIAV gp45. As observed with the antibodies contained in immune sera from experimentally infected horses, the binding of glycoprotein-specific monoclonal antibodies does not appear to depend significantly on protein conformation, as binding efficiencies are not altered significantly by changing detergent, salt, or reducing conditions.

Competitive binding studies between the monoclonal antibodies and polyclonal equine immune sera demonstrate that all of the antigenic sites defined by monoclonal antibodies are also recognized by equine antibodies produced during persistent infections. However, it appears that the equine antibodies recognize some additional glycoprotein sites not identified by our panel of monoclonal antibodies (Payne et al., 1989).

To further localize the antibody binding sites of EIAV gp90 and gp45, we have assayed the antigenicity of selected segments of the EIAV glycoproteins produced by recombinant expression systems in E. coli (Payne et al., 1989) and by synthetic peptide methodologies (Ball et al., 1989). A summary of epitope localizations is presented in Figure 4. The results of these studies indicate that epitope 90A is localized to the amino terminal of gp90, while epitopes 90B, 90C, and 90D, and 90F localize to the middle one-third of the glycoprotein; epitope 90E has yet to be localized to a specific peptide region. One of the two identified gp45 epitopes, 45A, was localized to the amino terminal end of gp45, while epitope 45B remains to be defined. These localizations by recombinant and synthetic peptide methodologies are in excellent agreement with the results of competitive ELISA assays that indicated a tight clustering of 90B, 90C, 90D and 90F monoclonal antibody binding sites (Hussain et al., 1988), in a pattern similar to that observed for HIV gp120 epitopes (Rusche et al., 1988; Palker et al., 1988; Takahashi et al., 1988). In addition, the 45A epitope is analogous to the HIV gp41 peptide that has been shown to be broadly reactive with AIDS patient sera, and, thus, of use as a diagnostic substrate (Gnann et al., 1988). Finally, in agreement with the data cited above, the reactivities of the monoclonal antibodies with recombinant and synthetic peptides lacking oligosaccharide confirm the peptide-specific nature of EIAV glycoprotein epitopes.

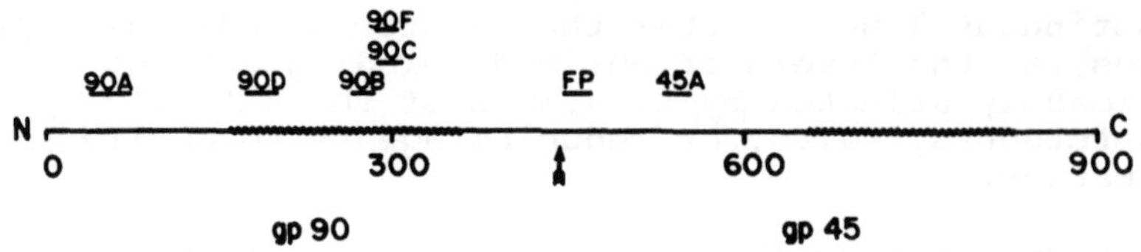

Figure 4. Schematic representation of EIAV env glycoprotein epitopes as defined by binding of monoclonal antibodies to specific recombinant and symthetic peptides (Payne et al., 1989; Ball et al., 1989). The solid line indicates conserved glycoprotein sequences and the wavy line represents variable amino acid sequences. The arrow locates the cleavage site between gp90 and gp45. Epitopes 90C and 90D are neutralizing, while the remaining epitopes are not involved in neutralization. The FP peptide constitutes the fusion peptide of EIAV which is reactive with sera from infected horses.

The recombinant and synthetic peptide antigens were also assayed for reactivity against a panel of sera from horses naturally infected with unknown strains of EIAV and from ponies experimentally infected with strains of EIAV derived from prototype virus (Payne et al., 1989; Ball et al., 1989). The observed reactivity patterns indicate that monoclonal antibody epitopes 90A and 45A are strongly recognized by all test sera, whether from experimentally or naturally infected animals. Also strongly recognized by sera from all infected animals is a recombinant peptide encompassing the carboxyl terminal one-third of gp90, to which no epitopes where defined by monoclonal antibodies. In contrast, the remaining epitopes were only recognized by certain test sera, with the highest frequency ofreactivity among experimentally infected animals and the lowest recognition among naturally infected animals. These variations in epitope reactivities appear to correlate with the "conserved" and "variable" glycoprotein gene regions previously deduced from nucleic acid sequences of related EIAV isolates (Payne et al., 1987), as discussed below.

The epitope localizations described here for EIAV constitute the most detailed dissection of a lentivirus glycoprotein currently available. This data provides important information and resources for analyzing the role of structural variations on EIAV antigenic properties and for analyzing in detail the nature of immune response to EIAV during the course of a persistent infection and eventual immunologic control.

VARIATION DURING PERSISTENT EIAV INFECTIONS

The unique periodic nature of chronic EIA can evidently be attributed to the sequential evolution of antigenic variants of EIAV that temporarily escape established humoral and cellular immune responses (Figure 5). The initial evidence for EIAV antigenic variation is based on comparative neutralization assays employing virus isolates and immune sera

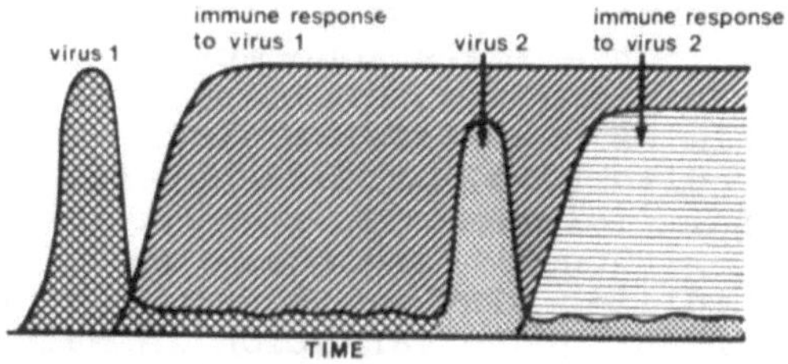

Figure 5. Model for the recurrent episodes of EIA during the chronic stage of disease. The infecting virus replicates in the horse producing a viremia and associated clinical signs of EIA. This viremia is brought under control by immune responses from the infected animal, and clinical manifestations of the disease disappear. However, mutations occurring during the replication of the infecting virus produce novel antigenic stains of virus which are able to circumvent established host immune responses, resulting in a subsequent viremia and recurrence of disease. This process can repeat itself.

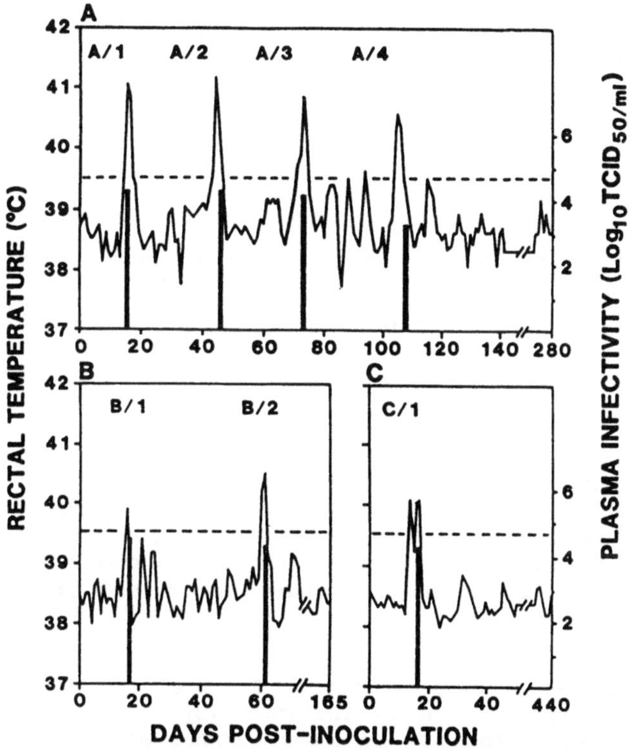

Figure 6. Clinical profile of three Shetland ponies infected with a standard inoculum of EIAV. All sustained rectal temperatures above 39.5°C (dashed line) are considered EIA-associated febrile episodes. The infectivity titers in tissue culture of plasma obtained during febrile episodes are shown as log TCID 50/ml (vertical bars). Viral isolates recovered during the respective febrile episodes are shown as A1, A2, etc. Panel A-pony 135F, panel B-pony 135P, panel C-pony 130. Adapted from Rwambo et al., 1989.

recovered from chronically infected horses during various stages of disease. For example Kono et al. (1973) demonstrated over 15 years ago that sera from infected horses were usually capable of neutralizing virus isolated prior to the time of serum collection, but had no effects on the infectivity of viral isolates recovered at intervals after the serum time point. Although the actual role of neutralizing antibodies in controlling EIAV replication is uncertain (see below), the role of antigenic variation in maintaining persistence appears to be well established. More recently a detailed characterization of EIAV antigenic variation has been made possible by the development of an animal model system in which Shetland ponies, inoculated with a standard strain of EIAV, reproducible develop chronic EIA during which EIAV isolates can be efficiently recovered during sequential cycles of disease (Orrego et al., 1982).

Examples of several experimental infections are presented in Figure 6. The clinical profiles demonstrate that although each animal responds uniquely to the standard inoculum, each animal brought virus replication and clinical symptoms under

Table 1. Neutralization of EIAV isolates by sera collected from an experimentally infected pony on different days post-inoculation.

	SERUM COLLECTION DAY[b]				
EIAV ISOLATE[a]	35	67	90	203	279
P3.3-1 (16)	1.5[c]	3.5	3.5	3.0	3.5
P3.3-2 (46)	1.5	3.0	3.0	2.0	2.0
P3.3-3 (75)	1.5	1.5	2.5	2.5	2.0
P3.3-4 (107)	1.0	1.0	1.5	2.5	2.5

[a]EIAV isolates recovered from pony 135F (c.f. Fig. 6) during sequential disease cycles. The numbers in parentheses indicate the time of virus isolation in days post inoculation.
[b]Days post inoculation on which serum was collected.
[c]Log_{10} neutralization index; values above 1.7 are considered positive.

control after 1-4 disease cycles. A panel of over 20 viral isolates from experimentally infected animals has been characterized by a variety of immunologic and biochemical procedures to ascertain the nature of antigenic variation in EIAV. The major conclusions of this survey are exemplified by the properties of a panel of viral isolates recovered during febrile episodes in experimentally infected ponies (Payne et al., 1987 a,b; Rwambo et al., 1989). Each virus isolate was initially examined in homologous neutralization assays using serum samples collected during sequential afebrile periods in the respective infected ponies. The results of these assays reveal several important properties of persistent EIAV infections. First, within each group of viral isolates from a single pony it can be demonstrated that each sequentially recovered isolate can only be inactivated by serum samples taken after the clinical episode of origin; none of the viral isolates can be neutralized by serum samples taken prior to the disease cycle of origin (Table 1). Second, cross neutralization assays with a panel of immune sera from experimentally and naturally infected horses indicate that each EIAV isolate is antigenically distinct, although limited cross neutralization can be observed (Rwambo et al., 1989).

The antigenic uniqueness of EIAV strains can also be demonstrated by assaying the reactivity of each virus in Western blots using a panel of monoclonal antibodies specific for gp90 and gp45, respectively (Table 2). These binding assays (Hussain et al., 1987) demonstrate variation at individual epitopes defined by the panel of monoclonal antibodies. However it should also be noted that variation of EIAV can result in altered glycoprotein structures that continue to bind a neutralizing monoclonal antibody, but which are no longer neutralized as a result of antibody binding (Hussain et al., 1987). Thus, structural variations may result in changes in function, without complete abrogation of the antibody binding properties of the glycoprotein.

These antigenic analyses indicate that a distinct antigenic variant is associated with each febrile episode in an experimentally infected animal and that the evolution of a new predominant strain of virus can take as little as two weeks or as long as several months. There is no detectable common pattern of variant evolution in the parallel infections in different ponies.

The biochemical nature of antigenic variation among the panel of EIAV isolates has been examined by peptide and glycopeptide mapping procedures to compare viral proteins and by oligonucleotide fingerprint analysis to compare genomic RNA (Payne et al., 1984, 1987a, b; Salinovich et al., 1986). The peptide/glycopeptide variations are specific for the viral envelope glycoproteins (gp90 and gp45); no structural alterations are observed among the internal viral proteins of the respective viral isolates. The combination of peptide and glycopeptide mapping indicates that each of the viral isolates contain structurally unique gp90 and gp45 components, as no two maps were identical in pairwise comparisons. Moreover, analysis of the variant peptides detected in sequential viral isolates from each pony reveals that peptide alterations appearing in one isolate are not necessarily retained in the next or in successive variants from that pony. Similarly, the oligonucleotide fingerprints differentiate among all of the viral isolates and once again demonstrate a noncumulative pattern of variation among sequential isolates. Pairwise comparisons of oligonucleotide fingerprints typically identify 3-4 variant oligonucleotides out of a total of over 50 oligonucleotides resolved for each genomic RNA. The degree of protein and genomic variation detected by these procedures suggests that the antigenic variation observed in EIAV is the result of random point mutations in the viral RNA, some of which result in altered glycoprotein amino acid sequences. Alterations in peptide maps and oligonucleotide fingerprints have not been detected during passage of the viral isolates in

Table 2. Comparison of epitope composition of various sequential EIAV isolates recovered from Shetland ponies experimentally infected with a standard virus inoculum.[a]

	EIAV Isolates	Isolation Day	Epitopes gp90	Epitopes gp45
Pony 91:	P3.1-1	12	ABCD	A
	P3.1-2	32	ABCDE	A
	P3.1-3	58	ABF	A
	P3.1-4	198	AE	A
Pony 127:	P3.2-1	13	AD	AB
	P3.2-2	41	ABCD	AB
	P3.2-3	99	ABCDF	AB
	P3.2-4	137	AE	AB
	P3.2-5	158	AE	AB
Pony 135F	P3.3-1	16	ACD	A
	P3.3-2	46	ABCDF	AB
	P3.3-3	75	ABCDF	AB
	P3.3-4	107	A	AB
Pony 135P	P3.4-1	18	ABCDF	AB
	P3.4-2	61	AB	AB
Pony 130	P3.5-1	17	ABCDF	AB

[a]Adapted from Hussain et al., (1987) and Rwambo et al., (1989). Epitope constitutions were determined by ELISA binding assays between samples of purified virus and selected monoclonal antibodies defining specific epitopes.

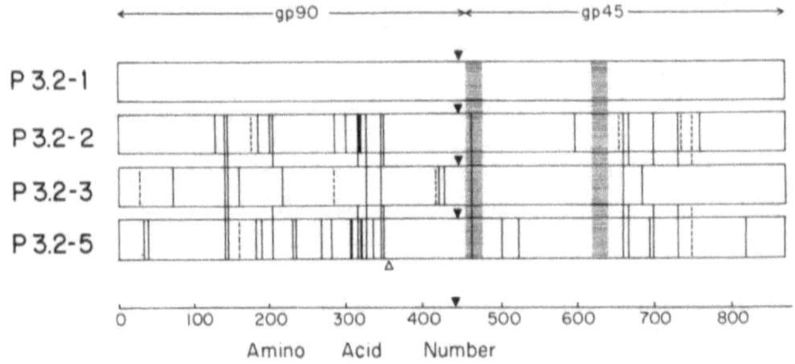

Figure 7. Schematic comparison of <u>env</u> gene sequences of four EIAV isolates, recovered during sequential febrile episodes from an experimentally infected pony. Nucleotide sequences for isolates P3.2-2, P3.2-3, and P3.2-5 are compared to the sequence for P3.2-1. Base substitutions not resulting in amino acid changes are indicated by dashed lines, while changes resulting in amino acid changes are represented by solid lines. Continuous lines represent nucleotide changes common to two or more viral isolates. The open triangle represents a three-base insertion in isolate P3.2-5. The solid triangles represent the precursor cleavage site. Shaded areas represent hydrophobic regions within gp45.

tissue culture, indicating the importance of immune selection in generating variants in horses. In these aspects, EIAV antigenic variation closely resembles the variation reported in other lentiviruses. However, the apparently noncumulative and completely random nature of EIAV variation, although similar to HIV, is in contrast to the cumulative, common patterns of variation described for visna virus.

To directly relate the observed antigenic and structural variations to nucleotide sequence variation in EIAV, the env gene sequences of four distinct virus isolates obtained from a single pony were determined and compared (Payne et al., 1977a). Figure 7 schematically depicts the patterns of sequence variation observed in the env genes of these isolates. All but one difference in the nucleotide sequence among the four viral isolates arise from single base substitutions; a single 3 base insertion was observed in isolate P3.2-5. Overall, the nucleotide divergence between any pair of viral isolates ranges from 16-37 bases out of a total of 2583 bases in the env gene. Approximately 75% of all nucleotide substitutions result in amino acid replacements, hence amino acid divergence ranges from 11-29 residues out of 860 total. In light of the extent of variation observed in peptide mapping, the relatively low level of nucleotide (0.6-1.4%) and amino acid (1.3-3.4%) divergence is somewhat surprising. However, the majority of all amino acid substitutions are nonconservative, predicting probable changes in structural and antigenic properties.

Variation is not evenly distributed throughout the env gene. Two to three times more amino acid substitutions occur in the exterior gp90 compared to the transmembrane gp45. Further, within gp90 two conserved and one hypervariable region can be identified. The first conserved region within gp90 includes the N-terminal 21 amino acid leader peptide and the following 110 amino acids, while the second conserved region encompasses amino acid residues 370 to 445 at the carboxyl terminus. In these regions, there are few substitutions, and all potential N-linked glycosylation sites and cysteine residues are absolutely conserved among the four isolates.

In contrast, a variable region of gp90 spans amino acid residues 143 to 366, in which pairs of viral isolates differ by 3% to 9% in amino acid sequence. A hypervariable region of 35 amino acids (residues 306-338) is located within the gp90 variable region. Within this hypervariable region, amino acid divergence between any two isolates ranges from 3% to 20%. EIAV gp90 is highly glycosylated and contains numerous potential N-linked glycosylation sites (N-X-S/T) within the variable region of the protein. Since about 40% of amino acid substitutions in the env gene involve the loss or gain of an asparagine residue, it is perhaps not surprising that the number of potential glycosylation sites within the variable region of gp90 range from 6-10 in the different isolates. In contrast to gp90, EIAV gp45 is more highly conserved among the four viral isolates, with amino acid substitutions between isolate pairs ranging from 0.7% to 2% (3-9/415 residues). Within gp45, the amino terminal half is more highly conserved than the carboxyl terminal half of the molecule, including two hydrophobic (putative transmembrane) regions (residues 448-473 and 617-636) and four N-linked glycosylation sites that are conserved among all isolates.

The sequence data also permits an estimation of the mutation rate of the EIAV env gene during a persistent infection. This calculation uses the formula R=D/2T (Gojobori et al., 1985), where R is the number of nucleotide substitutions per site per year, D is the divergence calculated from the proportion of nucleotide differences between pairs of isolates, and T is the divergence time represented here as the time between the isolation of the first isolate (P3.2-1) and the later isolates. Using these parameters, mutation rates of 1×10^{-1}, 2×10^{-2}, and 1×10^{-2} were obtained for isolates P3.2-2, P3.2-3, and P3.2-5 respectively. These mutation rates are about 10-fold higher than those reported for HIV (Hahn et al., 1986; Li et al., 1988) and 1000-fold greater than observed for other RNA viruses (Holland et al., 1982).

The preceding experiments characterize both antigenic and structural variations in EIAV glycoproteins that occur during replication of the virus. It remains necessary, however, to establish a correct perspective on the exact relationship between amino acid sequence variations and observed differences in protein antigencity and immunogenicity. These experiments are now in progress.

IMMUNE RESPONSES DURING INFECTION BY EIAV

Practical and specific diagnostic tests for EIAV-infected horses have been available since 1970, making this the first retrovirus diagnostic to be commercially available and to gain widespread usage among veterinarians (Issel and Coggins, 1979). It was also recognized by veterinarians in the late 1960's and early 1970's that immune responses to EIAV may play a role in pathogenesis. In particular, it appears that the characteristic anemia caused by EIAV is actually the result of lysis or phagocytosis of EIAV-coated erythrocytes mediated by viral specific antibodies produced by the horse; there does not appear to be any direct immunosuppression by EIAV. Another early observation was that infected horses during their viremic stages frequently contain antibody-coated EIAV that remains infectious (McGuire et al., 1972). Taken together, the initial deduction was that immune responses to EIAV were at best not beneficial and, most likely, actually harmful! Only recently has it been recognized that the eventual management of EIAV replication by immune responses in persistently infected horses represents a uniquely successful system in controlling retrovirus replication in general and lentivirus replication in particular.

Thus a major focus of our current research effort is to define the kinetics and specificities of immune responses

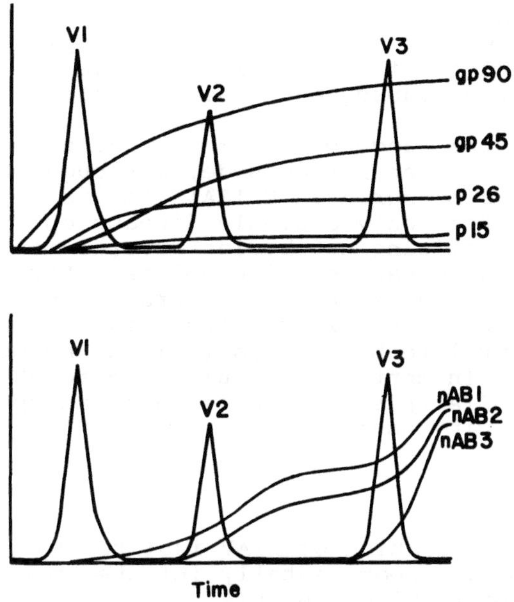

Figure 8. Schematic representation of the kinetics of antibody responses to EIAV during experimental persistent infections. Panel A summarizes antibody responses to individual viral antigens as determined in ELISA and immunoblot assays. Panel B depicts the temporal development of type specific neutralizing antibodies during the course of infection as measured in *in vitro* neutralization assays.

produced against sequential variants evolved during the chronic stage of EIA induced by experimental infections of Shetland ponies with our standard virus inoculum (Rwambo et al., 1989). In these experimentally infected animals, the first clinical episode is typically detected 15-20 days post infection, and viremia levels usually peak coincident with the first febrile episode. A schematic representation of the kinetics of humoral immune responses to EIAV antigens during persistent infections in ponies is presented in Figure 8. Antibodies to the p26 antigen and to apparently group-specific determinants of gp90 are first detectable within several days following the first clinical episode. It is interesting to note that these early group-specific gp90 antibodies bind to the gp90 component of a wide variety of EIAV strains assayed in immunoblot procedures. However, these early gp90 specific antibodies frequently fail to neutralize even the homologous EIAV strain, i.e., the virus predominating during the first viremia (c.f. Table 1). Thus, it appears that neutralizing antibodies, at least as defined by our current *in vitro* assays, do not play a significant role in the cessation of the initial viremia. Instead, neutralizing antibodies to EIAV are detected later in infection, are type specific, and appear coincident with detectable levels of gp45 specific antibody.

Following a second clinical episode, type specific neutralizing antibodies are detected more rapidly than observed following the first febrile episode and this pattern repeats itself after subsequent disease cycles.

In general, the range of serum neutralization increases with each subsequent febrile episode, although the total antibody titer (as measured in ELISA) remains unchanged. Even animals that manifest only a single disease episode display more broadly neutralizing antibodies several months after infection compared to earlier time points. For example, the neutralization data in Table 3 demonstrates that early serum from Pony C (Figure 6) two weeks after its initial EIA episode is unable to neutralize homologous (c/1) or heterologous strains of EIAV. In contrast, serum taken roughly one year after the single EIA episode revealed a broad spectrum of neutralizing activity against homologous and heterologous EIAV isolates.

Table 3. Neutralization of EIAV isolates by serum collected from an experimentally infected pony that had only one clinical episode of EIA.

	SERUM COLLECTION DAY[(b)]	
VIRUS ISOLATE[(a)]	35	438
Prototype	1.5[(c)]	3.0
P3.3-1	1.5	3.0
P3.3-2	1.5	2.0
P3.3-3	1.5	2.0
P3.3-4	1.0	1.5
P3.4-1	1.5	2.5
P3.4-2	1.5	2.0
P3.5-1	1.0	3.5

(a) Virus isolates are the same as described in Table 2. The prototype strain is the cell adapted laboratory strain of virus.

(b) Serum collected from pony 130 at the indicated days post inoculation.

(c) Log_{10} neutralization index. Values above 1.7 are considered positive. Data adapted from Rwambo et al., 1989.

Taken together, these initial studies suggest a "maturation" of EIAV-specific immune responses as the infected horse is exposed to a series of variant virus strains. Whether this broadening of antibody specificity is due to an accumulation of type specific responses to variable epitopes or to the strengthening of group specific responses to conserved epitopes remains to be determined. In most infected horses, the sequential appearance of EIAV variants and the evolution of neutralizing antibody response are demarcated by the cyclic disease episodes characteristic of chronic EIA. In certain animals, however, the neutralizing antibody response broadens in the absence of apparent recurrent disease episodes, perhaps resulting from long term exposure to low levels of virus replication.

Neutralizing antibodies are undoubtedly a narrow window to view the myriad of immune responses generated in response to EIAV infection. Several nonneutralizing epitopes, recognized by mice and horses, have been identified in EIAV gp90 and gp45 (Hussain et al., 1987). In addition, lymphocytes from EIAV-infected horses can be stimulated *in vitro* with EIAV antigens (Issel and Coggins, 1979). Thus, nonneutralizing humoral and cell mediated immune responses must also be examined in the immunologic management of EIAV infections. For example, cellular immune responses against the EIAV core proteins must be carefully analyzed, as CTL responses against the "internal" proteins of other viruses have proven to be an important factor in establishing a protective immune response.

EIAV AS A VACCINE MODEL

EIAV offers a number of advantages as a model for testing different approaches to developing an effective lentivirus vaccine. Large quantities of virus and purified viral antigens and *env* gene clones for antigen expression are available from a number of well characterized EIAV strains. The experimental infection system is highly reproducible and results in disease within 2-3 weeks, thus providing a timely challenge to determine the efficacy of any immunization protocol. Moreover, viral challenges can be mediated using horse fly transfers of blood, thus ensuring a biologically significant level of infecting virus.

Most importantly, EIAV infection in horses still appears unique among lentiviral infections in the eventual management of the infecting virus by the host immune responses. Although the exact mechanisms by which EIAV-infected horses attain this immunologic control are uncertain, the protected status of these long term inapparently infected animals is clear. For example, we have recently demonstrated that ponies experimentally infected with various strains of EIAV and having remained clinically asymptomatic for at least several months display no signs of disease when challenged with a variety of heterologous, virulent strains of EIAV (unpublished observations). Experiments are currently in progress to determine if the challenged animals resisted additional viral infections as well as prevented any apparent disease symptoms. Regardless of the outcome of this latter question, the ability of any candidate lentivirus vaccine to prevent disease upon challenge by a different antigenic variant of virus would be a welcome first step. The prevention of viral infection by any lentivirus vaccine will undoubtedly be the more difficult challenge.

Kono and his colleagues nearly twenty years ago demonstrated that immunization of horses with inactivated preparations of purified EIAV provided protection against infection by homologous, but not heterologous, strains of EIAV (Kono et al., 1970). However, there were no subsequent experiments to determine if a more rigorous immunization protocol might provide protection against heterologous challenge. For example, experiments employing serial immunizations with murine leukemia virus gp70 indicate the sequential appearance of type, group, and interspecies specific antibodies. We are currently evaluating a whole virus EIAV vaccine in ponies receiving from 3 to 9 immunization doses. The initial results of these studies indicate that all of the vaccinated ponies are protected against infection by homologous virus challenge, regardless of the particular immunization regimen. The animals are now being challenged with heterologous strains of EIAV to test the breadth of protection afforded by each vaccine protocol. If an effective whole virus vaccine can be developed, subsequent generations of vaccines utilizing subunit immunogens consisting of purified viral proteins, recombinant viral antigens, or synthetic peptides can be evaluated in the EIAV model system.

ACKNOWLEDGEMENTS

This work was supported in part by funds from the Louisiana Agricultural Experiment Station and by Public Health Service grant AI-25850 from the National Institutes of Health.

REFERENCES

Ball, J. , Henry, L., Issel, C., and Montelaro, R., Localization of EIAV glycoprotein epitopes using synthetic peptide methodologies, J. Virol., submitted (1989).

Gojobori, T. and Yokoyama, S., Rates of evolution of the retroviral oncogene of Moloney murine sarcoma virus and of its cellular homologues, proc. Natl. Acad. Sci. USA, 82:4198-4201 (1985).

Gnann, J., Schwimmbeck, P, Nelson, J., Traux, A., and Oldstone, M., Diagnosis of AIDS by using a 12 amino acid peptide representing an immunodominant epitope of HIV, J. Infect. Dis., 156: 261-265 (1987).

Hahn, B., Shaw, G., Taylor, M., Redfield, R., Markham, P., Salahuddin, S., Wong-Staal, F., Gallo, R., Parks, E., and Parks, W., Genetic variation in HTLV-III/LAV over time in patients with AIDS or risk for AIDS, Science, 232:1548-1553, 1986.

Holland, J., Spindler, F., Horodyski, E., Grabau, E., Nichol, S., and Van DePol, S., Rapid evolution of RNA genomes, Science, 215:1577-1585 (1982).

Hussain, K.A., Issel, C.J., Schnorr, K.L., and Montelaro, R.C., Antigenic analysis of equine infectious anemia virus

(EIAV) variants using monoclonal antibodies: Epitopes of glycoprotein 90 (gp90) of EIAV stimulate neutralizing antibodies, J. Virol., 61:2956-2961 (1987).
Hussain, K.A., Issel, C.J., Schnorr, K.L., West, M.D., Rwambo, P., and Montelaro, R.C., Epitope mapping of the envelope proteins of EIAV, Arch. Virol., 98:213-224 (1988).
Issel, C.J. and Coggins, L., Equine infectious anemia:. Current knowledge, J. Am. Vet. Med. Assoc., 174:727-733 (1979).
Kono, Y., Hirasawa, K., Fukunaga, Y., and Taniguchi, T., Recrudescence of EIA by treatment with immunosuppressive drugs, Natl. Inst. Anim. Health Q. (Tokyo), 16:8-15 (1976).
Kono, Y., Kobayasi, K., and Fukunaga, Y., Antigenic drift of equine infectious anemia virus in chronically infected horses, Arch. Virol., 41:1-10 (1973).
Kono, Y., Kobayasi, K., and Fukunaga, Y., Immunization of horses against equine infectious anemia (EIA) with an attenuated EIA virus, Natl. Inst. Anim. Health. Q. (Tokyo), 10:113-122 (1970).
Leis, J., Baltimore, D., Bishop, J., Coffin, J., Fleissner, E., Goff, S., Oroszlan, S., Robinson, H., Skalka, A., Temin, H., and Vogt, V., Standardized and simplified nomenclature for proteins common to all retroviruses, J. Virol., 62:1808-1809 (1988).
Li, W., Tanimura, M., and Sharp, P., Rates and dates of divergence between AIDS virus nucleotide sequences, Mol. Biol. Evol., 5:313-330 (1988).
McGuire, T., Crawford, T., and Henson, J., EIA: Detection of infectious virus-antibody complexes in the serum, Immunol. Commun., 1:545-551 (1972).
McGuire, T.C., Norton, L.K., O'Rourke, K.I., and Cheevers, W.P., Antigenic variation of neutralization sensitive epitopes of CAEV during persistent arthritis, J. Virol., 62:3488-3492 (1988).
Montelaro, R.C., Parekh, B., Orrego, A., and Issel, C.J., Antigenic variation during persistent infection by equine infectious anemia virus, a retrovirus, J. Biol. Chem., 259:10539-10544 (1984a).
Montelaro, R.C., West, M.D., and Issel, C.J., Antigenic reactivity of the major glycoprotein of equine infectious anemia virus, a retrovirus, Virol., 136:368-374 (1984b).
Narayan, O., Griffin, D.E., and Chase, J., Antigenic shift of visna virus in persistently infected sheep, Science, 197:376-378 (1977)
Narayan, O., Griffin, D.E., and Clements, J.E., Virus mutation during "slow infection." Temporal development and characterization of mutants of visna virus recovered from sheep, J. Gen. Virol., 41:343-352 (1978).
Orrego, A., Issel, C.J., Montelaro, R.C., and Adams, W.V., Jr., Virulence and in vitro growth of a cell-adapted strain of equine infectious anemia virus after serial passage in ponies, Am. J. Vet. Res., 43:1556-1560 (1982).
Ozel, M., Pauli, G., and Gelderblom, H., The organization of the envelope projections of the surface of HIG, Arch. Virol., 100:255-266 (1988).

Palker, T., Clark, M. Langlois A. Matthews T. Weinhold, K., Randall R. Bolognesi D. and Haynes B. Type-specific neutralization of HIV with antibodies to env - encloded synthetic peptides, Proc. Natl. Acad. Sci. USA, 84:2479-2483.

Parekh, B., Issel, C.J., and Montelaro, R.C., Equine infectious anemia virus, a putative lentivirus, contains polypeptides analogous to prototype-C oncornaviruses, Virology, 107:520-525 (1980).

Payne, S.L., Fang, F.D., Lui, C.P., Dhruva, B., Rwambo, P., Issel, C.J., and Montelaro, R.C., Antigentic variation and lentivirus persistence: Variations in envelope gene sequences during EIAV infection resemble changes reported for sequential isolates of HIV, Virology, 161:321-331 (1987a).

Payne, S., Parekh, B., Montelaro, R.C., and Issel, C.J., Genomic alterations associated with persistent infections by equine infectious anemia virus, a retrovirus, J. Gen. Virol., 65:1395-1399 (1984).

Payne, S.L., Rushlow, K., Issel, C.J., and Montelaro, R.C., Expression of EIAV env gene fragments in E. coli: Epitope mapping and serological analyses of constant and variable env gene regions, J. Virol., in press, (1989).

Payne, S.L., Salinovich, O., Montelaro, R.C., Issel, C.J., and Nauman, S.M., Course and extent of variation of equine infectious anemia virus during parallel persistent infections, J. Virol., 61:1266-1270 (1987b).

Perryman, L., O'Rourke, K., and McGuire, T., Immune responses are required to terminate viremia in EIAV infection, J. Virol., 62:3073-3076 (1988).

Preston, B.D., Poiest, B., and Loeb, L.A., Fidelity of HIV-1 reverse transcriptase, Science 242:1168-1171 (1988).

Roberts, J.D., Bebenek, K., and Kunkel, T.A., The accuracy of reverse transcriptase from HIV-1, Science 242:1171-1173 (1988).

Rusche, J., Javaherian, K., McDanal, C., Petro, J., Lynn, D., Grimaila, R., Langlois, A., Gallo, R., Arthur, L., Fischinger, P., and Bologinesi, D., Antibodies that inhibit fusion of human immunodeficiency virus-infected cells bind a 24-amino acid sequence of the virual envelope, gp120, Proc. Natl. Acad. Sci. USA, 85:3198-3202 (1987).

Rushlow, K., Olson, K., Stiegler, G., Payne, S.L., Montelaro, R.C., and Issel, C.J., Lentivirus genomic organization: The complete nucleotide sequence of the env gene region of equine infectious anemia virus, Virol., 155:309-321 (1986).

Rwambo, P M., Issel, C.J., Adams, W.V., Hussain, K.A., Miller, M., and Montelaro, R.C., Equine infectious anemia virus (EIAV): Humoral responses of recipient ponies and antigenic variation during persistent infection, Arch. Virol., in press (1989).

ANTIGENIC STRUCTURE OF THE FUSION GLYCOPROTEIN OF RESPIRATORY SYNCYTIAL VIRUS

Peter R. Paradiso, Branda Hu, Stephen Hildreth, Antonia Martin and Rasappa Arumugham

Praxis Biologics, Inc.
Rochester, NY 14623

INTRODUCTION

Respiratory syncytial virus (RSV) is the major cause of lower respiratory disease in infants and young children (McIntosh and Chanock, 1985). The virus is a member of the family Paramyxoviridae, and is an enveloped single stranded RNA virus. RSV, like other paramyxoviruses, has an envelop protein called the fusion (F) protein which is responsible for fusion of the viral and cell membranes during the infectious process. Once a cell is infected, F protein in the membrane can cause the fusion of adjacent cells resulting giant cells or syncytia. Antibody to the F protein is able to block the fusion process and thereby prevent the infection of cells by virus and spread of infection between cells (Walsh and Hruska, 1983; Walsh et al., 1985).

The RSV F protein is a 70,000 dalton glycoprotein which is proteolytically processed into two disulfide linked subunits designated F2 and F1 (Collins et al., 1984). This protein can be purified from RSV as well as virus infected cells and exists predominantly as a dimer in solution (Walsh et al., 1985). Monoclonal antibodies have been identified which have both the capacity to neutralize virus and inhibit fusion in vitro (Walsh and Hruska, 1983) and passively protect cotton rats from live virus challenge (Walsh et al. 1984).

A subunit vaccine for RSV, based on purified fusion glycoprotein, has been developed and shown to be highly immunogenic and protective in both a cotton rat and baboon model (Paradiso, unpublished data). As part of this development program we have characterized the nature of the immune response to the F protein both after infection and after vaccination with purified protein. We have further mapped a neutralizing and fusion inhibiting epitope based on the ability to bind functional antibody.

RESULTS AND DISCUSSION

Some of the important structural characteristics of the RSV fusion glycoprotein are presented in Figure 1.

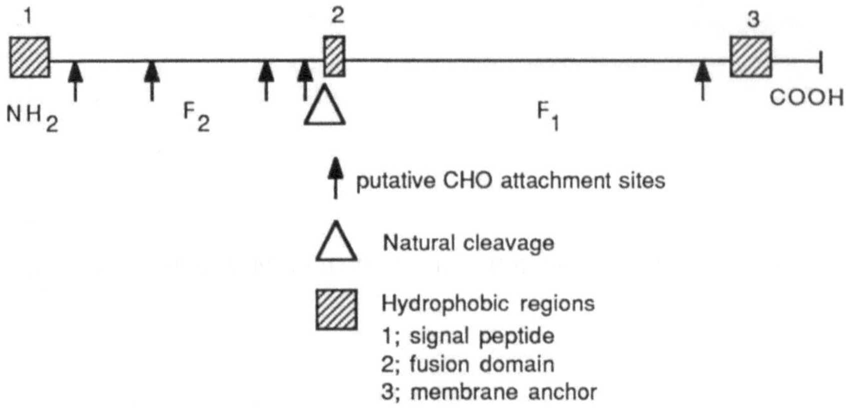

Figure 1. Linear diagram of the F protein.

The most significant features of this structure are common to the known paramyxovirus fusion proteins. Specifically, the protein is cleaved post-translationally at a trypsin sensitive site forming a new amino terminus for the F1 subunit at amino acid 137. The three hydrophobic domains function as signal sequence (amino terminus of F2), fusion related hydrophobic domain (amino terminus of F1) and membrane anchor sequence (carboxy terminus of F1). The potential sites of asparagine linked sugars are indicated.

Walsh et al. (Walsh and Hruska, 1983; Walsh et al., 1986) have described two monoclonal antibodies to the F protein, designated L4 and A5, which both neutralize virus and inhibit fusion. These monoclonal antibodies did not recognize identical regions of the protein but were partially overlapping based on competitive ELISA analysis. The L4 monoclonal is able to recognize the F1 subunit on western blot analysis and defines a linear epitope. The A5 monoclonal antibody is only very weakly reactive with the F1 subunit, while binding strongly to the native protein, and defines a much more conformationally dependent epitope. We were interested in defining the relative contribution of these two epitopes to the antibody repertoire induced either by natural infection, or by vaccination with purified protein.

The type of antibody elicited as a result of natural RSV infection was investigated using a pooled sample of human adult serum (PHS) which had a high titer to the F glycoprotein as a result of numerous natural exposures to RSV. The ability of the L4 and A5 monoclonal antibodies to block the binding of the human polyclonal IgG to F protein on solid phase was analyzed in a competition ELISA (Table 1). To minimize the potential effects of steric hindrance, Fab fragments of each monoclonal antibody were used.

The L4 was able to block 18% of the binding and A5 was able to block 27% of the binding of PHS. The two together blocked up to 38% of the binding which is less than the sum of the two and most likely reflects the overlap in the binding site of these two antibodies. Clearly these two epitopes represent immunodominant regions of the protein.

Table 1. Competition of Fab fragments of the L4 and A5 monoclonal antibodies with rabbit serum for F protein on solid support.

ng/well blocking antibody	% blocking L4	A5	L4+A5
0	0	0	0
6.25	0	2.1	9.4
25	2.9	11.7	22.6
100	11.4	19.6	38.3
400	17.2	26.1	30.3
1600	18.2	27.0	37.4

In similar experiments, rabbits were hyperimmunized with purified F protein in Freund's adjuvant. After three immunizations of 20 ug, the animals were bled and the serum used for competitive binding ELISA studies.

Table 2. Competition of Fab fragments of the L4 and A5 monoclonal antibodies with pooled human serum for F protein on solid support.

ng/well blocking antibody	% blocking L4	A5	L4+A5
0	0	0	0
6.25	13.9	12.5	29.2
25	35.6	36.7	68.3
100	50.8	58.1	78.2
400	52.4	65.7	81.0
1600	56.9	68.7	81.6

Competition against both monoclonal antibodies (Table 2) was very strong giving values of 57% for L4 and 69% for A5. Together these two functional epitopes accounted for over 80% of the total antibody produced after vaccination. Similar data was obtained with serum from animals immunized with F protein either adsorbed to aluminum hydroxide or without any adjuvant.

Since these two monoclonals clearly represent major neutralizing epitopes we were interested in mapping the region of the protein to which they bound. For these studies, the L4 monoclonal antibody was used since it was able to recognize linearized regions of the F protein. Preliminary experiments in which the F protein was hydolyzed with a variety of proteases including trypsin, chymotrypsin and S. Aureus V8 protease demonstrated that fragments of under 10K daltons were difficult to detect by L4 in either dot blot or western blot. To circumvent this problem, larger fragments were mapped using monospecific polyclonal antibodies to five synthetic peptides which spanned the whole of the F1 subunit. Cleavages were performed with either trypsin, endoproteinase lys-C or endoproteinase arg-C to define the carboxy-terminal specificity of each fragment. Defined fragments were then tested for their reactivity with the monoclonal antibody.

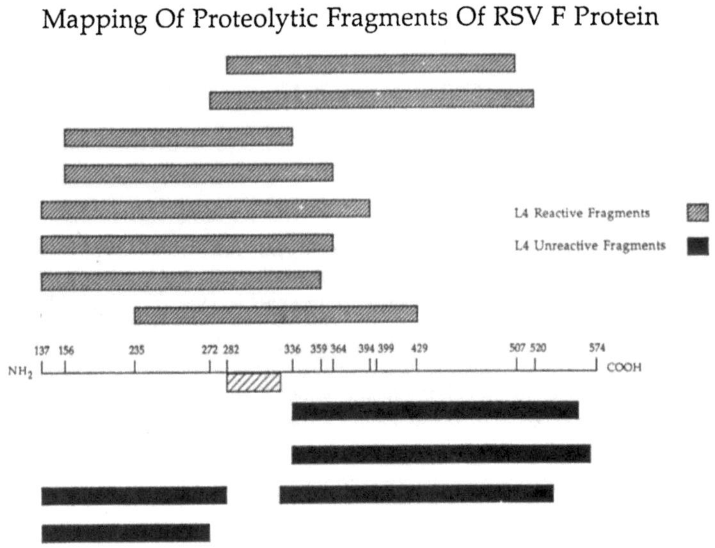

Figure 2. Mapping of L4 reactive and unreactive proteolytic fragments from the F1 subunit.

Figure 2 shows the series of peptides which were reactive with L4 (above the line) and those which were not (below the line). The hatched area defines the region which was included in all of the L4 reactive peptides but excluded from all of the non-reactive peptides. The boundaries defined were between amino acids 283 and 328.

To confirm the identification of this region, a peptide was synthesized from residue 283 to residue 315. The ability of this peptide to bind to the L4 monoclonal antibody was tested by ELISA with the peptide on solid phase. A non-specific peptide derived from residues 155-169 served as a control.

Figure 3 shows the dose dependent binding of peptide 283-315 to the L4 monoclonal antibody confirming the identification of this region as a neutralizing and fusion inhibiting epitope.

In conclusion we have demonstrated that epitopes defined by the monoclonal antibodies L4 and A5 constitute immunodominant epitopes on the RSV fusion glycoprotein. This is most clearly seen by the fact that more than 80% of the antibody response in rabbits to the purified protein is directed

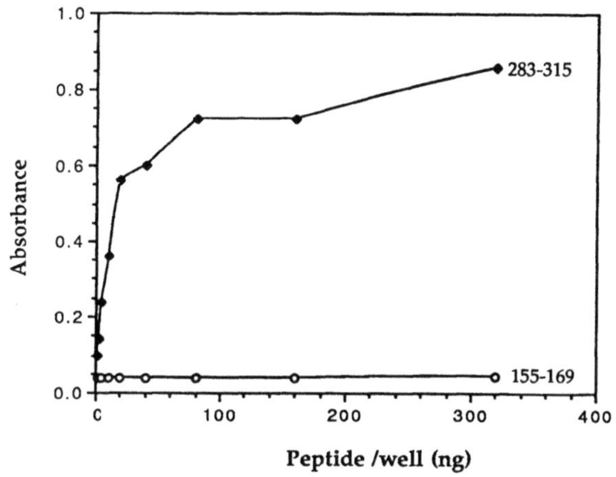

Figure 3. Binding of the L4 monoclonal antibody to peptide 283-315.

to these epitopes. Although the pooled human serum had a lower percentage of A5/L4. activity, the value of 38% is still a major contribution to the immune response. The latter value may reflect an increase in the antibodies directed at less dominant epitopes which are boosted after many exposures to the virus. Similar serological analysis with sera from infants or toddlers who have had few RSV episodes may be necessary to prove this hypothesis. The L4 epitope maps to amino acids 283-315 within the F1 subunit. The A5 epitope will be more difficult to map due to its conformational nature but appears to at least overlap the L4 epitope in the three-dimensional structure of the protein. The dominance of these biologically functional epitopes in the purified fusion protein underscores the potential of this protein as an RSV vaccine.

References

Collins, P., Huang, Y., and Wertz, G. (1984) Nucleotide sequence of the gene encoding the fusion (F) glycoprotein of human respiratory syncytial virus. Proc. Natl. Acad. Sci. U.S.A. 81:7683-7687

McIntosh, K. and Chanock, R. (1985) Respiratory Syncytial Virus. In Virology, B. Fields (ed.), Raven Press, New York.

Walsh, E., Brandriss, M., Schlesinger, J. (1985) Purification and characterization of respiratory syncytial virus fusion protein. J. Gen. Virol. 66:409-415.

Walsh, E., Cote, P., Fernie, B., Schlesinger, J., and Brandriss, M., (1986). Analysis of the respiratory syncytial virus fusion protein using monoclonal and polyclonal antibodies. J. Gen. Virol. 67:505-513.

Walsh, E., and Hruska, J., (1983). Monoclonal Antibodies to respiratory syncytial virus proteins: identification of the fusion protein. J. Virol. 47:171-177.

Walsh, E., Schlesinger, J. and Brandriss, M., (1984) Protection from respiratory syncytial virus infection in cotton rates by passive transfer of monoclonal antibodies. Infection and Immunity, 43: 756-758.

SIV INFECTED RHESUS MACAQUES: AN AIDS MODEL FOR IMMUNOPREVENTION AND IMMUNOTHERAPY

Murray B. Gardner

Department of Medical Pathology
University of California, School of Medicine
MSI-A
Davis, CA 95616

Introduction

The current worldwide epidemic of the acquired immunodeficiency syndrome (AIDS) caused by the human immunodeficiency lentivirus (HIV) presents an unprecedented challenge for vaccine prevention and immunotherapy. Never before have we have been faced with an organism which so successfully persists and evades the host defenses while going about its insidious but apparently inevitable destruction of the immune system. Contrary to more conventional viruses, neutralizing antibodies to the external envelope glycoprotein provide only partial, incomplete protection against HIV infection and, under some circumstances, may even enhance infection. Virus envelope mutations lead to changes in neutralization epitopes and cell tropism and may cause an increased virulence with disease progression. Escape from immune surveillance by hiding within cells may also pose a major hurdle to elimination of the virus via natural or vaccine-induced immunity. At the present time the target antigen(s) for vaccine induction of protective immunity against HIV or any other lentivirus is simply unknown. Presumably, cell mediated immunity against cell-associated virus core and envelope antigens plays an important role, both in protection and, possibly, in causing immunopathology. To help us sort our way through this biologic maze and develop successful vaccines and antiviral therapies, we need animal lentivirus models that resemble AIDS.

Fortunately, such a model has recently been discovered, namely, simian immunodeficiency virus (SIV) infection of captive macaques. This lentivirus is different from the simian Type D retroviruses which are related to the prototype

Mason Pfizer monkey virus, and which also cause a fatal immunodeficiency syndrome in macaques (For Review, see Gardner et al, 1988). This paper will review the major features of the SIV model of simian AIDS (SAIDS) and will describe the current status of vaccine research using this model at the California Primate Research Center (CPRC) at UC Davis.

SIV: Origin and Natural History

SIV consists of a heterogeneous group of lentiviruses which are the closest known relatives to HIV (For Reviews, Gardner et al, 1988 and Gardner and Luciw, 1988, in press). SIV and HIV have a similar genomic organization of their structural and functional genes, similar protein composition, similar reverse transcriptase activity, similar antigens and similar serologic reactivity. Both viruses use the CD4 receptor on T-helper lymphocytes and both cause profound lymphoid depletion and eventual death from acquired immunodeficiency in their respective hosts. Overall, the macaque SIV (SIVmac) is about 40% related in nucleotide sequences to HIV-1, the pandemic strain, and about 75% related to HIV-2, the West African, less pathogenic strain of HIV. SIV is found as an apparently nonpathogenic, persistent infection in the natural host which are various species of non-human primates from Africa, e.g. African green monkey (*Cercopithecus aethiops*), sooty mangabey (*Cereocebus atys*) and mandril (*Papio sphinx*). The prevalence of SIV in various monkey species throughout Africa is now being investigated. The route of natural transmission is unknown but sexual contact is suspected. SIV-naturally infected monkeys make an antibody response to the virus core and envelope proteins similar to that seen in healthy HIV seropositive humans. The lack of associated immunodeficiency in the naturally infected African monkeys is unexplained. Genetic analysis of recent SIV isolates from African green monkeys and a mandril indicate considerable heterogeneity with some isolates showing a closer similarity to HIV-1 than HIV-2 and others more similarity to HIV-2 than HIV-1 (Ohta, 1988; Fukasawa et al, 1988; Daniel et al, 1988a). The genetic dissimilarity between SIV isolates from different African monkey species suggests that SIV has existed as a heterogeneous family of lentiviruses in African non-human primates for a long time. It remains undetermined whether HIV-1 or HIV-2 was acquired at some time in the distant past by transpecies spread from monkeys to humans but the degree of genetic difference between SIV and HIV-1 suggests that this has not occurred in recent times.

In striking contrast to the asymptomatic infection of African monkey species with SIV, a fatal immunodeficiency syndrome with many parallels to human AIDS (Table 1) results when Asian macaques, e.g. rhesus, which are not the natural host, are infected with this virus (Letvin et al, 1985). Spontaneous infection of macaques with SIV has been found at 5 of the 7 US Primate Centers (New England, Delta, Yerkes, Washington, California) (Gardner et al, 1988). At Delta and

Yerkes Centers this infection was shown to be the result of cross-species spread via doctors' needles from naturally infected, healthy sooty mangabeys (Fultz et al, 1986; Murphey-Corb et al, 1986). SIV was also isolated at the California Center from healthy sooty mangabeys housed at a midwestern zoo (Lowenstine et al, 1986). Both of these SIV isolates from sooty mangabeys are pathogenic for rhesus and other macaques. Isolates of SIV from macaques at the New England, Washington and California primate centers in the past decade presumably were the result of inadvertent transmission from African monkeys via direct contact or needles. These SIV macaque isolates were obtained from rhesus macaques (M. mulatta) at the New England (Daniel et al, 1985) and Delta Primate Centers (Murphey-Corb et al, 1986), from a pigtailed macaque (M. nemestrina) at the Washington Primate Center (Benveniste et al, 1986) and from a stumptailed macaque (M. arctoides) at the California Primate Center (Lowenstine et al, 1988). Each of these isolates causes SAIDS when inoculated into naive macaques. The M. arctoides SIV caused a naturally occurring epizootic of SAIDS in the mid 1970s in a colony of stumptailed macaques at the CPRC. The prototype SIVmac isolate (strain 251) from the New England Primate Center has been most thoroughly characterized, including pathogenesis, molecular cloning and complete sequencing (Chakrabarti et al, 1986; Franchini et al, 1987; Hirsch et al, 1987; Kestler et al, 1988). However, none of the molecular clones of SIVmac, although infectious in vitro and in vivo, have as yet induced disease in monkeys. We have used the prototype SIVmac strain 251 (Daniel 1988a; Kannagi et al, 1986) for most of the initial pathogenesis and vaccine studies which are summarized in this report.

SIV Infection of Rhesus Macaques

Many similarities between SIVmac infection of rhesus monkeys and HIV-1 infection of humans are listed in Table 1. Features shared in common with all lentivirus infections of animals and humans are shown in Table 2. The characteristics of SIV infection of rhesus monkeys which make this an ideal model for AIDS vaccine studies are: 1) reproducibility of AIDS-like disease (SAIDS): rhesus monkeys are highly susceptible to infection with SIVmac. Virtually 100% of monkeys inoculated intravenously (IV) become viremic by 1 month post-inoculation (pi) and remain persistently infected until death (Daniel et al, 1987). 2) Availability of a well characterized pathogenic virus stock of SIVmac: Strain 251 of SIVmac has reliably induced SAIDS at several different primate centers. 3) Similarity of pathology and pathogenesis of SIV induced SAIDS and HIV induced AIDS: These similarities include T-helper cell tropism via the CD4 receptor, the serologic response to core and envelope antigens, the histopathology, lymphoid depletion, opportunistic infections, tumors (lymphomas), and fatal outcome (Letvin et al, 1985). Compared to HIV infected humans, SIV infected macaques have less pneumocystis carinii pneumonia, no Kaposi's sarcoma and more adenovirus infection. 4) Availability of macaques: Rhesus and other species of

Table 1 - SIV Infection of Macaques: Major Features[1]

I.	Lentivirus: Same genetic organization and 40-75% sequence homology with HIV-1 and HIV-2, respectively. Non-pathogenic in certain species of African monkeys (Cercopithecus). Exogenous. Origin unknown.
II.	Persistent infection of all recipients: Dose independent
III.	Susceptible to experimental infection via bloodstream and genital mucosa
IV.	Long incubation period: months-years
V.	100% mortality: 3 mo - 3 years
VI.	Vigorous humoral immune response to core and envelope antigens: Immune response affects survival period
VII.	Decrease in core antibody precedes clinical decline
VIII.	Cytopathic in vitro: Syncytia
IX.	Restricted cell tropism: T4 cells, macrophages
X.	Restricted virus expression in vitro
XI.	Similar tissue distribution of virus
XII.	Similar pathology including neuropathology. Syncytia
XIII.	Similar opportunistic infections and B cell lymphomas
XIV.	Strain variation, mostly in envelope

Table 2 - Properties of Lentiviruses

1. Long incubation period: healthy carrier state
2. Wide disease spectrum
3. Persistence in face of vigorous immune response
4. Viral regulatory genes control latency and activation
5. Restricted virus expression
6. Cytopathic effect. Syncytia: Non-oncogenic
7. Neurotropism
8. Macrophage tropism, "trojan horse"
9. High rate of envelope antigenic variation
10. Transmission by close physical contact

macaques breed well in captivity and established breeding colonies exist at several locations in the USA. Importation of these animals from abroad is not necessary. This animal resource is thus readily available at the US regional primate centers, at NIH and at several other primate research facilities (e.g. Southwest Foundation for Biomedical Research in San Antonio, Texas), most of which are closely affiliated with nearby academic institutions.

The disadvantages of this model system are: 1) lack of standardized reagents and assay systems. Because SIVmac was discovered only 3 years ago and only a limited number of rhesus can be housed in currently available biocontainment facilities, there has not elapsed sufficient time for carrying out of infectivity titrations by different inoculation routes or by using cell free vs cell associated virus. Standard assay systems including monoclonal antibodies for detecting and quantitating virus, standardized procedures for measuring neutralizing and other antibodies and assays for cell mediated immunity against SIV antigens (e.g. antibody dependent cell cytotoxicity, T cell proliferation and cytotoxicity) are all still under development. 2) Extreme susceptibility of macaques to SIVmac infection and disease. The in vivo susceptibility of rhesus monkeys to SIVmac infection has proven to be several logs greater than the in vitro susceptibility of rhesus cells. Minimal infectious doses for in vivo infection by different routes of inoculation e.g. IV vs genital mucosa, are just now being established. Therefore, the initial experimental transmission studies using SIV by the IV route often overwhelmed the animal and led to early deaths from SAIDS after only several months. This telescoping of pathogenic events may facilitate pathogenesis studies but it hinders the ability to study the therapeutic effect of various antiviral interventions and "stacks the deck" against vaccine protection. 3) Individual variability in susceptibility to SAIDS. Not unexpectedly, outbred rhesus show considerable variation in response to experimental infection with SIVmac. The length of the incubation period before onset of SAIDS is several months to about 3 years depending on the dose, strain and route of inoculation and strength of the individual animal's immune response to SIV (Kannagi et al 1986). The mean life span is about 9-10 months. About 1/4 of the SIVmac-infected monkeys fail to make a significant immune response and die of SAIDS within 2-4 months. 4) Lack of adequate biocontainment facilities for housing SIV-infected macaques. The major limiting factor to exploiting this simian model of AIDS is not so much for lack of monkeys or research grants as it is for lack of adequate housing for infected monkeys. Development of appropriate space to appropriately house more SIV infected macaques is now underway. The advantages of this system far outweigh the disadvantages. The SIV-infected macaque is an excellent AIDS model for studies of viral pathogenesis and for testing and development of antiviral therapies and preventive vaccines. If an experimental drug or vaccine works well in the SIV model one can feel confident that it will work against HIV and AIDS! We will now summarize the status of our early, ongoing efforts at the CPRC, UC Davis to use this simian AIDS model system for active immunization and post-infectious immunotherapy.

Active Immunization

Because we do not know the nature of protective immunity in AIDS or SAIDS and, because neutralizing antibody to native or recombinant viral envelope antigens alone has already been shown ineffective in protecting chimpanzees against HIV challenge (For Summary, see Gardner et al 1988), we have chosen to start our SIV (and HIV) vaccine research using relatively crude whole killed virus preparations. Obviously, such a whole virus approach to an AIDS vaccine is not practical nor likely to be approved by the Federal Drug Administration, unless of course, it works. It is a logical first step, however, to begin with this classical approach in AIDS vaccine development, and if successful, to then dissect the mechanisms of protection and define the protective epitopes for ultimate synthesis of peptides or production of genetically engineered subunit vaccines. SIVmac, grown in a human T cell line (HUT 78), was pelleted at 1000x G and inactivated by two 5 minute exposures to psoralin and UV light (Sutjipto et al, unpublished data). The inactivated virus was shown free of residual infectious virus by in vitro assay. Immunoblots of the concentrated virus, before and after inactivation, using antiserum from an infected rhesus, showed that the preparation contained mainly core antigens. The envelope of SIV and HIV is very fragile and easily stripped off by any purification procedure. Eight rhesus monkeys, free of antibodies to SIV or Type D virus, were inoculated with 0.5 mg intramuscularly (IM) of this material using threonyl muramyl dipeptide (MDP-Syntex), 240 ug, as an adjuvant. Little or no antibody was detectable in any of the animals by Western blot 2 months after this primary immunization. After a boost 2 months later with the same dose of immunogen and adjuvant a low level core antibody was detected in several monkeys. After a 2nd boost of this immunogen and adjuvant, given 2 months after the 1st boost, a strong antibody response was detected by immunoblot in all animals. This revealed primarily core antibody although several animals showed some envelope antibody. Western blot analysis typically reveals envelope antibody less well than does radioimmunoprecipitation which was not done. Using a newly developed colorimetric assay technique (MTT assay) (Pedersen and Marx, unpublished) each of these immune sera was shown to contain SIV neutralizing antibody, although in relatively low titers (1:64 - 1:1064). The immunized monkeys stayed free of infectious SIV, thus confirming the thorough inactivation of the SIV immunogen. The relatively low immunogenicity of this material was evidenced by the fact that it took 22 weeks and a total of 1.5 mg viral protein given as a primary immunization and two boosters to generate a reasonably strong humoral immune response. We plan to boost these monkeys one more time before challenge with SIVmac. Meanwhile infectivity titrations of this virus using exposure by the IV and genital mucosal routes are being completed (Miller et al, 1988).

Initial SIV vaccine studies at the New England Primate Center using formalin-inactivated SIVmac and MDP adjuvant failed to protect rhesus vs SIVmac challenge despite induction of high levels of binding antibody and moderate levels of neutralizing antibody (Letvin et al, 1987). The reason for this failure was thought to be that the immunogenic dose of SIV was too small (225 ug total) and the IV dose of challenge virus was too high, thus overwhelming the antiviral immunity. Subsequent vaccine experiments done there using a minimal IV infectious dose of challenge virus and after inducing stronger immunity with a larger dose of viral immunogen (2 mg total) have suggested a measure of protection against infection (Daniel 1988b).

Post-Infectious Immunotherapy

Salk proposed a novel strategy for the possible prevention of AIDS or delay in its onset by the immunization of healthy seropositive individuals with a whole killed HIV-1 immunogen (Salk, 1987). The idea is that HIV infected individuals, during the long incubation period (mean of 8 years) before onset of disease, may remain sufficiently immunocompetent to respond with an anamnestic or booster response to an inactivated whole HIV immunogen. If such a booster response could be elicited it might lower or even eliminate the amount of infectious virus and thereby reduce the contagiousness and delay or prevent the onset of AIDS. We have begun to test this hypothesis in SIV infected rhesus. Four months after onset of infection with SIVmac, 5 seropositive healthy rhesus were inoculated with 100 µg IM of a gamma irradiated whole SIVmac immunogen mixed with incomplete Freund's adjuvant. Five seropositive rhesus controls received the adjuvant alone. The vaccine preparation was immunogenic, noninfectious and safe in uninfected rhesus in which it induced a primary and secondary humoral immune response, mainly to core antigen as measured by Western blot and enzyme linked immunoassay (EIA) (Fig. 1). However, when given to SIV infected rhesus this SIV immunogen induced no detectable humoral or cellular anamnestic immune response after 28 days. There was no change in Western blot patterns (Fig. 2a,b), no increase or decrease in amount of antibody as measured by EIA against an HIV-2 whole virus lysate (Table 3a) or an SIVmac/HIV-2 specific transmembrane envelope peptide (Table 3b) (Norrby et al, 1987). However, only 2 of these immunized monkeys and 3 controls had normal levels of CD4 helper lymphocytes (>600/mm^3) at the time of receiving the SIV immunogen or adjuvant alone. All animals remained viremic. The data are summarized in Table 4. Thus, despite appearing healthy 4 months after infection, except for generalized lymphadenopathy, 5 of the 9 monkeys were already showing evidence of a compromised immune system making it difficult to assess their booster response to either the SIV immunogen or the adjuvant alone. Neutralizing antibody titers on these animals are now pending. In both vaccine and control groups SIV specific T cell proliferation in response to SIV pulsed autologous monocytes showed a general correlation with presence of normal CD4 cell counts but showed no evident response to the treatment. No adverse reactions were noted systemically or locally at the inoculation site. Whether or not this treatment will have any effect on the disease course remains to be seen but it appears at this one time point - 28 days after receiving the vaccine or adjuvant only - that neither treatment had any immediate beneficial or adverse effect on the infected monkeys (Gardner et al, 1988b). By 6 weeks after treatment one of the monkeys that received the SIV immunogen and two IFA controls had died with SIV induced SAIDS. Thus these treatments had no apparent effect on the natural course of SIV disease.

This approach to post infectious immunization needs to be further tested in seropositive macaques that are still fully immunocompetent, perhaps in conjunction with passive immunization. In view of the marked susceptibility of macaques to SIVmac this window of opportunity may be fairly short, perhaps no more than a few months depending on virus strain, dose and route of inoculation, and individual immune response capacity. However, these initial results in no way distract from the value of this model for carrying out further vaccine trials. The results reinforce our belief, that, if any AIDS vaccine or drug works successfully in this highly susceptible animal model system, its efficaciousness in humans is virtually guaranteed.

Summary

SIV infection of macaques constitutes a valuable model for development of AIDS vaccines and antiviral therapies. Initial results using inactivated whole SIV immunogens for immunoprevention and post-infectious immunotherapy in the SIV-rhesus monkey system are summarized.

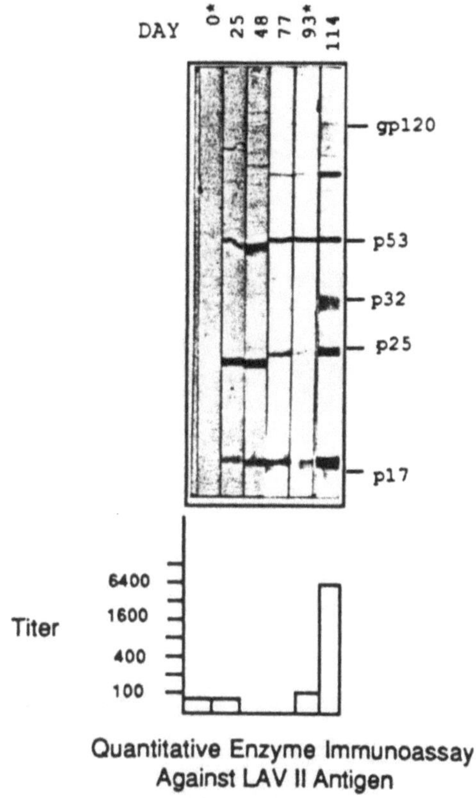

FIGURE 1. Anti-viral antibody response in a rhesus monkey immunized with whole inactivated SIVmac. Antigen for the immunoblot (upper panel) consists of purified SIV. This monkey (#23233) was immunized with 100 micrograms of whole inactivated SIVmac in incomplete Freund's adjuvant by intramuscular route on day 0 and boosted with the same material on day 93. Serum samples were collected on the days indicated at the top of each gel lane. Molecular weights for the major viral proteins are shown. gp120 is the external envelope glycoprotein, p53 is the core (gag) precursor polypeptide, p32 is the endonuclease, and p25 and p17 are the mature core proteins. The lower panel shows antibody titers using a commercial ELISA for detecting antibodies to HIV-2 and SIVmac.

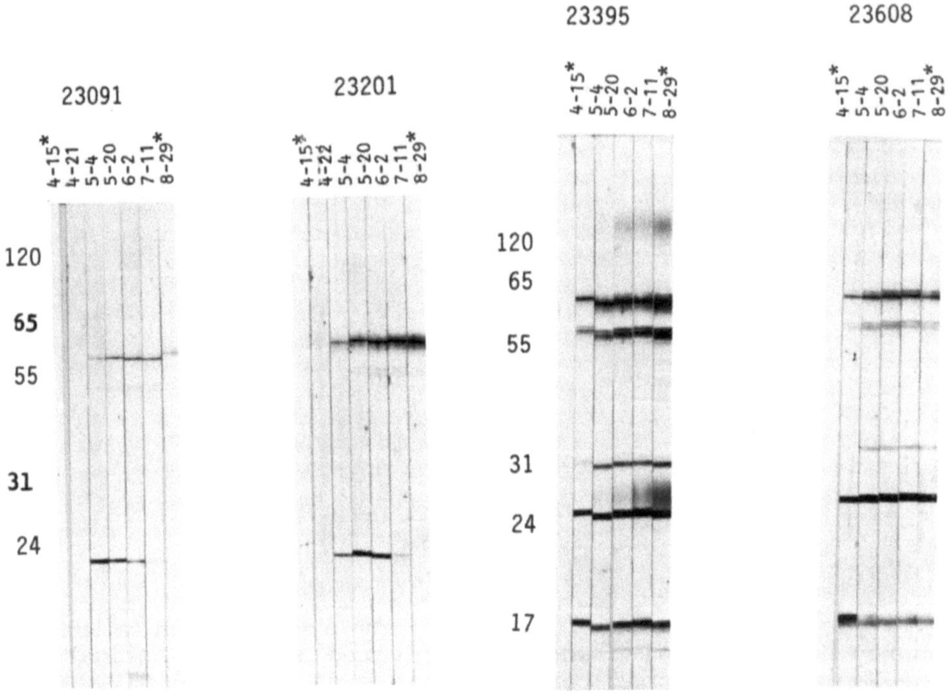

FIGURE 2a,b. Antiviral response as measured by immunoblot in representative rhesus monkeys infected with SIVmac. Antigen for the immunoblot consists of SIVmac. The animals were infected with SIV by intravenous inoculation on 4/1/88 (1st asterisk). Monkeys #23201 and #23608 received the inactivated SIV immunogen and monkeys #23091 and #23395 received only the incomplete Freund's adjuvant by the intramuscular route on 8/1/88 (2nd asterisk). Serum samples were collected on the days indicated at the top of each gel lane. Immunoblot patterns indicating relatively weak or strong antibody response in these SIV infected monkeys are shown in Fig. 2a and b, respectively. Molecular weights for the major viral proteins are shown. Gp120 is the external envelope glycoprotein, p65 is the reverse transcriptase, p55 is the core (gag) precursor polypeptide, p31 is the endonuclease, the broad band between p24 and p31 is the transmembrane glycoprotein, and p24 and p17 are the mature core proteins.

Table 3a

HIB II (Viral Lysate) EIA O.D.[1,2]

		4 *		Date			5 *	
Monkey	3-22	4-15	4-21	5-4	5-20	6-2	7-11	8-29
23091 Control	.024	.022	.031	.429	.570	.510	.463	.371
23201 Immunized	.019	.023	.080	.470	.578	.531	.489	.547
23395 Control	.021	.021	.071	.875	1.311	1.682	1.587	2.178
23608 Immunized	.010	.019	.030	.836	1.203	.593	1.541	1.494

Table 3b

SIVmac Synthetic Peptide EIA O.D.[1,3]

		4 *		Date			5 *	
Monkey	3-22	4-15	4-21	5-4	5-20	6-2	7-11	8-29
23091 Control	.062	.032	.052	1.008	1.095	1.292	1.050	.520
23201 Immunized	.032	.023	.422	1.380	1.906	1.800	.947	1.036
23395 Control	.029	.024	.203	1.512	1.945	2.033	1.274	1.618
23608	.033	.184	.034	.046	1.560	1.812	1.892	1.638

1. EIA = enzyme linked immunoassay, O.D. = optical density at 1:100 serum dilution.
2. HIV-2 is highly cross reactive with SIVmac.
3. This transmembrane envelope peptide is highly specific for SIVmac and HIV-2 (Norrby et al, 1987).
4. Monkeys were infected with SIVmac on 4/1/88.
5. Monkeys were inoculated with SIV immunogen on 8/1/88.

Table 4

Attempted Immunotherapy of SIV Infected Rhesus with an SIV Immunogen

	No. Treated	No. Immunocompetent[3] Before Immun.	No. Immunocompetent[3] After Immun.	No. with Booster Response to SIV Immunogen or IFA[4]	No. with Adverse Reactions after Treatment[5]
SIV Immunogen[1,2]	5	2	1	0	0
IFA	4	3	2	0	0

1. Given 4 mo after induction of experimental infection.

2. 100ug IM of gamma irradiated whole SIV mixed with Incomplete Freund's Adjuvant (IFA).

3. Based on absolute CD4 cell count (normal = 930 $\pm$ 364/mm^3).

4. Measured by immunoblot, EIA and SIV-specific T-cell proliferation and virus isolation 28 days after SIV immunogen or IFA only.

5. Measured by clinical status, total lymphocyte count, CD4 cell count and virus isolation 28 days after treatment. At 6 weeks after treatment one monkey that received the SIV immunogen and two IFA controls had died.

REFERENCES

Benveniste, R.E., Arthur, L.O., Tsai, C-C., Sowder, R., Copeland, T.D., Henderson, L.E., and Oroszlan, S. Isolation of a lentivirus from a macaque with lymphoma: Comparison with HTLV-III/LAV and other lentiviruses. J. Virol. 60:483-490, 1986.

Chakrabarti, S., Robert-Guroff, M., Wong-Staal, F., Gallo, R.C. and Moss, B. Expression of the HTLV-III envelope gene by a recombinant vaccinia virus. Nature 320:535-537, 1986.

Daniel, M.D., Letvin, N.L., King, N.W., Kannagi, M., Sehgal, P.K., Hunt, R.D., Kanki, P.J., Essex, M. and Desrosiers, R.C. Isolation of T-cell tropic HTLV-III-like retrovirus from macaques. Science 228:1201-1204, 1985.

Daniel, M.D., Letvin, N.L., Sehgal, P.K., Hunsmann, G., Schmidt, D.K., King, N.W., and Desrosiers, R.C. Long-term persistent infection of macaque monkeys with the simian immunodeficiency virus. J. Gen. Virol. 68:3183-3189, 1987.

Daniel, M.D., Li, Y., Naidu, Y.M., Durda, P.J., Schmidt, D.K., Finger, C.D., Silva, D.P., MacKey, J.J., Kestler, H.W. III, Sehgal, P.K., King, N.W., Hayami, M. and Desrosiers, R.C. Simian immunodeficiency viruses from African green monkeys, J. Virol, in press 1988a.

Daniel, M., Sehgal, P., King, N., Wyand, M., Schmidt, D. and Desrosiers, R. Protection of macaques against simian immunodeficiency virus infection by inactivated whole virus. IV International Conference on AIDS, Book I, p. 235, 1988c.

Franchini, G., Gurgo, H.G., Guo, R., Gallo, R.C., Collati, E., Fargnoli, K.A., Hall, L.F., Wong-Staal, F. and Reitz, M.S. Jr. Sequence of simian immunodeficiency virus and its relationship to the human immunodeficiency viruses. Nature 328:539-543, 1987.

Fukasawa, M., Miura, T., Hasegawa, A., Morikawa, S., Tsujimoto, H., Miki, K., Kitamura, T. and Hayami, M. Sequence of simian immunodeficiency virus from African green monkey, a new member of the HIV/SIV group. Nature 333:457-461, 1988.

Fultz, P.N., McClure, H.M., Anderson, D.C., Swensen, R.B., Anand, R. and Srinivasan, A. Isolation of a T-lymphotropic retrovirus from naturally infected sooty mangabey monkeys (Cercocebus atys). Proc. Natl. Acad. Sci. USA 83:5286-5290, 1986.

Gardner, M.B., Luciw, P., Lerche, N., and Marx, P. Non-human primate retrovirus isolates and AIDS. In K. Perk (ed.) Immunodeficiency Disorders and Retroviruses, Academic Press, 1988, in press.

Gardner, M.B. and Luciw, P. Simian immunodeficiency viruses and their relationship to the human immunodeficiency viruses. In M. Adler, J.Gold and J. Weber (eds.) AIDS '88, Gower Academic Journals, 1988, in press.

Gardner, M., Luciw, P., Marx, P., Lerche, N., McGraw, T., Jennings, M., Carlson, J. and Pedersen, N. An Approach to Immunotherapy of SIV-Infected Rhesus with an Inactivated Whole SIV Immunogen. Cold Spring Harbor Laboratory, in press, 1988.

Hirsch, V., Riedel, N. and Mullins, J.I. The genome organization of STLV-3 is similar to that of the AIDS virus except for a truncated transmembrane protein. Cell 49:307-39, 1987.

Kannagi, M., Kiyotaki, M., Desrosiers, R.C., Reimann, K.A., King, M.W., Waldron, L.M., and Letvin, N.L. Humoral immune responses to T cell tropic retrovirus simian T lymphotropic virus type III in monkeys with experimentally induced acquired immune deficiency-like syndrome. J. Clin. Invset. 78:1229-1236, 1986.

Kestler, H.W. III, Li, Y., Naidu, Y.M., Butler, C.V., Ochs, M.F., Jaenel, G., King, N.W., Daniel, M.D. and Desrosiers, R.C. Comparison of simian immunodeficiency virus isolates. Nature 331:619-621, 1988.

Letvin, N.L., Daniel, M.D., Sehgal, P.K., Desrosiers, R.C., Hunt, R.D., Waldron, L.M., MacKey, J.J., Schmidt, D.K., Chalifoux, L.V., and King, N.W. Induction of AIDS-like disease in macaque monkeys with T-cell tropic retrovirus STLV-III. Science 230:71-73, 1985.

Letvin, N.L., Daniel, M.D., King, N.W., Kiyataki, M., Kannagi, M., Chalifaux, L.V., Sehgal, P.K., Desrosiers, R.C., Arthur, L.O., and Allison, A. In Vaccines 87: Modern Approaches to New Vaccines, Cold Spring Harbor Laboratory, pp. 209-213, 1987.

Lowenstine, L.J., Pedersen, N.C., Higgins, J., Pallis, K.C., Uyeda, A., Marx, P., Lerche, N.W. and Gardner, M.B. Seroepidemiologic survey of captive old-world primates for antibodies to human and simian retroviruses and isolation of a lentivirus from sooty mangabeys (Cercocebus atys). Int. J. Cancer 38:563-574, 1986.

Lowenstine, L., Lerche, N., Marx, P., Gardner, M. and Pedersen, N. An epizootic of simian AIDS caused by SIV in captive macaques in the 1970s. In M. Girard and L. Vallette (eds.) Retroviruses of Human AIDS and Related Animal Viruses. Pasteur Vaccines, Paris, pp. 174-176, 1988.

Miller, C.J., Alexander, N., Jennings, M., Hendrickx, A.G., and Marx, P.A. Transmission of simian immunodeficiency virus (SIV) across the genital mucosas of male and female rhesus macaques. IV International Conference on AIDS, Stockholm, Sweden, Book II, p. 121, 1988.

Murphey-Corb, M., Martin, L.N., Rangan, S.R.S., Baskin, G.B., Gormus, B.J., Wolf, R.H., Andes, W.A., West, M. and Montelaro, R.C. Isolation of an HTLV-III-related retrovirus from macaques with simian AIDS and its possible origin in asymptomatic mangabeys. Nature 321:435-437, 1986.

Norrby, E., Biberfeld, G., Chiodi, F., von Gegerfeldt, A., Naucler, A., Parks, E. and Lerner, R. Discrimination between antibodies to HIV and to related retroviruses using site-directed serology. Nature 329:248-250, 1987.

Ohta, Y., Masuda, T., Tsujimoto, H., Ishikawa, K., Morikawa, S., Nakai, M., Honjo, S. and Hayami, M. Isolation of simian immunodeficiency virus from African green monkeys and seroepidemiologic survey of the virus in various non-human primates. Int. J. Cancer 41:115-122, 1988.

Peterson, N., Marx, P., and Gardner, M., unpublished data.

Salk, J. Prospects for the control of AIDS by immunizing seropositive individuals. Nature 327:473-476, 1987.

Sutjipto, S., Marx, P., and Gardner, M. Unpublished data.

THE DEVELOPMENT OF MOLECULAR VACCINES AGAINST MALARIA SPOROZOITES

W. Ripley Ballou*, James F. Young†,
Stanley J. Cryz††, and Jerald C. Sadoff**

Departments of Immunology* and Bacterial Diseases**
Walter Reed Army Institute of Research, Washington, D.C.
Molecular Genetics†, Smith Kline & French Laboratories
King of Prussia, PA
Swiss Serum and Vaccine Institute††, Berne, Switzerland

INTRODUCTION

The successful development of an efficacious molecular vaccine against malaria has long been a major research goal of the Walter Reed Army Institute of Research, the World Health Organization and major international public health agencies. Although this goal has not yet been achieved, immense progress has been made in the area of sporozoite vaccines since the gene encoding the Plasmodium falciparum circumsporozoite (CS) protein was cloned and sequenced in 1984 (Dame et al., 1984). The basis for a subunit approach to sporozoite immunity was established in the early 1970's in a series of landmark studies demonstrating that mice, primates, and humans could be protected against sporozoite challenge by immunization with radiation-attenuated sporozoites (Nussenzweig et al. 1969; Clyde et al., 1975; Rieckman et al., 1979). These studies demonstrated that protective immunity was species and stage specific, but since large numbers of sporozoites were required for its induction, it was not feasible to develop live attenuated sporozoite vaccines for general use. Moreover, the mechanism of immunity to irradiated sporozoites is complex, and includes the induction of both humoral and cellular immune responses. Vaccines designed to induce one or both of these types of immunity are currently being developed in our laboratories and those of our collaborators and will be discussed in this review.

HUMORAL IMMUNITY AGAINST SPOROZOITES

The major neutralizing epitopes on the surface of P. falciparum sporozoites are thought to be expressed in the 51kD CS membrane protein (Zavala et al., 1983). This protein is representative of a family of more than a dozen CS proteins from other Plasmodia whose structures have been determined to date. It consists of a centrally located repeat sequence having a species-specific 40-44 tetrapeptides [(NANP)x] that are flanked by regions containing short sequences of highly conserved amino acids (Dame et al., 1984). Studies using synthetic peptides have demonstrated that the vast majority of antibodies raised against sporozoite antigens and having neutralizing activity in vitro recognize a limited repertoire of epitopes situated within the CS repeat region (Zavala et al., 1983; Ballou et al., 1985). A significant majority of humans immunized or infected naturally by mosquitoes have been shown to induce antibody responses against these epitopes, thus supporting the rationale for developing vaccines targeting CS repeat epitopes (Zavala et al., 1985). The biological significance of the conserved regions remains unclear, but important T cell epitopes relevant to boosting by natural exposure to sporozoites have been identified in nonconserved flanking sequences, and may thus play an important role in future sporozoite vaccines (Good et al., 1987). As determined by immunofluorescence (IFA), antibodies against repeat epitopes bind to the membrane uniformly over the surface of the sporozoite (Nardin et al., 1979), neutralize sporozoite infectivity for liver cells (ISI assay) (Hollingdale et al., 1984), and precipitate CS protein from the surface of the sporozoite (CSP reaction) (Vanderberg et al., 1969). These in vitro characteristics have been used with varying degrees of success to monitor responses to subunit sporozoite vaccines, but none of them has yet been shown to accurately predict protection from experimental sporozoite challenges. Currently, the only method of determining vaccine efficacy involves sporozoite challenge of human volunteers, thus the identification of an assay predictive of protection remains a significant challenge to malaria sporozoite vaccine efforts. Together with our collaborators at the National Institutes of Health, Smith Kline and French Laboratories, and the Swiss Serum and Vaccine Institute, we have completed four separate sporozoite vaccine trials as of this writing. These studies are part of an ongoing systematic evaluation of vaccine candidates in the search for optimally immunogenic and efficacious sporozoite vaccines.

A Recombinant Sporozoite Vaccine, FSV-1

The first subunit *P. falciparum* sporozoite vaccine tested in human volunteers, FSV-1, consisted of alum-adjuvanted R32tet32, a recombinant CS protein expressed in *Escherichia coli* (Young et al., 1985). R32tet32 contains the CS sequences $[(NANP)_{15}(NVDP)_1]_2$ encoded by two tandem XhoII restriction fragments from the malaria gene fused to 32 amino acids derived from the tetracycline resistance gene of the vector plasmid. This molecule was selected as a clinical candidate following unsuccessful attempts to express and purify full length CS protein in *E. coli*. Preclinical studies in rodents and rhesus monkeys established its safety and immunogenicity, and indicated that antibodies raised against the recombinant molecule were biologically active against falciparum sporozoites (Wirtz et al., 1987). The vaccine was administered to 15 U.S. volunteers at doses of 10 to 800 ug at 4 week intervals for a total of three injections (Ballou et al., 1987). FSV-1 was safe, but it was a surprisingly poor immunogen for human subjects (Figure 1). Although 12 of 15 volunteers seroconverted, the mean antibody rise was small, and only one volunteer (who received the 800 ug dose) showed evidence of boosting upon repeated immunization. Among those volunteers that did respond, titers were typically higher after the initial dose, and declined following subsequent immunizations. Six of the volunteers went on to received a fourth dose of vaccine approximately nine months after the primary injection, and along with two non-immunized controls, were challenged with a chloroquine-sensitive strain of *P. falciparum* by allowing five mosquitoes with heavy sporozoite burdens to feed on them. Only 3 of 6 immunized volunteers had any detectable CS antibodies at the time of challenge and only the two receiving 800 ug doses were positive by IFA. The volunteer with the highest ELISA and IFA titer, and the only subject with significant prechallenge ISI activity, was completely protected from challenge. All other volunteers developed patent parasitemia within 9 to 13 days from challenge. The two volunteers having lower levels of CS antibodies had prepatent periods of 48 to 72 hours longer than nonimmunized controls. These results paralleled those of a simultaneous but separate vaccine efficacy trial using a synthetic peptide $(NANP)_{12}$ conjugated to tetanus toxoid (Herrington et al., 1987). These investigators found that seroconversion required a high dose of conjugate, occurred in only 60% of the volunteers, and as was seen with the recombinant vaccine, repeated dosing failed to boost

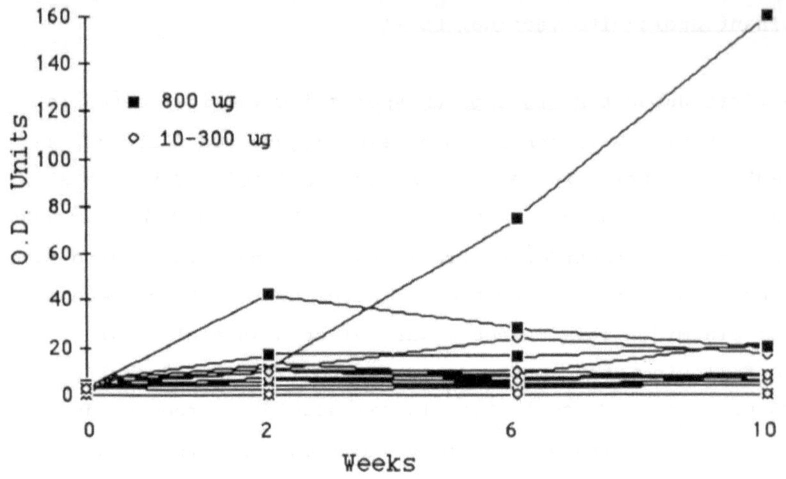

Fig. 1. Antibody responses to FSV-1 in 15 U.S.volunteers as measured by ELISA using R32LR as antigen. Sera were assayed two weeks after each of three doses of vaccine. Data for the 800 ug dose are highlighted to illustrate the two patterns of responses observed.

antibody responses. Similarly, only 1 of 3 volunteers with relatively high CS antibodies was protected against challenge, while prolonged prepatent periods occurred in the other two seropositive volunteers. These critical studies established the feasibility of protecting human volunteers against sporozoite-induced malaria by immunization with subunit sporozoite vaccines, but indicated that the generation of higher levels of CS antibodies would be required.

Modifications of the Immunizing Regimen

To address whether the route of administration or the interval between doses might influence CS antibody responses, we conducted a separate vaccine trial comparing intradermal (i.d.) versus standard intramuscular (i.m.) administration of FSV-1. The i.d. route was selected because studies with both rabies and hepatitis B vaccines have shown that small amounts of antigen can be made immunogenic by intradermal administration (Granden et al., 1985; Redfield et al.,1986).

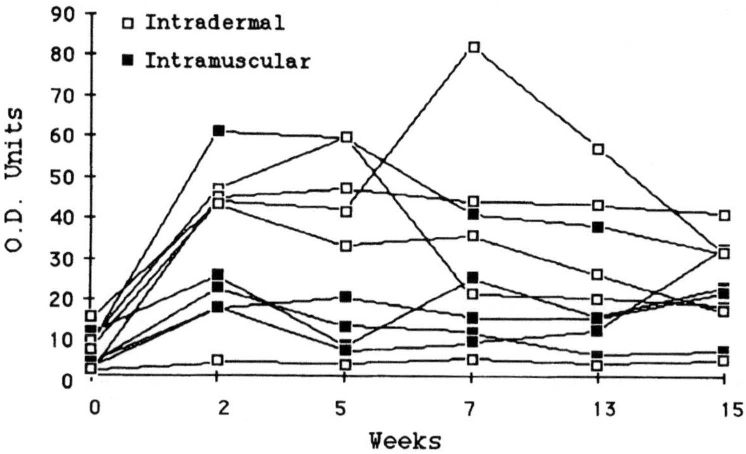

Fig. 2. Antibody responses to 800 ug of FSV-1 administered i.d. on weeks 0 and 5, or i.m. on weeks 0, 5, and 13. No significant difference between the two routes was detected when sera were assayed by ELISA 2 weeks after the second dose. Administration of a third i.m. dose 8 weeks later failed to result in a booster response.

Antigen processing and presentation by Langerhans cells are thought to play a role in enhanced antibody responses to antigens administered intradermally (Stingl et al., 1978). Thus, sporozoites that are inadvertently inoculated into the skin while the mosquito probes for a blood meal may contribute to the CS antibody responses seen in endemic populations. Ten volunteers were randomly assigned to receive 800 ug doses i.d. or i.m. at week 0. Each dose of i.d. vaccine was administered in five separate sites (0.1 ml/site) including over deltoids and across the upper back. Although no toxicity was observed with this regimen, volunteers receiving the i.d. injections complained of considerable local discomfort for up to twelve hours after immunization. One volunteer declined to receive subsequent doses. The other four i.d. volunteers, and all 5 i.m. recipients were boosted at week six. Local responses after the second i.d. dose were similar to the first. No adverse reactions occurred in the i.m. group, consistent with the results of the initial Phase I study. Antibody responses to either route of injection were indistinguishable, with 4/5 volunteers seroconverting in each group (Figure 2). The i.m. group was boosted 8 weeks later, but no significant rise in titer was observed. The results of these two studies indicated that the immune response to FSV-1 in naive U.S. volunteers could not be easily modulated by simple modifications of the immunization protocol.

The Issue of Genetic Restriction

The low immunogenicity of the recombinant and synthetic peptide-conjugate vaccines was not expected, but the results are consistent with the observation that 15 to 30 percent of the adult population of malaria-endemic regions fail to develop CS antibodies despite life-long exposure to malaria sporozoites (Hoffman et al., 1987; Good et al., 1988; Guidice et al., 1987). Studies using congenic mice have shown that antibody responses to the *P. falciparum* repeat epitopes present within R32tet32 are genetically restricted (Good et al., 1986; Hoffman et al., 1987b), and it has been postulated that some members of outbred human populations may be similarly unable to recognize T helper epitopes on sporozoite CS proteins. The induction of cognate help by T cells to enhance antibody responses requires that the immunogen be recognized and presented under control of Ir gene products. In this context it was expected that T helper epitopes on tetanus toxoid, if not within the much smaller tet32 region of FSV-1, might effectively overcome genetic restriction to *P. falciparum* CS repeats. The current results suggest that limits placed upon antibody responses to the repeat epitope by the human immune system may go beyond the issue of genetic restriction. Several T helper epitopes from nonconserved flanking regions of CS protein have been mapped in murine and human populations, and the inclusion of these epitopes in molecular sporozoite vaccines may enhance vaccine efficacy by promoting boosting of antibody levels with subsequent sporozoite exposure (Good et al., 1987). The failure of intact sporozoites to be universally immunogenic for human subjects is relevant to the recent observation that immunodominant T cell domains of the CS protein map to polymorphic regions of the circumsporozoite gene (Good et al, 1988), indicating that this approach to the generation of widespread antibody responses to sporozoite vaccines may be problematic.

$R32NS1_{81}$, a Recombinant Vaccine with Enhanced Immunogenicity

The generally poor human antibody response to R32tet32 has been attributed to the absence of appropriate helper T cell epitopes in the tet32 portion of the fusion protein. To address this issue, the vaccine was re-engineered to express the two tandem Xho II fragments encoding R32 fused to the amino terminus of an 81 amino acid sequence derived from the nonstructural protein 1 of influenza A (Egan et al., in preparation). Alum-adjuvanted $R32NS1_{81}$ (FSV-2) was shown to be safe and immunogenic in preclinical studies, and convalescent sera from patients recovering from influenza A did not react with the fusion protein. Because $NS1_{81}$ was a larger and more complex molecule than tet32, it was hoped that a fusion protein containing this molecule might enhance immunogenicity for human volunteers. FSV-2 was administered to 15 U.S. volunteers on weeks 0, 8, 16, and 21 at doses of 11.5, 115, or 1150 ug/dose. Like FSV-1, the vaccine was well tolerated, and no serious adverse reactions were noted during the study. FSV-2 was considerably more immunogenic than FSV-1 (IFA GMT = 13 for FSV-1 vs. 51 for FSV-2), and 14 of 15 volunteers seroconverted. One volunteer who received the 115 ug dose had a dramatic rise in antibody following the second boost (peak response > 300 ug/ml), but his titers began to decline shortly afterwards and did not increase after receiving a fourth dose some 5 weeks later. This volunteer and five other immunized subjects, and two nonimmunized control volunteers were challenged according to the same protocol as was used for the FSV-1 challenge. None of the volunteers was completely protected, although 24 - 48 hours delay in patency was seen in several immunized subjects. These results were not expected considering the high levels of CS antibodies ([Ig] ≈ 100 ug/ml) present in at least one volunteer at the time of challenge. A careful analysis of these sera and those from the FSV-1 study demonstrated similar isotype patterns (IgG1>IgG2>IgG3>IgG4), similar binding avidities using a sodium thiocyanate wash technique, and reactivity with intact by sporozoites by IFA. The only significant difference detected has been the failure of high titer day of challenge FSV-2 serum to inhibit sporozoite invasion in vitro (ISI) compared to strong inhibitory activity in analogous sera from the protected FSV-1 and tetanus toxoid conjugate volunteers. Further analysis of these results will be required to determine the significance of this finding on the design of future vaccine efficacy trials.

Novel Conjugate Vaccine Strategies

The possibility that carrier-mediated suppression in tetanus-immune U.S. volunteers receiving the tetanus conjugate sporozoite vaccine may have down-regulated CS antibody responses led us to evaluate a series of conjugate vaccines using carrier proteins with which volunteers had not previously been immunized. We selected cholera toxin (CT) and *Pseudomonas* toxin A (TA) as carrier molecules that could be suitably detoxified and prepared for use in human subjects and conjugated them to CS protein molecules (Sadoff et al., in preparation). To determine whether the number of CS repeats was an important factor for immunogenicity, we conjugated the carriers to short synthetic peptides, $(NANP)_3$ or $(NANP)_6$, or to a large recombinantly produced molecule R32LR (32 repeats). Conjugation strategies were designed to maximize CS protein:carrier ratios, however full substitution was difficult to achieve in all preparations (Table 1). Nevertheless, preclinical studies in rabbits demonstrated each of these vaccines to be highly immunogenic. Alum-adsorbed vaccines were administered to nonimmune volunteers (4-5 per

Table 1. Characteristics of *Plasmodium falciparum* CS Protein:Carrier Conjugate Vaccines

Vaccine	Molar ratio peptide:carrier	ug peptide/ 100 ug dose	Seroconversion (≥ 4x day 0)
R32TA	7:1	131	4/5
R32CT	6:1	96	0/4
R3CT	25:1	41	3/5
R6CT	33:1	110	0/4
R3TA	7:1	14	0/4

R3 = $(NANP)_3$; R6 = $(NANP)_6$; R32 = $[(NANP)_{15}(NVDP)_1]_2$

TA = Pseudomonas toxin A; CT = Cholera toxin

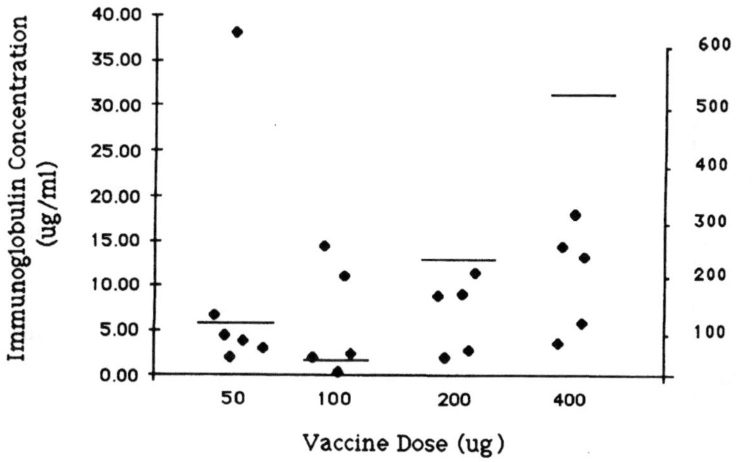

Fig. 3. Dose response study of P. falciparum sporozoite carrier:conjugate vaccine R32TA. Vaccine was administered at weeks 0 and 6, and sera were obtained at week 8. Horizontal lines indicate the geometric mean titer of sera when tested for immunofluorescence against intact sporozoites.

group) at weeks 0 and 6. Doses were administered on the basis of the weight of the carrier molecule. The vaccines R3CT, R6CT, R32CT, R3TA, and R32TA were administered at doses of 100 ug (containing 21-131 ug repeat protein) and were uniformly well tolerated with no significant local or systemic adverse reactions noted. All volunteers seroconverted to the carrier protein at levels consistent with previous results obtained using unconjugated carrier toxoid vaccines. In contrast, responses to CS epitopes were divergent, with the vaccines R3TA, R6CT, and R32CT being substantially less immunogenic than R3CT or R32TA (Table 1). Antibody responses did not correlate with the either the amount of CS protein, the conjugation ratio, or the carrier protein selected. Because the highest mean responses were seen with R32TA, it was selected for a separate dose response study. Volunteers (5-6 per group) received 50, 100, 200, or 400 ug of R32TA at weeks 0 and 6, and sera were obtained 2 weeks after each dose for assay. No adverse effects were seen, even at the highest dose of vaccine administered. After two doses, 21 of 22 volunteers had seroconverted (Figure 3). The only volunteer failing

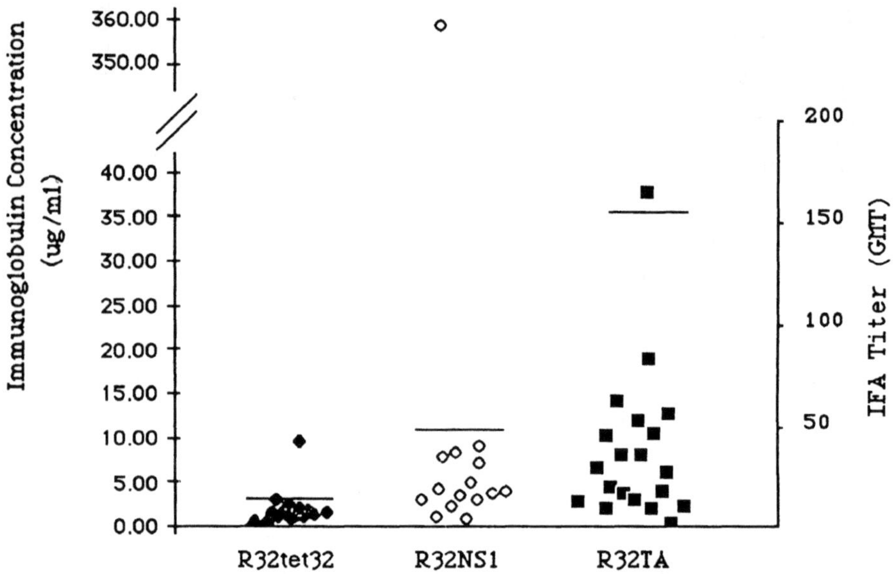

Fig. 4. Comparison of human Ig responses to three CS vaccines. Results are presented for three doses of R32tet32 (FSV-1) and R32NS1$_{81}$ (FSV-2), and two doses of R32TA. Horizontal lines indicate geometric mean IFA titers for all vaccinees in each group.

to respond received the 100 µg dose, and the volunteer with the greatest response received the 50 µg dose. Overall, the best responses occurred in the 400 ug dose group (mean, 11.00 ug/ml). These responses are in the range of those from protected volunteers (10-15 ug/ml), and IFA titers were 400 or greater in each volunteer at this dose.

Alternative Vaccine Designs for Enhanced Immunogenicity

The studies described above suggested that increased immunogenicity can be achieved by the use of better carrier molecules. Figure 4 clearly shows this trend, but challenge studies that we and others have conducted would predict that only a minority of the volunteers studied to date have achieved protective levels of CS antibodies. The approaches employed for these vaccines have been more or less "traditional", i.e., the vaccines were essentially soluble, nonparticulate proteins administered with alum as an adjuvant. A variety of other methods to enhance immunogenicity are now being studied to determine whether a greater proportion of vaccinees will respond maximally, including the use of particulate hybrid molecules containing an Xho II repeat fragment inserted into the Pre S2 region of hepatitis B and expressed in yeast (Rutgers et al., 1988), the incorporation of CS antigens into liposomes (Richards et al., 1988) or proteosomes (Lowell et al., 1988), the expression of complete CS protein in eukaryotic expression systems, and the use of alternative adjuvants (Clough et al., 1985). Clinical studies in human volunteers will be required to determine the usefulness of these alternative approaches.

CELL MEDIATED IMMUNITY AGAINST SPOROZOITES

The vaccines described thus far are designed to induce a CS protein-specific antibody response, but others have previously established that the immunity induced by irradiated sporozoites is strongly dependent upon cell-mediated immune responses to sporozoite or exoerythrocytic antigens (Chen et al., 1977; Verhave et al., 1978). We previously reported that protective immunity induced by irradiated P. berghei sporozoites is mediated by antigen-specific T cells (Egan et al., 1987). These T cell populations mayparticipate in cytotoxic or lymphokine-mediated killing of exoerythrocytic parasites in the liver (Ferreira et al., 1986). The effector arm of the immune response to sporozoites appears to be mediated by gamma-interferon and CD8+ T cells (Schofield et al., 1987; Weiss et al., 1988). Unlike live sporozoites, killed sporozoites are unable to induce protective immunity. Intracellular targeting of CS antigens may thus be necessary for induction of protective cell-mediated immune responses. We therefore investigated the use of attenuated Salmonella as vectors to present CS antigen. This transformant vaccine constitutively expressed CS antigen, and induced antigen-specific cell-mediated immune responses in mice and protection from sporozoite challenge in the absence of measurable antibodies to CS protein.

A plasmid containing the entire P. berghei CS protein gene was used

Table 2. Transformant Salmonella Vaccine Protects Mice from Challenge with 1500 *P. berghei* Sporozoites

Vaccine	Dose	Route	IFA	Infected / Challenged	%Protected
Wk4					
PbM206	10^4	s.c.	1+	5/9	44
M206	10^4	s.c.	-	5/5	0
Wk5					
PbM206	10^{10}	p.o.	-	4/10	60
PbM206	10^9	p.o.	-	3/10	70
PbM206	10^4	s.c.	-	12/18	33
M206	10^{10}	p.o.	-	5/5	0
M206	10^9	p.o.	-	5/5	0
M206	10^4	s.c.	-	10/10	0
Wk9					
PbM206	10^9	p.o.	-	2/10	80
M206	10^9	p.o.	-	7/10	30
Normal	NA	NA	-	7/8	13
SPZ	75K	i.v.	4+	0/5	100

to transform *S. typhimurium* M206, an avirulent strain with impaired ability to multiply within macrophages (Sadoff et al., 1988). Mice were immunized subcutaneously (s.c.) or orally (p.o.) and *Salmonella* were cultured from liver homogenates for up to 3 weeks after immunization. Mice immunized s.c. with the PbM206 vaccine developed low levels of antibodies to CS protein that were detected for less than 5 weeks while mice immunized p.o. with larger doses made no detectable antibodies. Mice immunized by either route were protected when challenged with sporozoites 4 to 9 weeks after immunization (Table 2).Overall, 64% of mice immunized with the vaccine PbM206 were protected, compared with 9% of mice immunized with the control ($p = 0.0001$). We further demonstrated that mice immunized orally with the vaccine developed DTH reactions to sporozoite lysates, as did

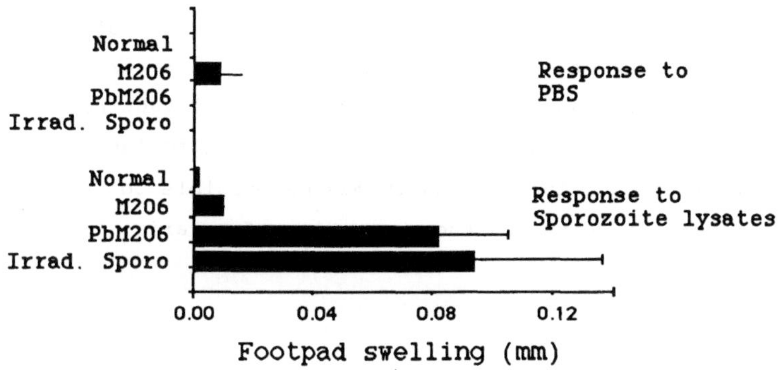

Fig. 5. DTH responses to P. berghei sporozoite antigens were determined by injection of 0.02 ml of a freeze-thawed lysate of sporozoites or PBS into hind footpads. Measurements were taken at 24 hours. Footpad swelling of mice immunized with the transformant vaccine PbM206 or irradiated sporozoites was significantly greater than normal controls or mice receiving nontransformed M206 ($p < 0.05$).

sporozoite-immunized mice (Figure 5), indicating that the vaccine induced a CS antigen-specific cell-mediated immune response. Protection in the absence of antibodies to sporozoites and the development of sporozoite-specific DTH responses suggested that the vaccine may have protected via cell-mediated mechanisms. Additional studies using partial CS genes have demonstrated that regions present on the amino and carboxy terminals of the gene are important in the protection mediated by these vaccines. Furthermore, concomitant administration of suboptimal doses of protective CS monoclonal antibodies enhances the efficacy of orally administered vaccines, suggesting a rationale for combined vaccination strategies in future clinical trials. Vaccine failures may have resulted from plasmid instability or variable levels of CS expression in the transformants, and current research efforts in our laboratories are directed at methods of producing stable, high expression constructs including cloning CS genes into the Salmonella chromosome.

These studies were the first to specifically identify the CS protein as a target antigen for protective cell-mediated immunity against sporozoites, and they demonstrate the feasibility of constructing analogous oral vaccines against human malarias. *S. typhi* mutants have been used safely in humans as oral vaccines and elicit cellular immunity against *S. typhi* antigens (Tagliabue et al.,1985; Levine et al., 1987). Thus it may be possible to develop an orally administered vaccine for the prevention of malaria.

ISSUES YET TO BE RESOLVED

Despite the immense efforts and progress that have been made in this field over the past decade, an efficacious vaccine for *P. falciparum* has yet to be identified. A major issue now confronting the field is how to optimally conduct efficacy trials. By any reckoning, the challenge models that have been employed to date have used an intense sporozoite challenge. Recent estimates by field workers have indicated that the sporozoite burden of a typical naturally infected Anopheline mosquito is on the order of 100 to 1000 sporozoites per gland (J. C. Beier, personal communications). Careful malariometric field studies in regions highly endemic for this disease indicate that nearly 100% of the population becomes infected at biting rates of approximately 1 infected bite per 24 hours (Hoffman et al., 1987). These rates must be contrasted with the experimental challenge in which five laboratory infected mosquitoes, each bearing sporozoite burdens of 1,000 to 100,000 per gland, to feed within a 5 to 20 minute period (Chulay et al., 1986). Protection of mice with monoclonal antibodies is clearly a function of sporozoite inoculum (Potocnjak et al., 1980), suggesting that the intense challenges employed in the vaccine trials enhance the risk of incorrectly rejecting vaccines that would be efficacious under more realistic settings. A related issue is the absence of an assay that is predictive of protection in human volunteers. Human challenge trials are expensive, time consuming, and potentially hazardous, and the identification of such an assay would be extremely useful. Among those assays currently available, the ELISA, CSP reaction, and IFA do not appear to be sufficiently discriminatory. While the ISI assay has shown high activity in the sera of both protected human volunteers, and low activities in unprotected volunteers (Hollingdale et al., in preparation), the assay is technically difficult and its validation as a predictive assay requires the accumulation of additional sera from immunized and protected volunteers. These issues can all be overcome, but they will require the continued commitment of both government and industry to malaria vaccine development programs.

REFERENCES

Ballou, W.R., Rothbard, J., Wirtz, R.A., Gordon, D.M., Williams, J.S., Gore, R.W., Schneider, I., Hollingdale, M.R., Beaudoin, R.L., Maloy, W.L., Miller, L.H., Hockmeyer, W.T. Immunogenicity of synthetic peptides from circumsporozoite protein of Plasmodium falciparum. Science. 228:996 (1985).

Ballou, W.R., Hoffman, S.L., Sherwood, J.A., Hollingdale, M.R., Neva, F.A., Hockmeyer, W.T., Gordon, D.M., Schneider, I., Wirtz, R.A., Young, J.F., Wasserman, G.F., Reeve, P., Diggs, C.L., Chulay, J.D. Safety and efficacy of a recombinant DNA Plasmodium falciparum sporozoite vaccine. Lancet. i:1277 (1987).

Chen, D.H., Tigelarr, R.E., Weinbaum, F.I. Immunity to sporozoite induced malaria infection in mice. I. The effect of immunization of T and B cell deficient mice. J Immunol. 188:1322(1977).

Chulay, J.D., Schneider I., Cosgriff T.M., Hoffman S.L., Ballou, W.R., Quakyi, I.A., Carter, R., Trosper, J.H., Hockmeyer, W.T. Malaria transmitted to humans by mosquitoes infected from cultured Plasmodium falciparum. Am J Trop Med Hyg. 35:66, 1986.

Clough, E.R., Audibert, F.M., Barnwell, J.W., Schlesinger, D.H., Arnon, R., Chedid, L.A. Biologically active antibodies elicited by a synthetic circumsporozoite peptide of Plasmodium knowlesi administered in saline with a muramyl dipeptide derivative. Inf Immun. 48:839, 1985.

Clyde, D.F., McCarthy, V.C., Miller, R.M., Woodward, W.E. Immunization of man against falciparum and vivax malaria by use of attenuated sporozoites. Am J Trop Med Hyg. 24:397 (1975).

Dame, J.B., Williams, J.L., McCutchan, T.F., Weber, J.L., Wirtz, R.A., Hockmeyer, W.T., Maloy, W.L., Haynes, J.D., Schneider, I., Roberts, D., Sanders, G.S., Reddy, E.P., Diggs, C.L., Miller, L.H.. Structure of the gene encoding the immunodominant surface antigen on the sporozoite of the human malaria parasite Plasmodium falciparum. Science. 225:593 (1984).

Egan, J.E., Weber, J.L., Ballou, W.R., Hollingdale, M.R., Majarian,W.R., Gordon, D.M., Maloy,W.L., Hoffman, S.L., Wirtz, R.A., Schneider, I., Woollett, G.R., Young, J.F., Hockmeyer, W.T. Efficacy of murine malaria sporozoite vaccines: Implications for human vaccine development. Science. 236:453 (1987).

Ferreira, A., Schofield, L.A., Enea, V., Schellekens, H., Meide, P., Collins, W.E., Nussenzweig, R.S., Nussenzweig, V. Inhibition of development of exoerythrocytic forms of malaria parasites by gamma interferon. Science. 232:881, 1986.

Good, M.F., Berzofsky, J.A., Maloy, W.L., Hayashi, Y., Fujii, N., Hockmeyer, W.T., Miller, L.H. Genetic control of the immune response in mice to Plasmodium falciparum sporozoite vaccine. Widespread nonresponsiveness to single malaria T epitope in highly repetitive vaccine. J Exp Med. 164:655 (1986).

Good, M.F., Maloy, W.L., Lunde, M.N., Margalit, H., Cornette, J.L., Smith, G.L., Moss, B, Miller, L.H., Berzofsky, J.A. Construction of synthetic immunogen: Use of new T-helper epitope on malaria circumsporozoite protein. Science. 235:1059 (1987).

Good, M.F., Pombo, D., Quakyi I.A., Riley, E.M., Houghten R.A., Menon, A., Alling, D.W., Berzofsky, J.A., Miller, L.H. Human T cell recognition of the circumsporozoite protein of Plasmodium falciparum: Immunodominant T cell domains map to the polymorphic regions of the molecule. Proc Natl Acad Sci, USA. 85:1199, 1988.

Guidice, G.D., Engers, H.D., Tounge, C., Biro, S.S., Weiss, N., Verdini, A.S., Pessi, A., Degremont, A.A., Freyvogel, T.A., Lambert, P.H., Tanner, M. Antibodies to the repetitive epitope of Plasmodium falciparum circumsporozoite protein in a rural Tanzanian community: a longitudinal study of 132 children. Am J Trop Med Hyg. 36:203, 1987.

Granden, M., Fridell, E., Kindmark, C.O. Intradermal immunization with reduced doses of human diploid cell strain rabies vaccine: Evaluation of antibody responses by ELISA and mixed hemadsorption test. Scand J Infect Dis. 17:173, 1985.

Herrington, D.A., Clyde, D.F., Losonsky, G., Cortesia, M., Murphy, J.R., Davis, J., Baqar, S., Felix, A.M., Heimer, E.P., Gillessen, D., Nardin, E., Nussenzweig, R.S., Nussenzweig, V., Hollingdale, M., Levine, M.M. Safety and immunogenicity in man of a synthetic peptide malaria vaccine against Plasmodium falciparum sporozoites. Nature. 328:257 (1987).

Hoffman, S.L., Oster, C.N., Plowe, C.V., Hollingdale, M.R., Beier, J.C., Chulay, J.D., Wirtz, R.A., Woollett G.R. Mugambi, M. Malaria not prevented by naturally acquired sporozoite antibodies: Implications for vaccine development. Science. 237:639 (1987).

Hoffman, S.L., Cannon, L.T., Berzofsky, J.A. Plasmodium falciparum: Sporozoite boosting of immunity due to a T cell epitope on a sporozoite vaccine. Exp Parasitol. 64:64 (1987).

Hollingdale, M.R., Nardin E.H., Tharavanij, S., Schwartz, A.L., Nussenzweig R.S. Inhibition of entry of Plasmodium falciparum and P. vivax sporozoites into cultured cells: an in vitro assay of protective antibodies. J Immunol. 132:909, 1984.

Levine, M.M., Herrington, D., Murphy, J.R., Morris, J.G., Losonsky, G., Tall, B., Lindberg, A.A., Svenson, S., Baqar, S., Edwards, M.F., Stocker, B. Safety, infectivity, immunogenicity, and in vivo stability of two attenuated auxotrophic mutant strains of Salmonella typhi, 541Ty and 543Ty, as live oral vaccines in humans. J Clin Invest. 79:888, 1987.

Lowell, G.H., Ballou, W.R., Smith, L.H., Wirtz, R.A., Zollinger, W.D., Hockmeyer, W.T. Proteosome-lipopeptide vaccines: enhancement of immunogenicity for malaria CS peptides. Science. 240:800, 1988.

Nardin, E.H., Gwadz R.W., Nussenzweig, R.S. Characterization of sporozoite surface antigens by indirect immunofluoresence: detection of stage and species-specific antimalarial antibodies. Bull WHO. 57:211, 1979.

Nussenzweig, R., Vanderberg, J., Most, H. Protective immunity produced by the injection of x-irradiated sporozoites of Plasmodium berghei. IV. Dose response, specificity and humoral immunity. Milit Med. 134:1176 (1969).

Potocnjak, P., Yoshida, N., Nussenzweig, R., Nussenzweig, V. Monovalent fragments (Fab) of monoclonal antibodies to a sporozoite surface antigen (Pb44) protect mice against malaria infection. J Exp Med. 151:1504 (1980).

Richards, R.L., Hayre, M.D., Hockmeyer, W.T., Alving, C.R. Liposomes, lipid A, and aluminum hydroxide enhance the immune response to a synthetic malaria sporozoite antigen. Inf Immun. 56:682, 1988.

Rieckmann, K.H., Beaudoin, R.L., Cassells, J.S., Sell, K.W. Use of attenuated sporozoites in the immunization of human volunteers against falciparum malaria. Bull WHO. 57:261 (1979).

Rutgers, T., Gordon, D., Gathoye, A.M., Hockmeyer, W.T., DeWilde, M., Rosenberg, M. Hepatitis B surface antigen as carrier matrix for the repetitive epitope of the circumsporozoite protein of Plasmodium falciparum. Biotechnology. (in press).

Sadoff, J.C., Ballou, W.R., Baron, L.S., Majarian, W.R., Brey, R.N., Hockmeyer, W.T., Young, J.F., Cryz, S.J., Ou, J., Lowell, G.H., Chulay, J.D. Oral Salmonella typhimurium vaccine expressing circumsporozoite protein protects against malaria. Science. 240:336 (1988).

Schofield, L., Villaquiran, J., Ferreira, A., Schellekens, H., Nussenzweig, R., Nussenzweig, V. g Interferon, CD8+ T cells and antibodies required for immunity to malaria sporozoites. Nature. 330:664 (1987).

Tagliabue, N.L., Caffarena A., Villa, L., Borasci, D., Cazzola, G., Cavalieri, S. Cellular immunity against Salmonella typhi after live oral vaccine. Clin Exp Immunol. 62:242, 1985.

Vanderberg, J., Nussenzweig, R., Most, H. Protective immunity produced by the injection of X-irradiated sporozoites of Plasmodium berghei. V. In vitro effects of immune serum on sporozoites. Milit Med. 134:1183 (1969).

Verhave, J.P., Strickland, G.T., Jaffe, H.A., Ahmed, A. Studies on the transfer of protective immunity with lymphoid cells from mice immune to malaria sporozoites. J Immunol. 121:1031 (1978).

Weiss, W.R., Sedegah, M., Beaudoin, R.L., Miller, L.H., Good, M.F. CD8+ T cells (cytotoxic/suppressors) are required for protection in mice immunized with malaria sporozoites. Proc Natl Acad Sci, USA. 85:573 (1988).

Wirtz, R.A., Ballou, W.R., Schneider, I., Chedid, L., Gross, M.J., Young, J.F., Hollingdale, M.R., Diggs, C.L., Hockmeyer, W.T. Plasmodium falciparum: Immunogenicity of circumsporozoite protein constructs produced in Escherichia coli. Exp Parasitol. 63:166 (1987).

Young, J.F., Hockmeyer, W.T., Gross, M.T., Ballou, W.R., Wirtz, R.A., Trosper, J.H., Beaudoin, R.L., Hollingdale, M.R., Miller, L.H., Diggs, C.L., Rosenberg, M. Expression of Plasmodium falciparum circumsporozoite proteins in Escherichia coli for potential use in a human malaria vaccine. Science. 228:958 (1985).

Zavala, F., Cochrane, A.H., Nardin, E.H., Nussenzweig, R.S., Nussenzweig, V. Circumsporozoite proteins of malaria parasites contain a single immuno-dominant region with two or more identical epitopes. J Exp Med. 157:1947 (1983).

DEVELOPING A GONOCOCCAL PROTEIN I VACCINE

M.S. Blake, L.M. Wetzler, E.C. Gotschlich, and P.A. Rice

Laboratory of Bacteriology and Immunology
The Rockefeller University
N.Y., N.Y.

The Maxwell Finland Laboratory for Infectious Diseases
Boston University
Boston, MA

Introduction

After many years of intensive research by numerous investigators, a safe and effective gonorrheal vaccine remains elusive. Our attention has focused on the outer membrane of the gonococcus and the potential advantages or disadvantages in the use of some of these gonococcal components for such a vaccine. The use of chromatographically purified products was selected over using boiled organisms (Braude et al., 1985) or delipidized gonococcal membrane vesicles (Arminjon et al., 1987; Buchanan and Arko, 1977) in order for us to monitor and assure the consistency of the vaccine. Reported here are some of our efforts on one such potential vaccine candidate, gonococcal protein I, and some of our studies on a minor but potentially hazardous contaminant, protein III.

Protein I

Protein I (PI) is the major outer membrane protein of the gonococcus and constitutes up to 60% of the membrane protein of the organism (Blake and Gotschlich, 1982). We have developed a method to extract and purify this protein from gonococci (Blake and Gotschlich, 1984), and have shown that when PI is purified in this manner and added to artificial lipid bilayers it produces an aqueous channel within these membranes. Thus, we believe that it forms a pore in the gonococcal outer membrane (Douglas et al., 1981; Lynch et al., 1983; Young et al., 1983) and as such seems extremely important for the survival of the organism. Such proteins in gram negative organisms are called porins. The studies of other gram negative porins (Garavito and Rosenbusch, 1980; Garavito et al., 1983) have demonstrated that they have a complex tertiary structure with a large majority of the protein containing a β-sheet conformation and that they associate into trimers. Our current data suggests that the gonococcal PI falls into the same category of proteins. It is a porin, forms trimers (Blake and Gotschlich, 1982) and is structurally related to the porins of *Escherichia coli* (Gotschlich et al., 1987).

Gonococcal PI has distinct advantages as a gonococcal vaccine. It is surface-exposed (Blake et al., 1981) and antibodies produced to PI epitopes have the potential of being protective. Both Joiner et al. (1985b) and Virji et al. (1986; 1987) have demonstrated that monoclonal antibodies (mAbs) to protein Ia (PIA) and protein Ib (PIB) can be bactericidal. Furthermore, mAbs to PI can be opsonic (Virji et al., 1986; Virji et al., 1987) and protect against gonococcal invasion in a cell line assay. Sarafian et al. (1983) demonstrated that Fab portions of PI mAbs can block opsonic activity of human sera. However, since PI most likely has the same complex conformation as other gram negative porins, the presentation of PI in a vaccine that will induce a significantly protective immune response against gonococcal disease needs to be carefully considered.

PI has a limited antigenic repertoire (Sandstrom and Buchanan, 1980; Sandstrom et al., 1982). The genes from each of the major subdivisions of protein I, PIA and PIB, have been cloned and their DNA sequences determined (Gotschlich et al., 1987; Carbonetti and Sparling, 1987). A comparison of the two anticipated amino acid sequences is shown in Figure 1. As can be seen, the similarities (shown by double dots between the two protein sequences) are quite remarkable. Two areas within the protein Ib sequence which are known to be surface-exposed are underlined. The region spanning amino acids 198 to 220 of protein Ib contains sites (shown by a "▾" above the sequence) which are cleaved by chymotrypsin and trypsin when whole cells are exposed to these two enzymes. Furthermore, we have raised monoclonal antibodies which are reactive with a synthetic peptide containing the PIB amino acid sequence between 120 to 134. These monoclonal antibodies cause aggregation of whole cells and are opsonic, suggesting that this area of the molecule is also surface-exposed.

To optimize an effective antibody response to protein I, we examined the combination of protein I with two different adjuvants. We investigated the effect of alum because its potentiation of the immune response had been shown for several different antigens and is currently the only adjuvant approved for human use. Liposomes were also studied because of their known adjuvant effects (Morein et al., 1984; Shek and Sabiston, 1982b; Shek and Sabiston, 1982a; Jiskoot et al., 1986) and because we had previously shown that protein I maintained both its structural and functional conformation within these vesicles (Young et al., 1983). As preparations of the PI-liposomes were made, they were examined to see if the protein I was oriented in the same manner as found in intact organisms. This was accomplished by exposing the liposomes to proteolytic enzymes known to cleave protein I on gonococci. The results

```
         10        20        30        40        50
MKKSLIALTLAALPVAAMADVTLYGTIKAGVETSRSVAHHGAQADRVKTATEIADLGSKI
::::::::::::::::::::::::: ::::: : ::: :       : :  :::: ::::
MKKSLIALTLAALPVAAMADVTLYGAIKAGVQTYRSVEHTDGKVSKVETGSEIADFGSKI
         10        20        30        40        50

         70        80        90       100       110
GFKGQEDLGNGLKAIWQLEQKAYVSGTDTGWGNRQSFIGLKGGFGKVRVGRLNSVLKDTG
:::::::::::::: ::::: : : :: ::::: ::: :::::::  : : ::: :: ::
GFKGQEDLGNGLKAVWQLEQGASVAGTNTGWGNKQSFVGLKGGFGTIRAGSLNSPLKNTG
         70        80        90       100       110

           130         140       150        160       170
GF-NPWE-GKSYYLGL--SNIAQPEERHVSVRYDSPEFAGFRA-VQYVPNDNSGKNHSES
   : :: ::     :  :  :: : :  ::::::::::::   ::: : :::: :  ::
ANVNAWESGKFTGNVLEISGMAQREHRYLSVRYDSPEFAGFSGSVQYAPKDNSGSNG-ES
         130       140       150       160       170

    180       190       200                        210
YHAGFNYKNSGFFVQYAGFYKRHSYTT-----------------EKHQVHRLVGGYDHDA
:: : :: ::::: ::::   :    : ▼        ▼      :: :::::::::: :
YHVGLNYQNSGFFAQYAGLFQRYGEGTKKIEYDDQTYSIPSLFVEKLQVHRLVGGYDNNA
180       190       200       210       220       230

 220       230        240       250       260       270
LYASVAVQQQDAKLTWRND-NSHNSQTEVAATAAYRFGNVTPRVSYAHGFKGSVYDADND
:: ::: :::::::      :::::::::::::::::::::::::::::::: :  :  :
LYVSVAAQQQDAKLYGAMSGNSHNSQTEVAATAAYRFGNVTPRVSYAHGFKGTVDSANHD
240       250       260       270       280       290

  280       290       300       310       320
NTYDQVVVGAEYDFSKRTSALVSAGWLQRGKGTEKFVATVGGVGLRHKF
:::::::::::::::::::::::::::: :::  : : :   : :::::
NTYDQVVVGAEYDFSKRTSALVSAGWLQGGKGADKIVSTASAVVLRHKF
300       310       320       330       340
```

Fig. 1. A comparison of the amino acid sequences of PIA (top) and PIB (bottom). The symbol : shows identical amino acids and . shows a conservative substitution.

showed that the orientation of protein I in the liposomes was identical to that in the gonococcal outer membrane. Different mixtures of each of the adjuvants were prepared and injected into rabbits. The sera from these different rabbits were examined for anti-protein I antibodies and for the ability of these antibodies to be bactericidal and/or opsonic. After approximately three injections of each of the preparations, high ELISA titers were raised to the protein I from both strain Pgh 3-2 (PIB) and strain UU1 (PIA) (Table I), the alum

TABLE 1

SERUM BACTERICIDAL TITERS

	Antisera	Bactericidal Titer	PI ELISA Titer
Pgh 3.2	Preimmune	1:320	0
	Alum 1:10	1:640	1/250,000
	Alum 1:40	1:640	1/240,000
	Liposome 0.10 mg	1:1280	1/220,000
	Liposome 0.20 mg	1:5120	1/180,000
	Freund's	1:5120	1/2,400,000
	Whole Org.	1:20,480	1/230,000
UU1	Preimmune	1:1280	0
	Alum no. 1	1:1280	1/200,000
	Alum no. 2	1:1280	1/94,000
	Liposome no. 1	1:5120	1/24,000
	Liposome no. 2	1:2560	1/24,000
	Freund's	1:1280	1/270,000

preparation being higher in each case. However, when these same sera were tested in either a bactericidal assay or an opsonophagocytosis assay, the liposome-derived antisera consistently gave increased bactericidal titers (Table 1) and opsonizing ability (Table 2) compared with the preimmune sera or the alum-derived antisera. This suggested to us that a greater proportion of antibodies generated with PI-liposomes were directed toward the surface-exposed epitopes of these molecules. To test this hypothesis, we used a synthetic peptide containing the PIB sequence 215 to 229, a known surface-exposed portion of PIB, in an ELISA assay. The reactivity of each of the PIB sera to this peptide was determined. The titers for the PIB-liposome generated sera were three times higher than those of the PIB-alum generated antibodies. Furthermore, in an ELISA inhibition assay, the ability of gonococci to remove the antibodies from each of the sera was examined. The antisera raised utilizing PIB-liposomes retained less than 45% of their reactivity after serial absorptions with strain Pgh 3-2 while the antisera raised utilizing PIB-alum retained greater than 75% of their reactivity. This could be interpreted to mean that 65% of the antibodies raised with PI in liposomes were reactive to exposed portions of PI in the gonococci, while of those antibodies raised using alum only 25% reacted with these epitopes. Similar experiments were performed using the sera raise against PIA from strain UU1 with similar results.

One possible explanation why the PI-alum antisera produced a lower percentage of antibodies to exposed epitopes of PI is that alum is ionic in nature. The alum might bind to the ionic, hydrophilic, exposed portions of PI and mask these epitopes from the immune system, whereas when PI is

TABLE 2

Opsonophagocytosis Survival

Antisera	Relative Survival
none	0.65
preimmune	0.67
mAb 9A1	0.05
alum-PI	1.39
Liposome-PI	0.05
Freunds-PI	0.07

incorporated into liposomes, it seems to mimic the orientation of the PI molecule in gonococci. The antibodies produced utilizing this method are therefore raised only to surface-exposed portions of the protein. Jiskoot et al. (1986) recently published the results of a study in which they raised antisera to PI utilizing alum, detergents and liposomes in various combinations. They raised the highest ELISA antibody titers to PI with alum, similar to what we observed. However, they did not characterize the functional activity of the antibodies.

Protein III

In keeping with our goal of using chromatographically pure reagents in the vaccine and in order to analyze its consistency, we felt it necessary to take into consideration another gonococcal outer membrane protein, protein III (PIII), which was sometimes present in our PI preparations. Although the PIII contamination was usually less than 0.5% of the total protein content, we elected to study this protein directly in order to more fully understand the consequences of such contamination of the protein I.

The PIII protein was first noted by McDade and Johnston (1980), who described its presence in all gonococci and its change in migration on SDS-PAGE upon treatment with a reducing agent. Other studies using surface cross-linking reagents showed that PIII was intimately associated with PI molecules in the membrane (McDade and Johnston, 1980; Newhall et al., 1980). Using peptide mapping studies, Judd showed that PIII was identical in all strains of gonococci (Judd, 1982a; Judd, 1982b). Surface-labelling and ability to react *in vitro* with monoclonal antibodies indicated that PIII had a surface-exposed domain (Swanson et al., 1982). Using a modification of our PI purification scheme, we developed a method for purifying PIII (Lytton and Blake, 1986). The amino terminal sequence was

```
proOmpA   MKKTAIAIAVALAGFATVAQAAPKDNTWYTGAKLGWSQYH
                  10        20        30        40
proOmpA   DTGFINNNGPTHENQLGAGAFGGYQVNPYVGFEMGYDWLG
                  50        60        70        80
proOmpA   RMPYKGSVENGAYKAQGVQLTAKLGYPITDDLDIYTRLGG
                  90       100       110       120
                      10        20        30
proPIII       MTKQLKLSALFVALLASGTAVAGEASVQGYTVSGQS

proOmpA   MVWRADTKSNVYGKNHDTGVSPVFAGGVEYAITPEIATRL
                 130       140       150       160
            40        50        60        70
proPIII   NEIVRNNYGECWKNAYFDKASQGRVECGDAVAVPEPEPAP
                                             :..:.
proOmpA   EYQWTNNIGDAHTIGTRPDNGMLSLGVSYRFGQGEAAPVV
                 170       180       190       200
            80        90       100       110
proPIII   VAVVEQAPQYVDETISLSAKTLFGFDKDSLRAEAQDNLKV
          ... ..::.  .. ..:....::.:.:..:..:.:..:..
proOmpA   APAPAPAPEVQTKHFTLKSDVLFNFNKATLKPEGQAALDQ
                 210       220       230       240
           120       130       140       150
proPIII   LAQRLSR--TNVQSVRVEGHTDFMGSEKYNQALSERRAYV
          :  .::.  ..  :: : :.:: .::. :::.::::::
proOmpA   LYSQLSNLDPKDGSVVVLGYTDRIGSDAYNQGLSERRAQS
                 250       260       270       280
           160       170       180       190
proPIII   VANNLVSNGVPASRISAVGLGESQAQMTQVCQAEVAKLGA
          :.. :.:.:.::..::: :.:::..  ...:
proOmpA   VVDYLISKGIPADKISARGMGESNPVTGNTC---------
                 290       300       310
           200       210       220       230
proPIII   KASKAKKREALIACIEPDRRVDVKIRSIVTRQVVPARNHH
            ...:.:.:::.:..:::::......:
proOmpA   --DNVKQRAALIDCLAPDRRVEIEVKGIKDVVTQPQA
                  320       330       340
proPIII   QH
```

Fig. 2. A comparison of the amino acid sequences of PIII and *E. coli* OmpA.

determined, polyclonal were sera raised to the protein, and the structural gene was cloned in a *lambda* gt11 expression vector (Gotschlich et al., 1986). The clone produced a full-sized, immunologically reactive PIII product upon western blot analysis and the typical increase in apparent MW in SDS-PAGE upon reduction. The PIII gene was sequenced by the chain termination method, and the translated results are shown in figure 2. Surprisingly, the calculated MW's for the complete protein and for the mature protein were 25,544 and 23,298 respectively. This was almost 8,000 less than expected from the migration on SDS-PAGE. However, a number of lines of evidence support the belief that PIII migrates aberrantly on SDS-PAGE. The sequence shown in figure 2 contains the complete sequence of the protein. In our attempts to confirm the above sequence, we tried to determine the carboxyl-terminal residue of gonococcal PIII by using several different methods of carboxypeptidase Y digestion. It was evident from these experiments that the carboxyl-terminus was blocked or unavailable for cleavage.

When the PIII sequence was compared to known sequences of other proteins, a homology to all enterobacterial OmpA proteins was found. The homology to *E. coli* OmpA is illustrated in figure 2. Note that as the similarity between OmpA and PIII begins there is a proline-rich region where OmpA has 5 proline residues interspersed with 6 valines or alanines. The same feature is seen in PIII except that in addition to the 6 valines or alanines the area also contains 3 glutamates and 1 glutamine. Thereafter the homology is very striking with the exception of the terminal 14 amino acids in PIII. Furthermore, we have compared the cyanogen bromide generated fragments from the purified cloned product to those generated by the purified gonococcal product. All the peptides had similar migrations with the exception of one in each case. The peptide from the gonococcal PIII was isolated using reverse phase HPLC on a diphenyl column and subjected to amino acid sequencing. This revealed that the sequence began at residue 183 and that this peptide contained the carboxyl-terminus. In addition to our failure to obtain a free residue upon carboxypeptidase Y digestion, the fact that this peptide fragment from the gonococcal PIII migrated approximately 2,000 daltons higher than did the fragment from the cloned product would suggest that an additional substance may be covalently bound to the carboxyl-terminus. Such a terminal addition does not seem to be the case with the *E. coli* OmpA protein. However, as one can see in figure 2, the carboxyl-terminus of PIII diverges at position 224 from the sequence of the *E. coli* OmpA. This might suggest that even though these two proteins have a high degree of structural homology, they may orient themselves differently in their respective outer membranes.

The surface-exposed regions of the OmpA of *E. coli* have been defined using several phages that use OmpA as a receptor (Datta et al., 1977; van Alphen et al., 1977). From these studies and from a comparison in the enzymatic cleavage of OmpA from intact organisms and cell envelopes (Chen et al., 1980), a model for the orientation of the OmpA protein within the membrane has been proposed. In this model, four surface-exposed loops are envisioned as starting very close to the amino terminal end of the molecule and terminating in the proline-rich area. The remainder of molecule containing the carboxyl-terminus forms a large periplasmic portion of the protein beginning at residue 198. This model would predict that the homology of the PIII sequence with that of OmpA begins where the OmpA molecule is within the periplasmic space. If the analogy and model of the two proteins is in fact correct, this would leave only residues 23 - 70 available for exposure on the surface of the gonococcus while the rest of the PIII molecule would be periplasmic. If, however, PIII were to be located in the membrane such that the homologous portions were to extend externally, then possible antibodies reactive to PIII might occur by prior exposure to other gram negative bacteria with an OmpA-like protein.

The function of PIII in gonococcal infection and\or outer membrane structure is unknown. However, it has been shown in very carefully controlled *in vitro* studies that human complement-fixing IgG antibodies to PIII were able to interfere with the bactericidal activity of IgG antibodies directed to other gonococcal surface antigens (Joiner et al., 1985a; Rice et al., 1985). Furthermore, using PIII affinity columns to remove and purify human PIII antibodies from a convalescent serum, we were able to show that the bactericidal effect of this serum upon the infecting strain could either be enhanced or decreased depending on the levels of PIII antibodies. Thus, this serum devoid of all PIII reactive antibodies was effective in killing the gonococci. However, as the antibodies to PIII increased the serum became less efficient. At a level equal to that in unadsorbed convalescent serum, the bactericidal activity of the serum was totally absent (Rice et al., 1986). These observations might explain why PIII is so highly conserved, in contrast to many other gonococcal surface proteins which exhibit considerable antigenic variation. Being highly conserved, PIII would elicit antibodies which would then decrease the potency of the bactericidal antibodies raised during the gonococcal infection. We are currently attempting to map the epitopes recognized by the anti-PIII blocking IgG antibodies. Our current data would suggest that many of these antibodies are directed towards PIII epitopes which are shared with OmpA. Using each of the cyanogen bromide derived fragments in an ELISA assay, we determined that many of the human anti-PIII antibodies in "blocking" sera were reactive to the fragment which contained the sequence between 139 to 180. As shown in figure 2, this area is quite homologous with the OmpA protein.

Through the generosity of Dr. Ulf Henning of The Max-Planck-Institut, we obtained some purified *E. coli* OmpA protein with which we constructed an affinity column. Human sera containing IgG "blocking" antibodies were passed through the column and reassessed for blocking activity. The preliminary data would suggest that most if not all of the "blocking" IgG remained attached to the OmpA column. These antibodies could be eluted and added back to the serum, which restored their "blocking" activity. These surprising results implied that the human "blocking" antibodies (1) were directed toward the OmpA-like regions on PIII and (2) did not conform to the proposed orientation of OmpA in *E. coli*; that is, in the OmpA model the carboxyl-terminus is located within the periplasm and <u>not</u> surface-exposed, whereas it seems that at least portions of the OmpA-like areas of PIII are available on the surface of the gonococci. To test this last hypothesis, we used a whole cell ELISA inhibition assay. Microtiter plates were sensitized with

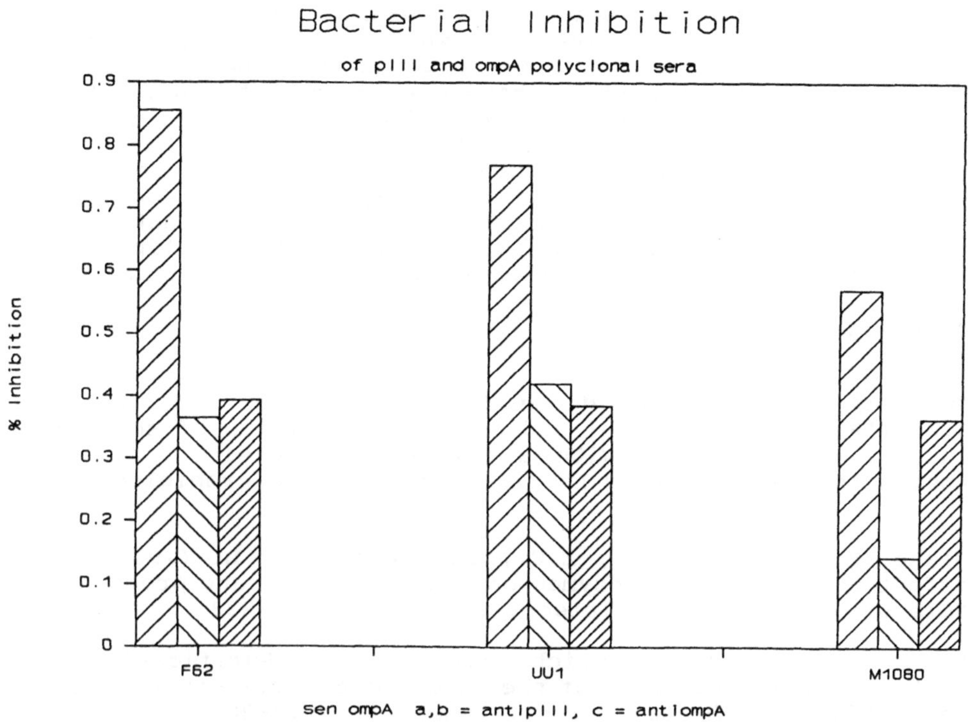

Fig. 3. Whole cell ELISA inhibition of rabbit polyclonal sera raised against purified PIII (a & b) and purified *E. coli* OmpA (c). F62 and UU1 are two gonococcal strains and M1080 is a meningococcal stain.

purified PIII and the polyclonal rabbit sera titered to give a half maximal reaction. At this dilution gonococci were added, incubated with the sera, and then removed by centrifugation. The supernate containing those antibodies which did not react with the intact gonococci were added to the pre-sensitized, washed plates. The ELISA assay was then performed as usual. The results of one of these assays is shown in figure 3. The data obtained using the anti-OmpA serum would suggest that approximately 40% of the OmpA cross-reactive epitopes on PIII are available on the surface of gonococcal organisms. With the use of monoclonal antibodies reactive to PIII, we are attempting to more clearly define the epitopes on PIII which are involved in this IgG "blocking" antibody activity. We are also interested in how these human antibodies are generated -- in particular, whether infections by *E. coli* or other Neisserial species might also elicit these "blocking" antibodies -- as well as in what effect these antibodies would have in opsonic tests and in the fallopian tube model of gonococcal pathogenesis.

After thorough examination of the data obtained from the PIII studies, we decided that even a 0.5% contamination of PIII in our protein I preparation was not tolerable. To eliminate PIII entirely from our preparation, we are in the process of constructing gonococcal mutants in which the gene for PIII has been deleted. At present, we have several mutants in which the β-lactamase gene has been inserted into the structural gene for PIII. This construction has been transformed back into gonococci. On SDS-PAGE western blot analysis of whole cell lysates of these mutants, no detectable PIII was observed. Thus, we believe protein I without the PIII contamination can be purified from these mutants, incorporated into liposomes, and potentially used as a safe, effective gonococcal vaccine.

References

Arminjon, P., Cadoz, M., Morse, S.A., Rock, J.P., and Sarafian, S.K. (1987). Bactericidal and opsonic activities of sera from individuals immunized with a gonococcal protein I vaccine. Abstracts Annual Meeting American Society of Microbiology 118.

Blake, M.S., Gotschlich, E.C., and Swanson, J.L. (1981). The effect of proteolytic enzymes on the outer membrane proteins of *Neisseria gonorrhoeae*. *Infect. Immun*. 33, 212-222.

Blake, M.S., and Gotschlich, E.C. (1982). Purification and partial characterization of the major outer membrane protein of *Neisseria gonorrhoeae*. *Infect. Immun*. 36, 277-283.

Blake, M.S., and Gotschlich, E.C. (1984). Purification and partial characterization of the opacity-associated proteins of *Neisseria gonorrhoeae*. *J. Exp. Med*. 159, 452-462.

Braude, A., Wunderlich, A., Chikami, G., Douglas, H., and McCutchan, J. (1985). Generation of adherence-blocking immunoglobin G (IgG) by vaccinating women with nonpiliated boiled gonococci: Observations on IgG binding to a high-molecular-weight protein in the outer membrane fraction. In The Pathogenic Neisseriae, G.K. Schoolnik, ed. (Washington D.C.: American Society of Microbiology), pp. 323-328.

Buchanan, T.M., and Arko, R.J. (1977). Immunity to gonococcal infection induced by vaccination with isolated outer membranes of *Neisseria gonorrhoeae* in guinea pigs. *J. Infect. Dis*. 135, 879-887.

Carbonetti, N.H., and Sparling, P.F. (1987). Molecular cloning and chracterization of the structural gene for protein I, the major outer membrane protein of *Neisseria gonorrhoeae*. *Proc. Nat. Acad. Sci. USA*. 84, 9084-9088.

Chen, R., Schmidmayr, W., Kraemer, C., Chen-Schmeisser, U., and Henning, U. (1980). Primary structure of major outer membrane protein II* (ompA protein) of *Escherichia coli* K-12. *Proc. Nat. Acad. Sci. USA*. 77, 4592-4596.

Datta, D.B., Arden, B., and Henning, U. (1977). Major proteins of the *Escherichia coli* outer cell envelope membrane as bacteriophage receptors. *J. Bacteriol.* 131, 821-829.

Douglas, J.T., Lee, M.D., and Nikaido, H. (1981). Protein I of *Neisseria gonorrhoeae* outer membrane is a porin. FEMS *Microbiol. Lett.* 12, 305-309.

Garavito, R.M., Jenkins, J., Jansonius, J.N., Karlsson, R., and Rosenbush, J.P. (1983). X-ray diffraction analysis of matrix porin, an integral membrane protein from *Escherichia coli* outer membranes. *Journal of Molecular Biology* 164, 313-327.

Garavito, R.M., and Rosenbusch, J.P. (1980). Three-dimensional crystals of an integral membrane protein: An initial X-ray analysis. *J. Cell Biol.* 86, 327-329.

Gotschlich, E.C., Blake, M.S., Koomey, J.M., Seiff, M., and Derman, A. (1986). Cloning of the structural genes of three H8 antigens and of protein III of *Neisseria gonorrhoeae*. *J. Exp. Med.* 164, 868-881.

Gotschlich, E.C., Seiff, M.E., Blake, M.S., and Koomey, M. (1987). Porin protein of *Neisseria gonorrhoeae*: cloning and gene structure. *Proc. Nat. Acad. Sci. USA.* 84, 8135-8139.

Jiskoot, W., Teerlink, T., Van Hoof, M.M.M., Bartels, K., Kanhai, V., Crommelin, D.J.A., and Beuvery, E.C. (1986). Immunogenic activity of gonocococcal protein I in mice with three different lipoidal adjuvants delivered in liposomes and in complexes. *Infect. Immun.* 54, 333-338.

Joiner, K.A., Warren, K.A., Frank, M.M., and Rice, P.A. (1985a). Blocking immunoglobulin G enhances complement consumption and deposition on *Neisseria gonorrhoeae*. In The Pathogenic Neisseriae, G.K. Schoolnik, ed. (Washington, D.C.: Amer. Soc. Microbiology), pp. 431-434.

Joiner, K.A., Warren, K.A., Tam, M., and Frank, M.M. (1985b). Monoclonal antibodies directed against gonococcal protein I vary in bactericidal activity. *The Journal of Immunology* 134, 3411-3419.

Judd, R.C. (1982a). ^{125}I-Peptide mapping of protein III isolated from four strains of *Neisseria gonorrhoeae*. *Infect. Immun.* 37, 622-631.

Judd, R.C. (1982b). Surface peptide mapping of protein I and protein III of four strains of *Neisseria gonorrhoeae*. *Infect. Immun.* 37, 632-641.

Lynch, E.C., Blake, M.S., Gotschlich, E.C., and Mauro, A. (1983). Reconstitution of a voltage-dependent anion-preferring porin from *Neisseria gonorrhoeae*. *Biophys. J.* 41, 62.

Lytton, E.J., and Blake, M.S. (1986). Isolation and partial characterization of the reduction-modifiable protein of *Neisseria gonorrhoeae*. *J. Exp. Med.* 164, 1749-1759.

McDade, R.L., Jr., and Johnston, K.H. (1980). Characterization of serologically dominant outer membrane proteins of *Neisseria gonorrhoeae*. *J. Bacteriol*. 141, 1183-1191.

Morein, B., Sundquist, B., Hoglund, S., Dalsgaard, K., and Osterhaus, A. (1984). Iscom, a novel stucture for antigenic presentationof membrane proteins from enveloped viruses. *Nature* 308, 457-460.

Newhall, W.J., Sawyer, W.D., and Haak, R.A. (1980). Cross-linking analysis of the outer membrane proteins of *Neisseria gonorrhoeae*. *Infect. Immun*. 28, 785-791.

Rice, P.A., Tam, M.R., and Blake, M.S. (1985). Immunoglobulin G antibodies in normal human serum directed against protein III block killing of serum-resistant *Neisseria gonorrhoeae* by immune human serum. In The Pathogenic Neisseriae, G.K. Schoolnik, ed. (Washington, D.C.: Amer. Soc. Microbiology), pp. 427-430.

Rice, P.A., Vayo, H.E., Tam, M.R., and Blake, M.S. (1986). Immunoglobin G antibodies directed against protein III block killing of serum resistant *Neisseria gonorrhoeae* by immune sera. *J. Exp. Med*. 164, 1735-1748.

Sandstrom, E., and Buchanan, T.M. (1980). Coagglutination class reagents identifies the same antigen as the principal outer membrane serotyping. In Genetics and immunobiology of pathogenic *Neisseria*, S. Normark, and D. Danielsson, eds. (Umea: University of Umea), pp. 67-71.

Sandstrom, E.G., Knapp, J.S., and Buchanan, T.M. (1982). Serology of *Neisseria gonorrhoeae*: W-antigen serogrouping by coagglutination and protein I serotyping by ELISA both detect protein I antigens. *Infect. Immun*. 35, 229-239.

Sarafian, S.K., Tam, M.R., and Morse, S.A. (1983). Gonococcal protein I specific opsonic IgG in normal human sera. The *Journal of Infectious Diseases* 148, 1025-1032.

Shek, P.N., and Sabiston, B.H. (1982a). Immune response mediated by liposome associated protein antigens: II. Comparison of the effectiveness of vesicle entrapped and surface associated antigen in immunopotentiation. *Immunology* 47, 627-632.

Shek, P.N., and Sabiston, B.H. (1982b). Immune response mediated by liposome associated protein antigens: I. Potentiation of the plaque-forming cell response. *Immunology* 45, 349-356.

Swanson, J.L., Mayer, L.W., and Tam, M.R. (1982). Antigenicity of _Neisseria_ _gonorrhoeae_ outer membrane protein(s) III detected by immunoprecipitation and western blot transfer with monoclonal antibody. *Infect. Immun*. 38, 668-672.

van Alphen, L., Havekes, L., and Lugtenberg, B. (1977). Major outer membrane protein d of _Escherichia coli_ K-12 Purification and _in vitro_ activity of bacteriophage K3 and F-pilus mediated conjugation. *FEBS Lett.* 75, 285-290.

Virji, M., Zak, K., and Heckels, J.E. (1986). Monoclonal antibodies to gonococcal outer membrane protein IB: use in investigations of the potential protective effect of antibodies directed against conserved and type-specific epitopes. *J. Gen. Microbiol.* 132, 1621-1629.

Virji, M., Fletcher, J.N., Zak, K., and Heckels, J.E. (1987). The potential protective effect of monoclonal antibodies to gonococcal outer membrane protein IA. *J. Gen. Microbiol.* 133, 2639-2646.

Young, J.D.E., Blake, M.S., Mauro, A., and Cohn, Z.A. (1983). Properties of the major outer membrane protein from *Neisseria gonorrhoeae* incorporated into model lipid membranes. *Biophysical J.* 80, 3831-3835.

THE GENETICS OF THE IMMUNE RESPONSE TO TRICHINELLA SPIRALIS ANTIGENS IN THE MOUSE

Michael Robinson and Chella S. David

Department of Immunology, Mayo Clinic, Rochester, MN 55905

Introduction

The production and successful utilization of a defined vaccine against a parasitic infection requires an understanding of the complex processes which occur during the development of most parasites, including the stage of the life cycle which can be effectively combatted by the host's immune system. For the activation of the host's immune system against any antigen there are many genetically controlled physical parameters involved, of which probably the most crucial are the interactions of the T cell receptor and the antigen/Major Histocompatibility Complex (MHC) molecules. In the case of many parasites, these particular molecular interactions have only recently been investigated. Although for several years it had been appreciated that there is a significant genetic element controlling the host's immune response to parasites, the actual nature of these controlling elements had not been elucidated (Wassom, 1985; Behnke and Robinson, 1985; Wakelin, 1985). However, recent appraisals of trends and patterns in the genetic make-up of hosts which can be categorized as good or bad responders to particular parasites has led us to realize that the known genotype of animals which respond to parasitic infections are not just epiphenomena but actually are an indication of controlling molecular properties. This is particularly true of the immune response to Trichinella spiralis in the mouse, the host/parasite system which has revealed the most data concerning the significance of the host genotype (Wassom et al., 1984).

T. spiralis is a nematode parasite which can form patent infections in most mammals (including man) and can therefore be easily investigated in common laboratory animals. However, it is the fact that the parasite forms a natural infection in the laboratory mouse which makes it such a useful model for the study of the immunogenetics of parasitic infections. The host usually acquires the parasite after eating the flesh of a previously infected animal and in the process of digesting the striated muscle, the encapsulated parasites are released into the lumen of the gut as mobile L1 larvae. These larvae then moult repeatedly to become adult worms, at which stage they are sexually mature and can immediately mate. The female parasites then burrow into the cytoplasm of the intestinal mucosa and release viviparous newborn larvae, which circulate the host's body via the lymphatic and venous system and eventually come to rest in the tissues. If the tissue is striated muscle, the larva coils around upon itself within the muscle fibre encysted by host derived material, in which position it will stay and remain viable in man for from six months to 30 years. The intestinal phase of the parasite is then normally rejected during a time span which varies greatly with, amongst other reasons, the number of previous infections and the host genotype. If the infected muscle is eaten by another suitable host, the larvae are released by the digestive secretions and the cycle is repeated. Because man normally eats very few carnivores, the most frequent source of infection is from eating infected meat of the domestic pig, although other animals on occasion, most notably bears, can act as an intermediate host. In laboratory infections, a host's muscle larvae can be released by in vitro digestion and the infective larvae collected and administered to another host. This can be done either orally in controlled quantities to produce a patent infection, or the larvae homogenized to form an antigenic solution, and injected or added to cell cultures during investigations of murine responses in vitro. As is the case with many parasitic organisms, the quantitative and qualitative course of the infection as an indication of the level of resistance of the host can be monitored in many ways. This variation in analysis can by itself be a source of discrepancy in the apparent resistance of the host and forms a marked divergence to the situation which exists when a purely synthetic antigen or one not normally infective, is used in immunological studies. In these situations for in vivo analysis, the quantity and isotype of antibody production is usually measured, while for in vitro experimentation, the main parameter to be measured is the proliferation of cultured lymphocytes. However, for a parasite such as T. spiralis with a relatively complex life cycle and stage specific immunogenicity, a number of methods of analysis of host resistance can be used. These include, but are not necessarily restricted to, LD50, antibody production in infected or immunized animals, in vitro proliferation of cells due to parasite derived antigens, the kinetics and quantitative analysis of rejection of adult worms and the level of larval deposition in the muscle. All of these factors, of course, vary with both the infective dose and the host genotype.

The Genetics of Regulatory Mechanisms in the Immune Response to T. spiralis

Because of these variable criteria, a confusion has arisen in the past over the genetic control of the immune response to T. spiralis in the mouse. Some authors have stated that only non-H-2 genes are involved in many of the observed immune responses to this parasite (Bell, McGregor and Adams, 1982), whilst other authors have postulated that non-H-2 genes control the rejection of adult worms but that genes within the H-2 region control the number of larvae found to encyst within the muscle (Wakelin and Donachie, 1981). Further analysis of data has now shown that both background and H-2 genes are involved in the overall immune response to T. spiralis and this control is reflected when resistance is analyzed by any of the parameters described above (Wassom, 1985). The important factor, however, is the level of infection of the mice. Only at low levels of infection (circa 100L1) can H-2 controlled immunity be readily demonstrated. At higher infection levels (c. 500L1) there is a dose dependent suppression of immunity which overrides the quantitative effect of the host protective response (Wassom et al., 1984). This ability of nematode parasites to immunosuppress the host has been well characterized for other species, most notably Nematospiroides dubius (=Heligmosomoides polygyrus) in the mouse, where it has been shown that overall resistance is a consequence of conflicting host protective immunity and parasite protective measures (Robinson, Behnke and Williams, 1988).

Actual identification of the role of the various genes discovered to be implicated in the immune response of mice to *T. spiralis* is still open to doubt in some cases, particularly the role of genes mapping outside of the H-2 complex. However, using congenic and H-2 recombinant strains of mice, the role of an MHC linked Ia gene (Ts-1) was clearly shown in the immune response to *T. spiralis* (Wassom, David and Gleich, 1979). A second gene, Ts-2, also modulates the hosts susceptibility to *T. spiralis* and has been mapped between the S and D regions (Wassom et al., 1983). Mice expressing the H-2^d haplotype at the D or L region have been shown to have a downregulated response to *T. spiralis* when compared to certain other haplotypes (Wakelin and Donachie, 1983, Wassom, David and Gleich, 1980). It is not known exactly how this Ts-2 gene exerts its effect, or indeed if there is more than one gene exerting a regulatory influence at the designated locus. Certainly there are genes of vital immunological importance which are known to map to this region, most notably those for TNFα and TNFβ, which although implicated in resistance to certain other parasites (Ruddle, 1987) have not been directly linked to resistance to *T. spiralis* infections. Our laboratory has made an interesting recent observation that there is also a hierarchical relationship between the titre of anti-*T. spiralis* muscle larva antiserum (as measured by RIA) and the haplotype possessed at the L region. A mouse possessing the L^k gene produces a greater anti-*T. spiralis* antibody titre than does a mouse which is L^q while a mouse which is L^b produces a higher specific antibody titre than one which is L^d (Table 1). This relationship applies to mice injected with *T. spiralis* muscle larvae homogenate in complete Freund's adjuvant. We are now investigating whether this pattern is repeated during an infection produced by oral inoculation of viable larvae. Whether the "L gene effect" is coincidental to other observed D end gene effects or whether all of these effects are a manifestation of the same gene has yet to be established. While there is no doubt that a clear relationship exists, a protective role for the antisera described has yet to be confirmed.

Table 1

RIA of B10 H-2 Recombinant Sera and *T. spiralis* Muscle Larvae Antigen

Strain	H-2 Genotype K	A	E	S	D	L	cpm*
B10.K	k	k	k	k	k	k	19,774 ± 207
B10.RKD1	k	k	k	d	d	d	14,856 ± 406
B10.AQR/K1	q	k	k	d	d	d	13,460 ± 340
B10.AKM	k	k	k	k	q	q	10,040 ± 102
B10.MBR	b	k	k	k	q	q	7,415 ± 261
B10.RKQ1	k	k	kb		q	q	8,504 ± 4
B10.Q	q	q	q	q	q	q	15,790 ± 560
B10.RQB1	q	q	q	q	b	b	19,363 ± 1,710
B10.RQDB	q	q	q	q	d	b	18,468 ± 644
B10.T(6R)	q	q	q	q	d	d	5,674 ± 16

*Sera dilution = 1/100
T. spiralis antigen at 1 µg/well

Of the Major Histocompatibility Complex (MHC) genes known to be involved in the immune response to *T. spiralis*, investigation of the Ia genes has revealed the most interesting results to date. From initial observations carried out several years ago (Wassom, David and Gleich, 1979), it became apparent that B10 congenic mice of the b, q, s and f haplotypes were better responders to *T. spiralis* infections than mice of other haplotypes (Wassom, 1985). These four haplotypes have genetic defects which result in an inability to express a surface I-E molecule (David, 1976). The I-E molecule has also been implicated in the modulation of the murine immune response to *N. dubius* (Wassom, Krco and David, 1987) and to *Leishmania donovani* (Blackwell and Roberts, 1987) as well as to other non-parasite antigens (Baxavanis et al., 1982; Oliveira et al., 1985).

How the presence of an I-E molecule can act to downregulate the specific immune response to antigens has not yet been established but there are several possibilities. The first of these is that the I-E positive animals produce specific T suppressor cells which inhibit the cellular and humoral response to the parasite (Wassom, Krco and David, 1987). The second possibility is that I-E positive mice produce autoreactive T cells when antigen is presented in the context of the I-E molecule, which can act in some way to inhibit or kill potentially host protective, I-A restricted cells (Quintans et al., 1986). It is known from work carried out previously in this laboratory that B10.BR mice, which are $A^k_\alpha A^k_\beta E^k_\beta E^k_\alpha$, preferentially produce T cell clones against *T. spiralis* which are $E^k_\beta E^k_\alpha$ restricted, although *T. spiralis* muscle larvae antigen is capable of being presented to T cells in the context of I-A by I-E negative mice (Krco et al., 1983) Nevertheless, *T. spiralis* antigen is preferentially presented in the context of the I-E rather than the I-A molecule and this could indicate that the actual method of antigen presentation serves to specifically downregulate the host's immune system.

One further intriguing mechanism by which this downregulation may occur is by the parasite being able to induce the host to produce T helper (Th) cells of either the Th1 type to follow the pathway of delayed type hypersensitivity and cellular cycotoxicity responses, or of the Th2 type to produce a response via antibodies and immediate hypersensitivity (Mossmann et al., 1986). It is entirely possible that antigens being presented to T cells by either I-E or I-A molecules could be the deciding factor as to whether the hosts immune system preferentially goes in the direction of a T cell inflammatory response (Th1) or a more humoral response with production of IgE and IgG1, and anti-parasite effects acting via mast cell inflammatory products (Th2) (Vitetta et al., 1985; Coffman et al., 1986). When originally postulated, Mossman and co-workers stated that for a gut living nematode, e.g. *Nippostrongylus brasiliensis* in mice, the ideal host protective immune response would be via a Th2 type (Coffman et al., 1988). This idea would concur with certain established data, in that good responders, e.g. H-2^q haplotype I-E^- mice, would produce Th2 cells preferentially and hence T cell clones from these strains would produce IL4 and IL5, but not IL2 or INFγ. It has been discovered in this laboratory that B10.Q (H-2^q) mice anti-*T. spiralis* T cell clones are difficult to produce and usually require the addition of exogenous IL2 to keep them growing. As Th2 cells require IL2 but produce little or none of this lymphokine, then this criterion would fit the classification (Powers, Abbas and Miller, 1988). Alternatively, it is equally possible that it is a Th1 type of response which is most effective against gut nematodes and certainly the role of large quantities of IgG1 and IgE in parasite rejection has been frequently questioned (Williams and Behnke, 1983; Miller, 1984), while many authors have accepted the fact that the effective immune response to *T. spiralis* is via a T cell inflammatory mechanism (Larsh and Weatherly, 1975; Almond and Parkhouse, 1985). At the present time we are analyzing culture supernatants from anti-T. spiralis T cell clones to attempt to resolve this situation, but either way there is little doubt now that Th cells form a heterogeneous population and that antigen presentation could well be a factor which influences the final quantities of each type. What has been obvious for a number of years to parasitologists is that there is a marked strain difference in the quantity and isotype of antibody levels in mice to T. spiralis infections. Differences in total and parasite specific immunoglobulin levels had been noted in a number of studies (Crandall and Crandall, 1972; Rivera-Ortiz and Nussenzweig, 1976) and of particular importance was the finding that levels of two Th2 regulated immunoglobulin isotypes namely IgG1 and IgE fluctuated greatly with the strain of the host (Rivera-Ortiz and Nussenzweig, 1976) and also in the case of IgG1 with the level of infection (Golinska and Bany, 1976). Although the significance of these differences was not fundamentally appreciated at the time, the evidence points to an antigen/MHC class II regulation of antibody responses via Th cells.

There is now increasing evidence of a pivotal role for the genes of the T cell receptor (Tcr) and the minor lymhocyte stimulating (Mls) antigens in the regulation of the immune response. The restricted use of certain Tcr genes in T cell clones for a specific antigen indicates that the actual response to antigens is orchestrated by Th cells in a regulated manner (Morel, Livingstone and Fathman, 1987; Winoto et al., 1986; Matis et al. 1987). As the use of such Tcr variable genes is so limited in these situations, it appears likely that the rearrangement of Tcr genes is linked to the effective immune response. Certainly the Tcr gene repertoire is partially controlled by the MHC molecules with which it comes into contact (Blackman et al., 1986), and this has been shown to be manifest in one example by the clonal deletion of $V_\beta 17a$ positive T cells from I-E positive mice (Kappler, Roehm and Marrack, 1987; Kappler et al., 1987). To further complicate matters, the Mls system appears to be intimately involved with MHC antigens and the Tcr gene repertoire. Certain MHC molecules can function as important regulators for T cell recognition of Mls determinants (Abe and Hodes, 1988) while the Mls^a locus has been shown to play a crucial role in the clonal deletion of $V_\beta 8.1$ Tcr gene positive (Kappler et al. 1988) and $V_\beta 6$ Tcr gene positive (MacDonald et al., 1988) cells from the host. Hence the Mls loci and the T cell receptor genes, both seem to be involved in MHC/antigen/T cell receptor interactions and therefore their role in the immune response to T. spiralis, while not yet elucidated, cannot be denied.

The crux of the issue as to whether or not the murine host produces a protective response against T. spiralis, or alternatively fails to activate the immune system to reject the parasite, appears to reside in the interaction between antigens, MHC Class II molecules, and presumably their recognition by the T cell receptor. As T. spiralis muscle larva antigens appear to be preferentially presented by I-E molecules, then it is likely that a component of the parasite antigen preferentially binds to the I-E molecule rather than the I-A. Although unproven, there is circumstantial evidence to suggest that this may be the case. Considerable data has accumulated to suggest that antigens are more readily presented by I-E rather than I-A (or vice versa) (Puri, Lonai and Friedman, 1986) and this preference is related to the structure of the functional agretope (Sette et al., 1988). Also Buss and co-workers (1987) have shown that a major feature determining the antigenicity of a polypeptide segment is the degree of affinity between the antigen and the Ia molecule. As mentioned previously for T. spiralis the degree of immunosuppression induced by the parasite is not only dependent upon the level of infection but also the host H-2 haplotype (Wassom et al., 1984). This leaves a working hypothesis that T. spiralis produces a particular antigen or set of antigens which at low concentrations would be presented extremely efficiently by I-E molecules but less well by I-A. When presented by I-E there is some means of downregulation of the host's immune system, or the specific host responses do not act to efficiently reject the parasite. This would result in the observed difference between I-E$^+$ and I-E$^-$ mice. However, at much higher concentrations of antigens this preferential affinity could be overcome and the antigen more readily presented by I-A and hence the difference between I-E$^+$ and I-E$^-$ mice diminishes. When B10 congenic and H-2 recombinant mice are injected with T. spiralis antigen and serum collected, and immunoprecipitated with ^{125}I-labelled T. spiralis antigen followed by electrophoresis and autoradiography, then I-E$^+$ mice strongly recognize an antigen which is not precipitated by serum from I-E$^-$ mice (Fig. 1).

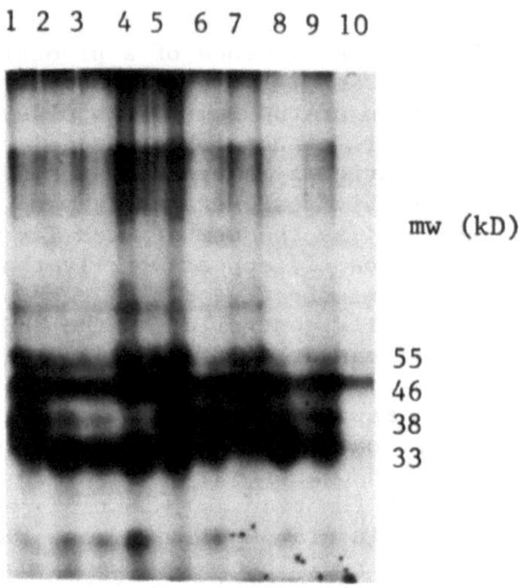

No.	Strain	K	A	E	S	D	L	I-E
		H-2 Genotype						
1	B10.K	k	k	k	k	k	k	+
2	B10.RQB1	q	q	q	q	b	b	−
3	B10.RQDB	q	q	q	q	d	b	−
4	B10.Q	q	q	q	q	q	q	−
5	B10.RKD1	k	k	k	d	d	d	+
6	B10.AQR/K1	q	k	k	d	d	d	+
7	B10.AKM	k	k	k	k	k	q	+
8	B10.RKQ1	k	k	kb		q	q	−
9	B10.MBR	b	k	k	k	q	q	+
10	B10.T(6R)	q	q	q	q	d	d	−

Figure 1. SDS-PAGE analysis of H-2 recombinant mouse sera immunoprecipitated with ^{125}I-labelled T. spiralis muscle larva antigen.

To further investigate the role of the I-E molecule in the immune response to T. spiralis we have begun to examine whether or not it is just the possession of the I-E molecule which is important or whether the degree of immunosuppression induced by I-E molecules is haplotype dependent. From in vivo results it is known that H-2^d mice are better responders, compared to H-2^k for example, and in general within the I-E^+ classification there are detectable H-2 haplotype determined differences in resistance to T. spiralis infections (Wassom et al., 1983). Preliminary data using E_α^k transgenic mice on an H-2^b background ($E_\beta^b E_\alpha^k$) suggest that in fact it is the E_β chain of the I-E heterodimer which is important in the presentation of T. spiralis muscle larvae antigen while the E_α chain has not yet been found to play a significant role (unpublished observations).

The Function and Identity of Antigens Implicated in the Immune Response to Trichinella spiralis

As might be expected for a complex metozoan organism such as T. spiralis, there are numerous antigens which have been identified, many of which are stage specific (Bell, McGregor and Despommier, 1979). Several of these antigens, found in infective L1 larvae, newborn L1 larvae and adults, have been isolated and used to elicit protective responses in mice, rats or pigs (Ortega, Pierres, Mackenzie and Parkhouse, 1984; Silberstein and Despommier, 1984, 1985; Marti, Murrell and Gamble, 1987; Chipman, 1957). However, the picture is far from complete and while many authors agree that newborn larvae are a target for host resistance (Maloney and Denham, 1979; James and Denham, 1974; Kazura and Grove, 1978), there appear to be other stages of the life cycle susceptible to immunological attack, as the rate of expulsion of adult worms varies with the number of the hosts previous exposures to the parasite (Larsh, 1963). This does not necessarily mean that it is the adults themselves which are affected by the hosts response, as damage to intestinal larval stages could result in defective adults with a shortened life span (Silberstein and Despommier, 1985).

The most characterized antigens to date have been those from infective L1 larvae. There are at least 20 of these antigens, many of which are known to be glycoproteins (Despommier and Lacetti, 1981). However, isolation of functional antigens has not been as straightforward as first imagined. For instance, Silberstein and Despommier (1984, 1985) and Gamble (1985) have identified three major antigens of muscle larva which, when injected, could give considerable resistance to a challenge infection in mice. Unfortunately, monoclonal antibodies to the most protective of the three antigens identified by Silberstein and Despommier (of approximate m.w. 48 kD) also bind to many other antigens (Robinson et al., manuscript in preparation). 48 kD protein was initially found to be localized in the cellular granules of a secretory organ called a stichosome, and along with certain other identified host protective antigens of 37 kD and a 50/55 kD complex has also been identified on the surface of infective L1 larvae (Silberstein and Despommier, 1984). It is likely, therefore, that the epitope recognized by anti-48 kD antibody also exists on other antigens and hence it is to be expected that even a monoclonal antibody would bind to more than one solubilized protein. This has led us to suspect that the actual epitopes recognized may not be polypeptides but in fact either a conserved oligosaccharide epitope or phosphorylcholine (Maizels et al., 1987; Ubeira et al., 1987). Initial experiments using periodate oxidation of carbohydrate moieties have tended to rule out an oligosaccharide epitope and the role of phosphorylcholine is still being examined. It is interesting to note, however, that while we have four monoclonal antibodies specific for this host protective 48 kD molecule, the overall pattern of antigens recognized appears individually different. Furthermore, affinity purification of specific muscle larva antigens recognized by the monoclonal antibodies results in a protein mixture which does give considerable protection in CD1 mice against a challenge infection but unfortunately the unbound fraction of muscle larvae homogenate is even more host protective and induces greater cell proliferation *in vitro* in primed mouse cells (unpublished observations). Therefore it appears that *T. spiralis* produces a conglomerate of antigens which may have host protective properties but it is not known which of these antigens are crucial to resistance in all strains of animals. However, studies carried out in this laboratory with antisera from mice, human and porcine hosts of *T. spiralis* indicate that all infected mammals examined to date are able to produce antibodies to three major antigens of approximate molecular weight of 41, 46, and 55 kD (Robinson et al., manuscript in preparation).

Conclusions

The study of the murine response to antigen of *T. spiralis* involves a realization of the complexities involved in devising a vaccine against a complex metazoan parasite. It is now being accepted that many parasites are able to manipulate the host's immune system, while at the same time being inherently immunogenic when in situ. It is a relatively straightforward procedure to partially immunize a host against many parasites but what has not yet been achieved is the development of any synthetic or defined antigen vaccine which will immunize a genetically diverse host population against a parasite, without inducing immunopathology or other adverse side effects. It is an interesting consideration that many parasites may have evolved to produce immunodominant antigens which manipulate the host's immune system to prevent an optimum response. It could well be in the parasite's best interests to induce a degree of immunity in the host which prevents parasite overcrowding but stops short of sterilizing immunity and this has led some authors to suggest that *T. spiralis* infections actually induce concomitant immunity (Almond and Parkhouse, 1985) in common with the totally unrelated trematode parasite, *Schistosoma mansoni* (Smithers and Terry, 1969).

We hypothesize that what is really required is an antigen or mixture of antigens which will be efficiently presented to T cells in such a way that the resultant immune response is directed to cause the maximum damage to the parasite burden. Continuing studies of the molecular profile of the hosts Major Histocompatibility Complex molecules will allow research in the near future to pick a polypeptide from known biochemical characteristics which will have an epitope able to most satisfactorily bind to the required MHC molecule and hence be presented to T cells to elicit the maximum response in the hosts defense, without inducing other immunological distractions.

References

Abe, R. and Hodes, R.J. (1988) T cell recognition of Mls^c. I. Influence of MHC gene products in Mls^c-specific T cell recognition. *J. Immunol.* 140:4132-4138.

Almond, N.M. and Parkhouse, R.M.E. (1985) Nematode Antigens. IN: *Current Topics in Microbiology and Immunology 120: Parasite Antigens in Protection, Diagnosis and Escape*. Springer-Verlag, Berlin,Heidelberg, pp. 173.

Baxavenis, C.N., Ishii, N., Nagy, Z.A., Klein, J. (1982) Role of the E^k molecule in the generation of suppressor T cells in response to LDH_β. Scand. J. Immunol. 16:25.
Behnke, J.M. and Robinson, M. (1985) Genetic control of immunity to Nematospiroides dubius: a nine-day anthelmintic abbreviated immunizing regime which separates weak and strong responder strains of mice. Parasite Immunol. 7:235.
Bell, R.G., McGregor, D.D., and Adams, L.S. (1982) Trichinella spiralis: genetic basis for differential expression of phase-specific intestinal immunity in mice. Exp. Parasitol. 53:315.
Bell, T.G., McGregor, D.D., and Despommier, D.D. (1979) Trichinella spiralis: mediation of the intestinal component of protective immunity in the rat by multiple phase-specific anti-parasite responses. Exp. Parasitol. 47:140.
Blackman, M., Yague, J., Kubo, R., Grey, D., Coleclough, C., Palmer, E. Kappler, J., and Marrack, P. (1986) The T cell repertoire may be biased in favor of MHC recognition. Cell 47:349.
Blackwell, J.M. and Roberts, M.B. (1987) Immunomodulation of murine visceral leichmaniasis by administration of monoclonal anti-Ia antibodies: differential effects of anti-I-A vs. anti-I-E antibodies. Eur. J. Immunol. 17:1669.
Buus, S., Sette, A., Colon, S.M., Miles, C. and Grey, H.M. (1987) The relation between major histocompatibility complex (MHC) restriction and the capacity of Ia to bind immunogenic peptides. Science 235:1353.
Chipman, P.B. (1957) The antigenic role of excretions and secretions of adult Trichinella spiralis in the production of immunity in mice. J. Parasitol. 43:593.
Coffman, R., Ohara, J., Bond, M., Carty, J., Zlotnik, A. and Paul, W.E. (1986) B cell stimulatory factor-1 enhances the IgE response of lipopolysaccharide activated B cells. J. Immunol. 136:4358.
Coffman, R.L., Seymour, B.W.P., Lebman, D.A., Hiraki, D.D., Christiansen, J.A., Shrader, B., Cherwinski, H.M., Savelkoul, H.F.J., Finkelman, F.D., Bond, M.W. and Mosmann, T.R. (1988) The role of helper T cell products in mouse B cell differentiation and isotype regulation. Immunol. Rev. 102:5.
Crandall, R.B. and Crandall, C.A. (1972) Trichinella spiralis: immunologic response to infection in mice. Exp. Parasitol. 31:378.
David, C.S. (1976) Serologic and genetic aspects of murine Ia antigens. Transplant. Rev. 30:299.
Despommier, D.D. and Laccetti, A. (1981) Trichinella spiralis: partial characterization of antigens isolated by immuno-affinity chromatography from the large particle fraction of the muscle larvae. J. Parasitol. 67(3):332.
Gamble, H.R. (1985) Trichinella spiralis: immunization of mice using monoclonal antibody affinity-isolated antigens. Exp. Parasitol. 59:398-404.
Golinska, Z., and Bany, J. (1976) Serum immunoglobulins in inbred mice strains infected with Trichinella spiralis and Trichinella pseudospiralis. IN: Trichinellosis, Kim, C., and Pawlowski, A., eds. University of New England Press, Hanover, NH, p. 151.
James, E.R., and Denham, D.A. (1974) The stage specificity of the immune response to Trichinella spiralis. IN: Trichinellosis, Kim, C., eds. Intent Publishing, New York, p. 345.
Jones, P.P., Murphy, D. B. and McDevitt, H.O. (1981) Variable synthesis and expression of E_α and A_e (E_β) Ia polypeptide chains in mice of different H-2 haplotypes. Immunogenetics 12:321.
Kappler, J.W., Roehm, N. and Marrack, P. (1987) T cell tolerance by clonal elimination in the thymus. Cell 49:273.

Kappler, J.W., Staertz, Y., White, J. and Marrack, P.C. (1988) Self-tolerance eliminates T cells specific for Mls-modified products of the major histocompatibility complex. Nature 332:35.

Kappler, J.W., Wade, I., White, J., Kushnir, E., Blackman, M., Bill, M., Roehm, N., and Marrack, P. (1987) A T cell receptor V_β segment that imparts reactivity to a class II major histocompatibility complex product. Cell 49:263.

Kazura, J.W. and Grove, D.I. (1978) Stage-specific antibody-dependent eosinophil-mediated destruction of Trichinella spiralis. Nature 274:588.

Krco, C.J., Wassom, D.L., Abramson, E.J. and David, C.S. (1983) Cloned T cells recognize Trichinella spiralis antigen in association with an $E^k_\beta E^k_\alpha$ restriction element. Immunogen. 18:435-444.

Larsh, J.E., Jr. (1963) Experimental trichinosis. Adv. Parasitol 1:213.

Larsh, J.E. and Weatherly, N.F. (1975) Cell mediated immunity against certain parasitic worms. Adv. Parasitol. 13:183.

MacDonald, H.R., Schneider, R., Lees, R.K., How, R.C., Acha-Orbea, H., Festenstein, H., Zinkernagel, R.M., and Hengartner, H. (1988) T cell receptor V_β use predicts reactivity and tolerance to Mls^a-encoded antigens. Nature 332:40.

Maizels, R.M., Kennedy, M.W., Meghji, M., Robertson, B.D. and Smith, H.V. (1987) Shared carbohydrate epitopes on distinct surface and secreted antigens of the parasitic nematode Tonorara canis. Journal of Immunology 139:207-214.

Maloney, A. and Denham, D.A. (1979) Effects of immune serum and cells on newborn larvae of Trichinella spiralis. Parasite Immunol. 1:3.

Marti, H.P., Murrell, K.D. and Gamble, H.R. (1987) Trichinella spiralis: immunization of pigs with newborn larval antigens. Exp. Parasitol. 63:68.

Mathis, D.J., Benoist, C., Williams, V.E., Kanter, M., and McDevitt, H.O. (1983) Several mechanisms can account for defective E_α gene expression in different mouse haplotypes. Proc. Natl. ACad. Sci. USA 80:273.

Matis, L.A., Sorger, S.B., McElligot, D.L., Fink, P.J. and Hedrick, S.M. (1987) The molecular basis of alloreactivity in antigen-specific, major histocompatibility complex-restricted T cell clones. Cell 51:59.

Miller, H.R.P. (1984) The protective mucosal response against gastrointestinal nematodes in ruminants and laboratory animals. Vet. Immunol. and Immunopathol. 6:167.

Morel, P.A., Livingstone, A.M., and Fathman, C.G. (1987) Correlation of T cell receptor V_β gene family with MHC restriction. J. Exp. Med. 166:583.

Mossmann, T.R., Sherwinski, H., Bond, M.W., Giedlin, M.A. and Coffman, R.L. (1986) Two types of murine helper T cell clones. 1) Definition according to profiles of lymphokine activities and secreted proteins. J. Immunol. 136:2348.

Murphy, D.B., Jones, P.P., Loken,M.R. and McDevitt, H.O. (1980) Interaction between I region loci influences the expression of a cell surface Ia antigen. Proc. Natl. Acad. Sci. USA 77:5405.

Oliveira, D.B.G., Blackwell, N., Virchis, A.E. and Axelrod, R.A. (1985) T helper and T suppressor cells are restricted by the A and E molecules, respectively, in the F antigen system. Immunogenetics 22:169.

Ortega-Pierres, G., Mackenzie, C.D. and Parkhouse, R.M.E. (1984) Protection against Trichinella spiralis induced by a monoclonal antibody that promotes killing of newborn larvae by granulocytes. Parasite Immunol. 6:275.

Powers, G.D., Abbas, A.K., and Miller, R. A. (1988) Frequencies of IL-2 and IL-3 secretory T cells in naive and antigen-stimulated lymophocyte populations. J. Immunol. 140:3352.

Puri, J., Lonai, P. and Friedman, V. (1986) Antigen-Ia interaction and the proteolytic processing of antigen: the structure of the antigen determines its restriction to the A or E molecule of the major histocompatibility complex. Eur. J. Immunol. 16:1093.

Quintans, J., Suzuki, H., Sosman, J.A. and Shah, P.D. (1986) Immunoregulation by T cells. 1) Characterization of the I-E^k-specific Lbd self-reactive T cell clone that helps, suppresses and contrasuppresses B cell responses. J. Immunol. 136:1974.

Rivera-Ortiz, G.I. and Nussenzweig, R. (1976) Trichinella spiralis: anaphylactic antibody formation and susceptibility in strains of inbred mice. Exp. Parasitol. 39:7.

Robinson, M., Benhke, J.M. and Williams, D.J.L. (1988) Immunity to Heligmosomoides polygyrus (Nematospiroides dubius): survival or rejection of adult worms following transplantation to mice refractory to larval challenge. J. Helminthol. (in press).

Ruddle, N.H. (1987) Tumor necrosis factor and related cytotoxins. Immunology Today 8:129.

Sette, A., Buus, S., Colon, S., Miles, C., and Grey, H. (1988) I-A^d binding peptides derived from unrelated protein antigens share a common structural motif. J. Immunol. 141:45.

Silberstein, D.S. and Despommier, D.D. (1984) Antigens from Trichinella spiralis that induce a protective response in the mouse. J. Immunol. 132:898.

Silberstein, D.S. and Despommier, D.D. (1985) Effects on Trichinella spiralis of host responses to purified antigens. Science 227:948.

Smithers, S.R. and Terry, R. J. (1969) Immunity in schistosomiasis. Ann. N.Y. Acad. Sci. 160:826.

Ubeira, F.M., Leiro, J., Santamarina, M.T., Villa, T.G. and SanMartin-Ovran, M.L. (1987) Immune response to Trichinella epitopes: the antiphosphorylcholine plaque-forming cell response during the biological cycle. Parasitology 94:543-553.

Vitetta, E.S., Ohara, J., Myers, C., Layton, J., Krammer, P.H. and Paul, W.E. (1985) Serological, biochemical and functional identity of B cell-stimulatory factor-1 and B cell differentiation factor for IgG1. J. Exp. Med. 162:1726.

Wakelin, D. (1985) Genetic control of immunity to helminth infections. Parasitology Today 1:17.

Wakelin, D., Donachie, A.M. (1981) Genetic control of immunity to Trichinella spiralis. Donor bone marrow cells determine responses to infection in mouse radiation chimaeras. Immunology 43:787.

Wakelin, D. and Donachie, A.M. (1983) Genetic control of immunity to Trichinella spiralis. Influence of H-2 linked genes on immunity to the intestinal phase of infection. Immunology 48:343.

Wassom, D.L. (1985) Genetic control of the host response to parasite helminth infections. IN: Genetic Control of Host Resistance to Infection and Malignancy (Skamene, E., ed.) Alan R. Liss, New York, p. 449.

Wassom, D.L., Brooks, B.O., Babish, J.G., and David, C. S. (1983) A gene mapping between the S and D regions of the H-2 complex influences resistance to Trichinella spiralis infections in mice. J. Immunogen. 10:371.

Wassom, D.L., Brooks, B.O. and Cypess, R.H. (1983) Trichinella spiralis: Role of non-H-2 genes in resistance to primary infection. Exp. Parasitol. 55:153.

Wassom, D.L., David, C.S. and Gleich, G.J. (1979) Genes within the major histocompatibility complex influence susceptibility to Trichinella spiralis in the mouse. Immunogen. 9:491.

Wassom, D.L., David, C.S. and Gleich, G.J. (1980) MHC linked genetic control of the immune response to parasites: Trichinella spiralis in the mouse. IN: Genetic Control of Natural Resistance to Infection and Malignancy (Eds. E. Skamene, P.A.L. Kongshown and M. Landy) Academic Press, Inc., New York, p. 75.

Wassom, D.L., Dougherty, D.A., Krco, C.J. and David, C.S. (1984) H-2 controlled, dose dependent suppression of the response that expels adult Trichinella spiralis from the small intestine of mice. Immunology 53:811.

Wassom, D.L., Krco, C.J., and David, C.S. (1987) I-E suppression and susceptibility to parasite infections. Immunology Today 2:39.

Wassom, D.L., Wakelin, D., Brooks, B.O., Krco, C.J. and David, C.S. (1984) Genetic control of immunity to Trichinella spiralis infections of mice. Hypothesis to explain the role of H-2 genes in primary and challenge infections. Immunology 51:625.

Williams, D.J.L. and Behnke, J.M. (1983) Host protective antibodies and serum immunoglobulin isotypes in mice chronically infected or repeatedly immunized with the nematode parasite Nematospiroides dubius. Immunology 348:34.

Winoto, A., Urgan, J.L., Lan, N.C., Goverman, J., Hood, L., and Hansburg, D. (1986) Predominant use of a V_{α} gene segment in mouse T-cell receptors for cytochrome C. Nature 324:679.

MODULATION OF ANTIBODY RESPONSES BY CONJUGATES OF ANTIGENS WITH MONOMETHOXYPOLYETHYLENE GLYCOL

Alec H. Sehon

MRC Group for Allergy Research
Department of Immunology
Faculty of Medicine
The University of Manitoba
Winnipeg, MB, Canada

ABSTRACT

The use of tolerogenic conjugates of monomethoxypolyethylene glycol (mPEG) and diverse antigens, including xenogeneic monoclonal immunoglobulins, is reviewed with respect to the induction of specific abrogation of antibody responses. Evidence is presented for activation of specific suppressor T (Ts) cells by antigen-mPEG conjugates and for the release of suppressor lymphokines (TsFs) from these cells; both Ts cells and TsFs are able to suppress antigen-specific immune responses on transfer to syngeneic mice. The therapeutic potential of tolerogenic mPEG conjugates is discussed with respect to IgE-mediated allergies and diseases which could be ameliorated by administration of xenogeneic monoclonal antibodies and of their respective immunotoxins, as would be the case for suppression of rejection of organ transplants, localization and destruction of tumors, and *in vivo* annihilation of the human immunodeficiency virus (HIV).

INTRODUCTION

In this paper are reviewed:

(i) the fundamental experimental results obtained in the author's laboratory (Lee and Sehon, 1977) in relation to the antigen-specific downregulation of IgE and IgG antibody responses by the use of *nonimmunogenic and tolerogenic conjugates of the antigens* in question with the synthetic water soluble polymer, mPEG,

(ii) the possible therapeutic applications of these conjugates to diverse fields, such as common forms of immediate hypersensitivity, organ transplantation, destruction of tumor cells, and annihilation of HIV and HIV-infected cells of asymptomatic carriers of HIV.

It was demonstrated (Abuchowski, 1976) that after covalent coupling of an enzyme to the chemically bifunctional polyethylene glycol (PEG), the resulting conjugate was devoid of the immunogenicity of the original unmodified enzyme, and that the clearance rate of the conjugate was significantly lower than that of the unmodified enzyme. In view of the long established facts that (i) administration of some antigens (e.g., xenogeneic immunoglobulins) in a deaggregated form resulted in specific tolerance (Dresser, 1961), and (ii) the clearance rate of the deaggregated antigens was significantly lower than that of the original immunogenic form of the antigens (Weigle, 1973), this author *hypothesized that mPEG conjugates of various antigens would be not only nonimmunogenic but also immunosuppressive*. Indeed, as documented by experimental results obtained in this and other laboratories, this hypothesis was confirmed (Lee and Sehon, 1977; Lee and Sehon, 1978; Savoca et al., 1984).

To avoid the possible cross-linking of protein antigens by PEG, the use of mPEG -- which has only one chemically functional, terminal hydroxyl group -- is preferred for the preparation of the corresponding conjugates. Thus, the mPEG conjugate of an antigen protein (Ag), represented by the formula $Ag(mPEG)_n$, consists of the respective protein onto which are grafted a number (n) of mPEG chains; the subscript n is defined as the average degree of conjugation. In most of the recent experiments, the molecular weight of the mPEG preparations used for conjugation was about 6,400 Da.

OVERVIEW OF EXPERIMENTAL FINDINGS

(a) Immunosuppressive Effects of Conjugates of mPEG with Ovalbumin (OA) on the Anti-OA Response in Mice and Rats

These studies were prompted by the quest for a new, effective, immunologically specific therapeutic procedure for IgE-mediated allergies, which would not have the disadvantages associated with the "time honoured" hyposensitization regimen among which one may cite:
(i) the risk of systemic anaphylaxis which dictates the administration of the allergenic preparations in only minute doses, (ii) as a result of the above limitation the course of the classical regimen requires many years, and hence it is time-consuming, inconvenient and expensive, and (iii) the effectiveness of this regimen is frequently at best marginal.

In the original experiments, injection into naive mice or rats of PEG conjugates of OA was shown to abrogate their ability to mount primary IgE and IgG responses to the subsequent injection of the unmodified OA (Lee and Sehon, 1977; Lee and Sehon, 1978); in these experiments the immunization was performed in the presence of $Al(OH)_3$ which served as adjuvant. Moreover, as a result of (i) the shielding of some of the epitopes of the antigen by the long PEG chains, and/or (ii) the conformational changes of the antigen induced by the amphipathic PEG chains, the conjugates were nonallergenic as demonstrated in passive cutaneous anaphylaxis tests and did not induce systemic anaphylaxis in presensitized rats.

It is also important to stress that mice tolerized by treatment with $OA(mPEG)_n$ were incapable of mounting subsequently an immune response not only to OA but also to a hapten, such as the 2,4-dinitrophenyl (DNP) group, when the latter was presented in the form of a conjugate with OA as the carrier. However, the same mice mounted a regular anti-DNP antibody response on immunization with DNP coupled to unrelated carriers, such as *Ascaris suum* antigens (Lee and Sehon, 1978). These findings suggested that the suppressogenic effect of mPEG conjugates was exerted on T helper (Th) cells specific for the carrier incorporated into the conjugate. Moreover, it was demonstrated that immunosuppressive $Ag(mPEG)_n$ conjugates injected into adult mice induced antigen-specific suppressor T (Ts) cells which exerted their tolerogenic effect on transfer into intact naive mice. It was also shown that these cells released an antigen-specific suppressor factor (TsF) which was capable of abrogating the immune response of its recipients to the antigen in question (Sehon, 1982).

In recent experiments IgE and IgG anti-OA antibodies were induced in both neonatal and adult mice in the absence of adjuvants (Athota and Sehon, unpublished). Thus, injection of $OA(mPEG)_{12}$ into B6D2F1 mice on days 1, 3 and 42 after birth, followed by injections of OA on days 43 and 71, resulted in a dramatic antigen-specific suppression (99%) of both IgE and IgG1 anti-OA antibody responses. A similar degree of suppression was induced with $OA(mPEG)_{10}$ conjugates in adult mice. Moreover, transfer of unfractionated spleen cells or of surface Ig negative (sIg^-) splenic cells of either of the above two groups of animals into adult, non-irradiated syngeneic recipients induced a profound (75-80%) suppression of the anti-OA IgE and IgG1 antibody responses compared to the antibody responses in recipients of sIg^- cells from donors treated with phosphate buffered saline (PBS) instead of $OA(mPEG)_{10}$ conjugates. Furthermore, the suppressogenic activity of these cells was impaired by their treatment with monoclonal antibodies to the Thy 1.2 markers in the presence of complement. All these results, taken together, strongly supported the earlier conclusion that the transfer of suppressogenic activity by spleen cells from $OA(mPEG)_n$ treated mice was mediated by Ts cells and not by a carry-over of $OA(mPEG)_n$ conjugates by the donors' spleen cells. It is also interesting to point out that injection of $OA(mPEG)_n$ into mice on days 1, 3 and 42 after birth induced a more potent Ts cell effect (measured in terms of the degree of suppression of the recipients of these cells) than the treatment with $OA(mPEG)_n$ in the neonatal stage only, or at day 42 only. These data may indicate that injection of the tolerogenic mPEG conjugates into neonates induced a population of Ts cells which were further amplified upon restimulation by the conjugate in adult life.

In other very recent experiments designed to isolate and characterize the TsF (Athota and Sehon, unpublished data), it was shown that the freeze/thaw (F/T) extract of sIg^- cells of $OA(mPEG)_{10}$ treated mice was also capable of specifically suppressing the anti-OA antibody response in the recipients of this extract, which were immunized one day later with OA. Most importantly, utilizing an immunosorbent consisting of the murine B16G monoclonal antibodies (MAbs) to an unrelated TsF (Steele et al., 1986), it was demonstrated that this immunosorbent depleted the F/T extracts of TsF. Moreover, the TsF could be recovered from the immunosorbent by elution without loss of biological activity, as substantiated on transfer into naive adult mice. This factor was also shown to bind specifically to an OA-immunosorbent and to be elutable from this immunosorbent. In a different experimental system, the TsF induced by tolerogenic mPEG conjugates was shown to be protease, but not RNase or DNase sensitive (Wilkinson, 1988). From all these results it was inferred that the TsFs consisted of at least two domains, i.e., one antigen-specific domain, and a second domain sharing a common epitope(s) which reacted with B16G Mabs. Hence, one may anticipate that the TsFs generated by tolerogenic antigen-mPEG conjugates are proteins and, as such, would be amenable to synthesis by recombinant DNA (rDNA) technology. It is obvious that if this prediction were validated, one may be in a position to generate specific TsFs for therapeutic purposes in man by a combination of antigen-mPEG and rDNA technologies.

(b) Conjugates of mPEG with Allergens for Use in Clinical Trials

As stated earlier, mPEG conjugates of antigens were essentially nonallergenic, i.e., they were essentially devoid of the capacity to trigger the release of vasoactive amines from mast cells to which the homologous IgE antibodies were attached. Hence, mPEG conjugates of a variety of allergens were tested in clinical trials in a number of medical centers in the U.S.A., Canada, Germany, Switzerland, Spain and Scandinavia -- conducted under the sponsorship of Pharmacia AB Uppsala (Sehon, 1986; Dreborg and Åkerblom, in press) -- for their ability to suppress *the ongoing production of the corresponding IgE antibodies in allergic patients*. For these clinical trials, mPEG conjugates of standard allergenic preparations containing multiple components were synthesized and, therefore, the resulting conjugates were complex mixtures of mPEG derivatives. To overcome this disadvantage, which is common to all hyposensitization therapies, work is underway in this laboratory for *synthesizing the major allergens of pollens in a pure state by rDNA methodology* (Hatton et al., 1989). Clearly, the availability of pure allergens in large amounts will make it possible to establish the epitopes responsible for common allergies and to synthesize well defined (i) tolerogenic mPEG conjugates of the whole allergenic molecules, and (ii) tolerogenic conjugates of their epitopes with polyvinyl alcohol (PVA), as had been achieved for the synthesis of tolerogenic PVA conjugates of a number of haptens, including the benzylpenicilloyl group which represents the main allergenic moiety responsible for penicillin allergy (Sehon, 1982).

As would be expected on the basis of the results obtained in the animal model system, the patients participating in the above mentioned clinical trials were able to tolerate much higher doses of mPEG derivatives of the allergens (expressed in terms of their protein content) than of the unmodified allergens. The beneficial effects of a short course of injections were similar to those obtained with standard extracts after a more extensive treatment, i.e., these patients showed fewer symptoms of rhinoconjunctivitis and required less medication during the pollination season. It is also worth noting that, in addition to these promising results, these clinical trials have not led to untoward physiological effects. However, surprisingly, the level of the ongoing IgE antibody responses in these patients did not drop markedly and even increased in some patients. Moreover, remarkably, the level of IgG antibodies of these patients increased somewhat faster than that of individuals receiving -- by necessity -- lower doses of the standard allergenic preparation; this effect would be obviously particularly beneficial in allergies such as honeybee venom allergy (Müller et al., 1985). The intriguing, diverging effects of these conjugates on the different isotypes (i.e., IgE vs IgG) -- depending on the state of immunity of the recipients (i.e., in naive mice versus presensitized patients) -- were confirmed in subsequent experiments in pre-immunized mice (Sehon and Lang, 1986). Thus, whereas administration of OA(mPEG)$_n$ conjugates to naive mice led to suppression of their IgE and IgG responses, injection of these conjugates into presensitized mice had the paradoxical dual potential of downregulating the ongoing (i.e., secondary) IgE response and of upregulating the formation of IgG antibodies. Clearly, further experiments are required for elucidating the difference in the effects of these conjugates on subpopulations of B cells in various states of differentiation.

(c) Conjugates of mPEG with Xenogeneic Monoclonal Immunoglobulins (xIg) and Immunotoxins (IT)

Therapeutic procedures involving injections into humans of xIg are becoming increasingly prevalent as exemplified by the administration of

(i) mouse MAbs to subsets of human T cells which are used in organ transplantation (Ortho Multicenter Group, 1985), (ii) IT also referred to as "magic bullets", which consist of toxins (Tx) coupled to murine or rat MAbs to tumor associated antigens and which are used for the selective *in vivo* destruction of tumors (Dillman, 1985; Frankel, 1988), and (iii) radio-labeled anti-tumor MAbs used for both the *in vivo* diagnostic localization of tumors by scintigraphy and their destruction (Frankel, 1988). However, as predicted (Sehon et al., 1983), most of the patients produce antibodies to the epitopes of murine MAbs and Tx (Frankel, 1988); in fact the toxins are even more immunogenic than xIg. It is also to be noted that, in addition to the fact that the diagnostic and therapeutic effectiveness of these immunoconjugates may be seriously counteracted by the patients' antibodies to Mabs on re-administration of these MAb preparations the patients may undergo anaphylactic reactions, such as bronchospasm, dyspnoea and hypotension (Dillman, 1985). Moreover, as recently demonstrated in patients receiving MAbs labeled with different radioactive elements for diagnostic imaging of tumors, the resulting immune complexes consisting of the patients' antibodies and of the MAbs were deposited primarily in the liver (Zimmer et al., 1987).

In spite of the recent elegant synthesis of tailor-made chimeric (Brüggemann et al., 1987) and "humanized" antibodies (Riechmann et al., 1988) by rDNA methodologies, even if these antibodies were nonimmunogenic in man, it is likely that the corresponding IT would be immunogenic because of the high immunogenicity of Tx. Consequently, the effectiveness of these IT, just as that of xenogeneic MAbs, will be undermined by their inherent immunogenicity. Therefore, on the basis of the results summarized in section (a), studies were recently undertaken in this laboratory for the synthesis of tolerogenic mPEG conjugates of xIg, in the expectation that *injection of these tolerogens prior to initiation of the treatment per se would induce specific immunosuppression* and that, consequently, it would be possible to administer subsequently the xIg without engendering anti-xIg antibodies (Wilkinson et al., 1987).

In the absence of the resources required to test the validity of this premise with mouse MAbs in nonhuman primates, the model used in the author's laboratory consisted of strains of syngeneic adult mice, human monoclonal (myeloma) immunoglobulins (HIgG) as a source of xIg, and heat aggregated HIgG (i.e., haHIgG) as the immunizing antigen. The HIgG was aggregated by heating at 63°C for 25 min in order to increase its immunogenicity and to use it without adjuvants. Clearly, if this system were to be adapted for use in humans, the nonpegylated xIg would not be heat aggregated.

A diversity of protocols, differing in the number of doses of a given conjugate and of haHIgG injected over differing lengths of time, as well as in the intervals between injections, were used for testing the capacity of the conjugates to induce long-term suppression (Maiti et al., 1988). However, in general, for screening the tolerogenic capacity of $HIgG(mPEG)_n$ conjugates with n values in the range of 20-35, the protocol consisted of the i.v. injection of 100 μg of each conjugate into a group of 3 mice each, followed 7 days later by an i.p. immunization of 20 μg of haHIgG. The control groups, as usual, received PBS in lieu of the mPEG conjugate and were treated otherwise in an identical manner as the test groups. Thus, for example, as illustrated in Table 1, a single i.v. injection of 100 μg of $HIgG(mPEG)_{31}$ administered on day 0 was capable of suppressing the IgG1 anti-HIgG response up to 133 days in animals which received three i.p. immunizing doses of 20 μg of haHIgG each on days 7, 49 and 119 (this experiment was not continued beyond day 133).

Table 1. Suppression of Primary, Secondary and Tertiary Responses by a Single I.V. Injection of 100 μg $HIgG(mPEG)_{31}$ on Day 0

Treatment[a]	Primary IgG1 ELISA Titres		Secondary IgG1 ELISA Titres		Tertiary IgG1 ELISA Titres	
	d23	d37	d63	d70	d126	d133
PBS	22,000	12,000	188,000	200,000	60,000	80,500
$HIgG(mPEG)_{31}$	800	600	800	800	900	1,290

[a]All mice received 3 i.p. injections of 20 ug of haHIgG on days 7, 49 and 119 and their anti-HIgG antibody titers were determined on the days indicated.

To test the kinetics of induction of suppression by $HIgG(mPEG)_{31}$, the tolerogen was injected into groups of mice 10, or 30, or 60 days prior to immunization with a single dose of 20 μg of haHIgG. With respect to the response of the control groups which received PBS, the IgG1 anti-HIgG antibody response of all the three test groups was reduced profoundly (>92%) even by one injection of the tolerogen at any of the above intervals prior to immunization with haHIgG (Maiti et al., 1988).

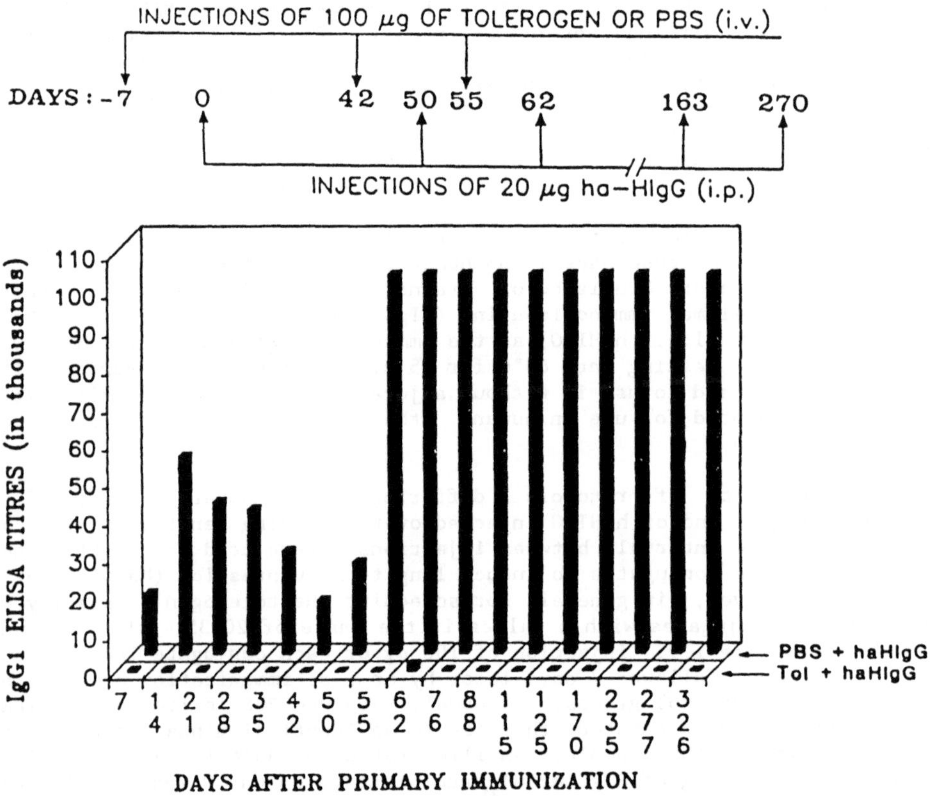

Fig. 1. Induction of tolerance to HIgG by treatment of mice with $HIgG(mPEG)_{22}$. The control group received PBS and the test group 3 injections of 100 μg $HIgG(mPEG)_{22}$ on days -7, 42 and 55. Both groups were immunized 5 times with 20 μg haHIgG on days 0, 50, 62, 163 and 270. (The days are to be read vertically.)

The results of the experiment shown in Fig. 1 demonstrate that injections of $HIgG(mPEG)_{22}$ on days -7, 42 and 55 resulted in suppression of the IgG1 anti-HIgG antibody response in excess of 95% for over 300 days in animals which had received 5 doses of the antigen over a period of 270 days.

In Fig. 2 is illustrated an experiment designed to test the specificity of the immunosuppressive effect of $HIgG(mPEG)_{22}$. Indeed, it was demonstrated that injection of an unrelated antigen (i.e., OA) on day 146 into mice, which had been immunosuppressed with regard to HIgG, induced an IgG1 anti-OA antibody response which was essentially identical to that induced in mice which had not been treated with $HIgG(mPEG)_{22}$. In summary, these results demonstrate that profound antigen-specific suppression (>90%), which lasted for at least 320 days, was achieved by $HIgG(mPEG)_n$ conjugates in spite of multiple injections of the immunogenic form of HIgG. Moreover, it is to be pointed out that the suppression induced by these tolerogenic conjugates encompassed multiple IgG subclasses, i.e., IgG1, IgG2a and IgG2b anti-HIgG antibodies.

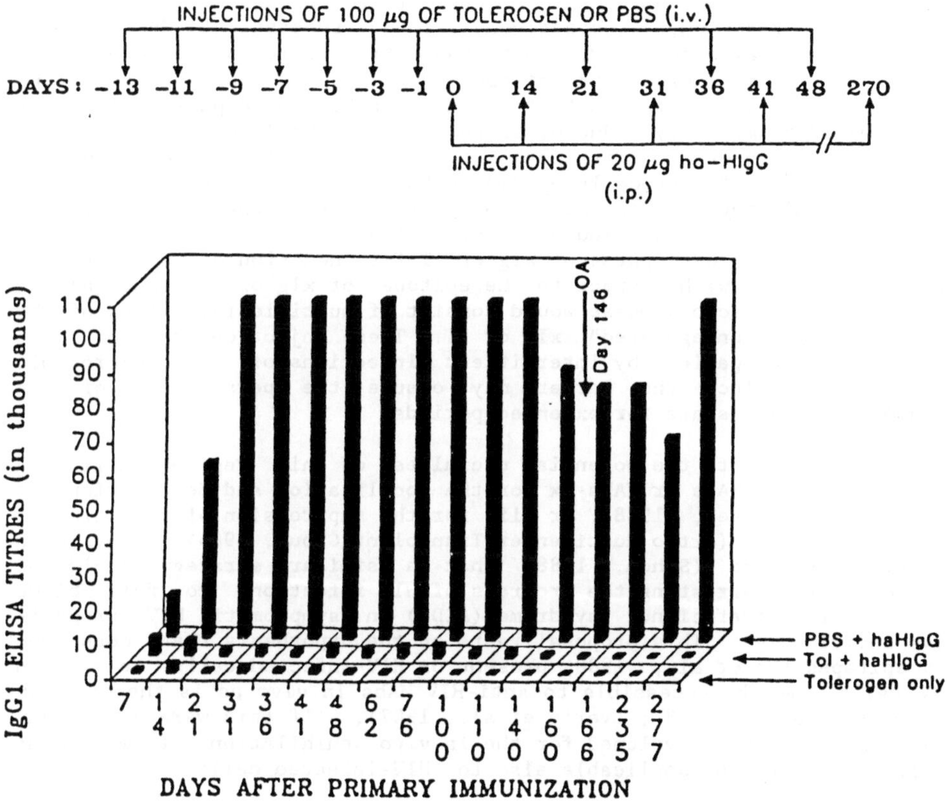

Fig. 2. Induction of long-lasting suppression by $HIgG(mPEG)_{22}$. Two test groups received 10 injections of 100 μg $HIgG(mPEG)_{22}$ and the control group received 10 injections of PBS. One of the test groups, as well as the control group, received 5 injections of 20 μg haHIgG on days 0, 14, 31, 41 and 270. (The days are to be read vertically.) For the specificity test, the 2 test groups and a group of untreated mice were immunized with 20 μg of aggregated OA on day 146; the anti-OA IgG_1 antibodies were determined 14 and 20 days later.

In support of the likelihood that a host tolerized to an xIg of a given isotype by the corresponding $xIg(mPEG)_n$ conjugate would be also tolerant to other xIgs of the same isotype but differing in their idiotypes, one may quote again the earlier findings in this laboratory demonstrating that tolerization of mice by $OA(mPEG)_n$ resulted in the abrogation of not only the anti-OA response, but also of an anti-hapten response, as long as the hapten in question was presented to the suppressed animals in the form of a covalent conjugate with OA (Lee and Sehon, 1978). This finding may be considered as an indication that the suppressor cells induced by tolerogenic mPEG conjugates exerted their downregulating effect not only on T helper cells recognizing the protein incorporated in the mPEG conjugate, but also on lymphocytes recognizing additional epitopes coupled to the same protein. Hence, experiments are in progress in this laboratory to establish whether or not cross-tolerization with respect to different idiotypes, and possibly even with respect to different isotypes, may be induced by a given tolerogenic $HIgG(mPEG)_n$ conjugate.

SUGGESTED NEW CLINICAL APPLICATIONS OF mPEG CONJUGATES

As stated in the Introduction, coupling of mPEG chains to an antigen or allergen resulted, respectively, in a substantial decrease in antigenicity or allergenicity, which was attributable to conformational changes and/or masking of the epitopes caused by the coupling reaction. By analogy, it is anticipated that the antigen-binding capacity of xIg or xIg-Tx would be markedly reduced, if not totally impaired, as a result of conjugation with mPEG. Therefore, to ensure the efficacy of the destruction of the target cells by "magic bullets", *it is suggested that the clinical regimen include two phases:* (a) *the first (i.e., immunosuppressive) phase* would consist of a series of injections of tolerogenic mPEG conjugates of xIg or IT for induction of a state of immunosuppression with respect to the epitopes of xIg or IT, and (b) *the second (i.e., effector) phase* would consist of multiple injections of the corresponding "nonpegylated" xIg or IT. These injections may be also preferably accompanied by intermittent injections of the tolerogenic derivatives, since the latter may ensure the persistence of the immunosuppressive state for extended periods.

In addition to the potential usefulness of this regimen utilizing appropriate (i) MAbs or MAbs-Tx for the localization and destruction of tumor cells (Frankel, 1988), or (ii) for the suppression of rejection of organ transplants (Ortho Multicenter Transplant Group, 1985), it has been recently proposed (Sehon, 1988) that a similar strategy could be developed for arresting the progress of HIV infection to full blown acquired immunodeficiency syndrome (AIDS) in asymptomatic HIV-seropositive individuals. This proposal rests on the hypothesis that conserved HIV epitopes (Ho et al., 1988) present on HIV-infected cells, e.g., on $CD4^+$ cells, may be accessible to anti-HIV MAbs *in vivo* as is the case *in vitro* (Fung et al., 1987; Lyerly et al., 1987). If this were confirmed, the armamentarium developed for the *in vivo* annihilation of malignant cells by IT might be applicable also to HIV-infected cells.

Since HIV-infected individuals are immunologically competent during their asymptomatic phase, to ensure the efficacy of the above treatment, it is suggested -- as recommended earlier -- that the therapeutic strategy for HIV-infected individuals consist also of two steps, i.e., *the first* would involve a series of injections of tolerogenic mPEG conjugates of anti-HIV MAbs or of the corresponding IT, and *the second* would consist of injections of the respective nonpegylated anti-HIV MAbs or of their Tx-tagged derivatives, accompanied at optimal intervals by injections of the corresponding mPEG conjugates. It is also visualized that the presence of the HIV genome in a cryptic, latent state in cells other than $CD4^+$ cells would not invalidate the proposed strategy, since the anti-HIV IT present in circulation would latch onto any viruses released from or appearing on these cells, and thus inactivate or destroy the virus and the corresponding cells. Clearly, the injection of both reagents, or of only the anti-HIV IT may have to be continued at regular intervals for as long as the HIV has not been totally eradicated, or the patient has not developed effective immunity for the inactivation of any residual virus. Moreover, even if all the HIVs could not be eliminated from all the cells of an infected individual, it is reasonable to anticipate that -- under the protective umbrella of passive immunization with anti-HIV IT -- the proposed regimen would result at least in the reduction of the viral load. This strategy would, therefore, slow down -- if not stop altogether -- the progression from the asymptomatic phase to full-blown AIDS. This regimen may thus provide protection during the critical period necessary for (i) the induction of a strong autologous anti-HIV response in patients by immunization with suitably designed vaccines, and/or (ii) their treatment with effective anti-viral agents when, and if, these become available.

CONCLUSIONS

As emphasized above, for all these potential therapeutic strategies, the pretreatment of patients with tolerogenic mPEG conjugates of MAbs, or more appropriately with mPEG conjugates of the corresponding IT, will be essential. Although these studies hold the promise that mPEG conjugates of xIg (and IT) may prove useful for clinical applications, it ought to be stressed that it still remains to be demonstrated that conjugates of xIg suppress also the immune response to the idiotypic determinants of cell-bound xIg (Benjamin and Waldmann, 1986; Jonker and DenBrok, 1987). Finally, it ought to be pointed out that administration of mPEG conjugates of immunogenic proteins, such as allergens and xenogeneic enzymes, has proven a safe therapeutic procedure in patients suffering from common allergies (Dreborg and Åkerblom, in press), cancer (Abuchowski et al., 1984) and adenosine deaminase deficiency in severe combined immunodeficiency diseases (Hershfield et al, 1987). Therefore, no *a prior* objection appears to exist for the use of mPEG conjugates of xIg in man, particularly in life-threatening diseases.

ACKNOWLEDGEMENTS

The work described in this paper was supported by grants from the Medical Research Council of Canada and the National Institute of Allergy and Infectious Diseases, National Institutes of Health, Bethesda, MD. (IA-14526-11). The author would also like to acknowledge the efficient and cheerful secretarial assistance of Gail Falkenberg and Shamina Singh.

REFERENCES

Abuchowski, A., 1976, Effect of covalently attached polyethylene glycol on the immunogenicity and activity of enzymes; Ph.D. Dissertation, Rutgers Univ., New Brunswick, N.J.

Abuchowski, A., Kazo, G.M., Verhoest, C.R., Van Es, T., Kafke Witz, D., Nucci, M.L., Viau, A.T., and Davis, F.F., 1984, Cancer therapy with chemically modified enzymes. I. Anti-tumor properties of polyethylene glycol-asparaginase conjugates, Cancer Biochem. Biophys., 7:175.

Athota, R. and Sehon A.H. (unpublished data).
Benjamin, R.J. and Waldmann, H., 1986, Induction of tolerance by monoclonal antibody therapy, Nature, 320:449.
Brüggemann, M., Williams, G.T., Bindon, C.I., Clark, M.R., Walker, M.R., Jefferis, R., Waldmann, H., and Neuberger, M.S., 1987, Comparison of the effector functions of human immunoglobulins using a matched set of chimeric antibodies, J. Exp. Med., 166:1351.
Dillman, R.O., 1985, Monoclonal antibodies in the treatment of cancer, CRC Crit. Rev. Oncol. Hematol., 1:357.
Dreborg, S., and Åkerblom, E., Immunotherapy with polyethylene glycol modified allergens, CRC Crit. Rev. Therapeutic Drug Carrier Systems, (in press).
Dresser, D.W., 1961, Acquired immunological tolerance to a fraction of bovine gamma globulin, Immunology, 4:13.
Frankel, A.E., 1988, "Immunotoxins," Kluwer Academic Publishers, Boston.
Fung, M.S.C., Sun, C., and Sun, N-C., 1987, Monoclonal antibodies that neutralize HIV-1 virions and inhibit syncytium formation by infected cells, Biotechnol., 5:940.
Hatton, T.W., Ekramoddoullah, A.K.M., Kisil, F.T., Sehon, A.H., and Hill, R.D., 1989, Identification of cDNA clones representing Group I pollen allergens from Kentucky Bluegrass, in: Advances in the Biosciences, Vol. 74, "Allergy and Molecular Biology", A.S. El Shami and T.G. Merrett, eds., Pergamon Press, New York.
Hershfield, M.S., Buckley, R.H., Greenberg, M.L., Melton, A.L., Schiff, R., Hatem, C., Kurtzberg, J., Markert, M.L., Kobayashi, R.H., Kobayashi, A.L., and Abuchowski, A., 1987, Treatment of adenosine deaminase deficiency with polyethylene glycol-modified adenosine deaminase, N. Eng. J. Med., 316:589.
Ho, D.D., Kaplan, J.C., Rackauskas, I.E., and Gurney, M.E., 1988, Second conserved domain of gp120 is important for HIV infectivity and antibody neutralization, Science, 239:1021.
Jonker, M. and DenBrok, J.H.A.M., 1987, Idiotype switching of CD4-specific monoclonal antibodies can prolong therapeutic effectiveness in spite of host anti-mouse IgG antibodies, Eur. J. Immunol., 17:1547.
Lee, W.Y. and Sehon, A.H., 1977, Abrogation of reaginic antibodies with modified allergens, Nature, London, 267:618.
Lee, W.Y. and Sehon, A.H., 1978, Suppression of reaginic antibodies with modified allergens. II. Abrogation of reaginic antibodies with allergens conjugated to polyethylene glycol, Int. Archs. Allergy appl. Immun., 56:196.
Lyerly, H.K., Matthews, T.J., Langlois, A.J., Bolognesi, D.P., and Weinhold, K.J., 1987, Human T-cell lymphotropic virus IIIb glycoprotein (gp120) bound to CD4 determinants on normal lymphocytes and expressed by infected cells serves as target for immune attack, Proc. Natl. Acad. Sci., 84:4601.
Maiti, P.K., Lang, G.M., and Sehon, A.H., 1988, Tolerogenic conjugates of xenogeneic monoclonal antibodies with monomethoxypolyethylene glycol: I. Induction of long-lasting tolerance to xenogeneic monoclonal antibodies, Int. J. Cancer, Supplement 3:17.
Müller, U., Lanner, A., Schmid, P., Bischof, M., Dreborg, S., and Hoigné, R., 1985, A double-blind study on immunotherapy with chemically modified honey bee venom: Monomethoxypolyethylene glycol-coupled versus crude honey bee venom, Int. Arch. Allergy appl. Immunol, 77:201.
Ortho Multicenter Transplant Study Group, 1985, A randomized clinical trial of OKT3 monoclonal antibody for acute rejection of cadaveric renal transplant, N. Engl. J. Med., 313:337.
Riechmann, L., Clark, M., Waldmann, H., and Winter, G., 1988, Reshaping human antibodies for therapy, Nature London, 332:323.

Savoca, K.V., Davis, F.F., and Palczuk, N.C., 1984, Induction of tolerance in mice by uricase and monomethoxypolyethylene glycol-modified uricase, Int. Arch. Allergy appl. Immunol., 75: 58.
Sehon, A.H., 1982, Suppression of IgE antibody responses with tolerogenic conjugates of allergens and haptens, Progr. Allergy, 32:161.
Sehon, A.H., 1986, Potential use of allergens modified with monomethoxypolyethylene glycol for immunotherapy of atopic patients, in Proc. XII Int. Congr. Allergol. and Clin. Immunol., The C.V. Mosby Company, St. Louis.
Sehon, A.H., 1988, Could Immunological Strategies for Cancer Treatment be Adapted for AIDS?, Clin. Immunol. Newsletter, 9:77.
Sehon, A.H., Lang, G.M., and Carter, B.G., 1983, Further insights into the mechanism underlying the downregulation of IgE antibodies by conjugates of allergens with monomethoxypolyethylene glycols (mPEGs). in: "Proc. XIth Int. Cong. Allergol. and Clin. Immunol., J.W. Kerr and M.A. Ganderton, eds., McMillan, London.
Sehon, A.H. and Lang, G.M., 1986, The use of nonionic, water soluble polymers for the synthesis of tolerogenic conjugates of antigens, in: "Mediators of Immune Regulation and Immunotherapy", S.K. Singal and T.L. Delovitch eds., Elsevier, New York.
Steele, J.K., Stammers, A.T., Chan, A., and Levy, J., 1986, Characterization of an anti-idiotypic T cell hybridoma involved in the regulation of the immune response to the P815 mastocytoma, J. Immunol., 137:3550.
Weigle, W., 1973, Immunological unresponsiveness, Advances in Immunology, 16:61.
Wilkinson, I., Jackson, C-J.C., Lang, G.M., Holford-Strevens, V., and Sehon, A,H., 1987, Tolerance induction in mice by conjugates of monoclonal immunoglobulins and monomethoxypolyethylene glycol. Transfer of tolerance by T cells and by T cell extracts, J. Immunol., 139:326.
Wilkinson, I., 1988, Tolerogenic monomethoxypolyethylene glycol derivatives of xenogeneic monoclonal immunoglobulins. Ph.D. Dissertation, The University of Manitoba.
Zimmer, A.M., Goldman-Lieken, R.E., and Kazikiew, J.M., 1987, Radioimmunotherapy retreatment of cutaneous T cell lymphoma: effect of plasmapheresis on human antimurine antibody titers, J. Nucl. Med., 28:603.

INDEX

MIX
Papier aus verantwortungsvollen Quellen
Paper from responsible sources
FSC® C105338

If you have any concerns about our products,
you can contact us on
ProductSafety@springernature.com

In case Publisher is established outside the EU,
the EU authorized representative is:
Springer Nature Customer Service Center GmbH
Europaplatz 3, 69115 Heidelberg, Germany

Printed by Libri Plureos GmbH
in Hamburg, Germany